Thermal and Optical Remote Sensing: Evaluating Urban Green Spaces and Urban Heat Islands in a Changing Climate

Thermal and Optical Remote Sensing: Evaluating Urban Green Spaces and Urban Heat Islands in a Changing Climate

Editors

John O. Odindi
Elhadi Adam
Elfatih M. Abdel-Rahman
Yuyu Zhou

MDPI • Basel • Beijing • Wuhan • Barcelona • Belgrade • Manchester • Tokyo • Cluj • Tianjin

Editors

John O. Odindi
University of KwaZulu-Natal
South Africa

Elhadi Adam
University of the
Witwatersrand
South Africa

Elfatih M. Abdel-Rahman
International Centre of Insect
Physiology and Ecology
(ICIPE)
Kenya

Yuyu Zhou
Iowa State University
USA

Editorial Office
MDPI
St. Alban-Anlage 66
4052 Basel, Switzerland

This is a reprint of articles from the Special Issue published online in the open access journal *Remote Sensing* (ISSN 2072-4292) (available at: https://www.mdpi.com/journal/remotesensing/special_issues/RS_Urban_Green_UHI).

For citation purposes, cite each article independently as indicated on the article page online and as indicated below:

LastName, A.A.; LastName, B.B.; LastName, C.C. Article Title. *Journal Name* **Year**, *Volume Number*, Page Range.

ISBN 978-3-0365-6275-9 (Hbk)
ISBN 978-3-0365-6276-6 (PDF)

Cover image courtesy of Getty Images

Contents

About the Editors

John O. Odindi

John O. Odindi holds a PhD in Environmental Geography with focus on terrestrial remote sensing. He is a full professor and former Head of Discipline of Geography and Environmental Sciences at the University of KwaZulu-Natal, South Africa. His main interests are urban green spaces and urban microclimate using optical, thermal and Radar datasets, remote sensing of above ground biomass, local and regional carbon modelling, remote sensing of rangeland nutritive value and crop yield, remote sensing of plant invasive species and crop disease modelling. His current projects include; Drone based high spatial resolution crop phenotyping, Data science for global health and climate change, Earth observation big data science, Ecosystem/urban heat island nexus using remotely sensed datasets, The integrated pest management strategy for food security in Eastern Africa (FAW-IPM) and the African reference laboratory (with satellite stations) for the management of pollinator bee diseases and pests for food security. Professor Odindi has been previously selected as one of the 52 African researchers with outstanding contribution to Geo-information and Earth Observation on the continent. He is a Brown International Advanced Research Institute (BIARI) fellow of the Watson Institute, Brown University, Rhode Island, USA.

Elhadi Adam

Elhadi Adam (Professor) holds a PhD in Environmental Science with a specific focus on hyperspectral remote sensing. His expertise lies in the applications of remote sensing in applied environmental science and precision agriculture. He is interested in vegetation mapping and monitoring by integrating field-based, airborne and space-borne hyperspectral remote sensors for detecting subtle patterns in natural vegetation and crops. He focuses on biochemical concentration, invasive species mapping, wetland health, herbaceous biomass distribution and characterization of heterogeneous landscapes. His current research project is developing a support information system to provide a broader range of agronomically-relevant information about crop biophysical status and leaf biochemistry using remote sensing techniques and machine learning algorithms. Prof Adam's current research also focuses on detecting and mapping disease on crops and quantifying forest fragmentation and its impact on biodiversity and ecosystem services.

Elfatih M. Abdel-Rahman

Elfatih M. Abdel-Rahman (Research Scientist) holds a PhD in Environmental Sciences with over 14 years of experience in working with multidisciplinary teams of Environmental Scientists, Agronomists, Entomologists, Plant Pathologists, Vector Scientists, Mathematical Modelers, Computer Scientists, Engineers, Geographers and Social Scientists. He uses remote sensing, Earth observation and geospatial modelling tools to improve the management of complex agro-natural and urban systems and to enhance crop risks monitoring, forecasting and early warning systems. This extensive experience, coupled with his background as an Agronomist and a university Associate Professor, gives him an excellent foundation to carry out various remote sensing and geospatial modelling tasks to develop advisory tools for improving the health of human, plant, animal, and environment. His current research and development projects focus on assessing the effect of landscape dynamics and climate change on agricultural pests, malaria and tsetse vectors, crop yield using satellite data, data science and machine learning algorithms. He is a member of the Editorial Advisory Board of ISPRS Journal of Photogrammetry and Remote Sensing and the Editorial Boards of the International

Journal of Tropical Sciences, and Frontiers in Multi- and Hyper-Spectral Imaging. Dr. Abdel-Rahman ranks among the best 10% scientists who apply remote sensing and GIS in agriculture in Africa. He is a Brown International Advanced Research Institute (BIARI) fellow of the Watson Institute, Brown University, Rhode Island, USA.

Yuyu Zhou

Yuyu Zhou is an Associate Professor in the Department of Geological and Atmospheric Sciences at Iowa State University, USA. He holds a Bachelor of Science degree in Geography and Master of Science degree in Remote Sensing from Beijing Normal University and a Ph.D. in Environmental Sciences from the University of Rhode Island. Dr Zhou's general focus is the quantification of spatiotemporal patterns of environmental change and developing modeling mechanisms to bridge the driving forces (both natural and socioeconomic factors) and consequences of environmental change so that the impacts of human activities on environment can be effectively measured, modeled, and evaluated. Specifically, his research focuses and interest include the application of remote sensing, GIS, Integrated Assessment Modelling, and spatial analysis in ecosystem, environmental, and social sciences, including interdisciplinary areas: Energy sources, demand, climate change impact, high spatial and temporal energy and emissions modelling, hydrology and ecosystem modelling. He was awarded a NASA Land-Cover Land-Use Change (LCLUC) Research Grant for Early Career Scientist to study global urban expansion in the context of climate change. He previously worked as a Geographical Scientist at the Joint Global Change Research Institute, a partnership of the Pacific Northwest National Laboratory (PNNL) and the University of Maryland.

Preface to "Thermal and Optical Remote Sensing: Evaluating Urban Green Spaces and Urban Heat Islands in a Changing Climate"

Urbanization, typified by land-use-land-cover transformation is a major cause of bio-physical, thermodynamic, surface energy and micro- and macro climate perturbations. These changes commonly result in environmental deterioration that in turn adversely affects bio-physical processes and quality of urban life. A major consequence of urbanization is the higher thermal values compared to the surrounding peri-urban and rural areas, causing the Urban Heat Island (UHI) effect. In recent decades, above-average heat during summer has become prevalent in global cities, a trend that is expected to continue. It is anticipated that the intensifying UHI effect, in concert with increasing anthropogenic activities, will exacerbate the vulnerability of urban landscapes to climate-related disasters such as floods and heatwaves. Hence, UHIs have become an invaluable theme in environmental research. A recent proliferation of optical and thermal remotely sensed datasets offers great potential for understanding the relationship between the urbanization processes and their respective bio-physical and climatic implications. This book focuses on the theoretical principles and practical adoption of remote sensing approaches and datasets in understanding the nexus between urbanization, natural landscapes, urban micro-climate, and climate change. This book provides a basis for understanding urban ecological and natural patterns, critical for the management of urban physical, ecological and social processes. Specifically, understanding past, current, and future Land Surface Temperature (LST) patterns and drivers is critical for, among others, urban environmental management, urban spatial planning, the optimal and sustainable use of urban landscapes and climate change mitigation. The book's first two chapters explore the potential of downscaling remotely sensed data and improved feature extraction to determine the effect of urban surface types on thermal characteristics. Chapter one adopts a Step-by-Step Random Forest Downscaling-Morphological (SSRFD-M) model to relate natural surfaces to LST, while chapter two proposes the absolute and relative indicators for the detailed derivation of landscape features and thermal values using Geofen 2 (GF-2) and Landsat 8 Thermal Infra-Red (TIR), respectively. Chapter three to five adopt the standardized Local Climate Zone (LCZ) typology to relate urban landscape feature types to thermal characteristics. Chapter three and four use the World Urban Database and Access Portal Tool (WUDAPT) and the LCZ framework in South Africa (Cape Town, Thohoyandou and East London) and Zimbabwe (Bulawayo), respectively, while chapter five relates seasonal LCZ to daytime LST in Riyadh, Saudi Arabia. Chapter six to nine investigate multi-city urban LULC and the contribution of the climate, urbanization and CO_2 to UHI in multiple cities. Chapter six proposes a Landsat imagery time series approach in Google Earth Engine platform to map built-up areas in 305 cities, while chapter seven compares Surface Urban Heat Island (SUHI) in relation to the SUHI fraction's key drivers in 305 Chinese cities. Chapter eight determines a critical competitive point of Artificial Surfaces (AS) and Urban Blue-Green Space (UBGS) on LST in 28 cities, while chapter nine relates the Global Land Surface Satellite (GLASS) Fractional Vegetation Cover (FVC) to CO_2, urbanization, and climate in 32 major cities. Chapter ten to twelve investigate a range of UHI-mitigation approaches. Chapter 10 explores the value of urban green spaces in mitigating UHI in Marrakesh, Morocco, while chapter eleven investigates the role roof colours in assimilating surface temperature. Using vegetation's morphological Spatial Patter Analysis (MSPA), chapter twelve adopts ArborCam^{TM} multispectral high-resolution imagery to determine the role of golf courses in assimilating urban LST. This book should be of interest to both specialists and generalists interested in, among others, urban planning, ecological conservation, the urban micro-climate, atmospheric science, environmental management, and climate change.

John O. Odindi, Elhadi Adam, Elfatih M. Abdel-Rahman, and Yuyu Zhou
Editors

Article

Step-By-Step Downscaling of Land Surface Temperature Considering Urban Spatial Morphological Parameters

Xiangyu Li [1], Guixin Zhang [2,*], Shanyou Zhu [1] and Yongming Xu [1]

1 School of Remote Sensing and Geomatics Engineering, Nanjing University of Information Science and Technology, Nanjing 210044, China; lxyxzs@nuist.edu.cn (X.L.); zsyzgx@nuist.edu.cn (S.Z.); xym30@nuist.edu.cn (Y.X.)
2 School of Geographical Sciences, Nanjing University of Information Science and Technology, Nanjing 210044, China
* Correspondence: 001631@nuist.edu.cn

Abstract: Land surface temperature (LST) is one of the most important parameters in urban thermal environmental studies. Compared to natural surfaces, the surface of urban areas is more complex, and the spatial variability of LST is higher. Therefore, it is important to obtain a high-spatial-resolution LST for urban thermal environmental research. At present, downscaling studies are mostly performed from a low spatial resolution directly to another high resolution, which often results in lower accuracy with a larger scale span. First, a step-by-step random forest downscaling LST model (SSRFD) is proposed in this study. In our work, the 900-m resolution Sentinel-3 LST was sequentially downscaled to 450 m, 150 m and 30 m by SSRFD. Then, urban spatial morphological parameters were introduced into SSRFD, abbreviated as SSRFD-M, to compensate for the deficiency of remote-sensing indices as driving factors in urban downscaling LST. The results showed that the RMSE value of the SSRFD results was reduced from 2.6 °C to 1.66 °C compared to the direct random forest downscaling model (DRFD); the RMSE value of the SSRFD-M results in built-up areas, such as Gulou and Qinhuai District, was reduced by approximately 0.5 °C. We also found that the underestimation of LST caused by considering only remote-sensing indices in places such as flowerbeds and streets was improved in the SSRFD-M results.

Keywords: step-by-step downscaling of LST; land surface temperature; urban spatial morphology

Citation: Li, X.; Zhang, G.; Zhu, S.; Xu, Y. Step-By-Step Downscaling of Land Surface Temperature Considering Urban Spatial Morphological Parameters. *Remote Sens.* **2022**, *14*, 3038. https://doi.org/10.3390/rs14133038

Academic Editors: Yuyu Zhou, Elhadi Adam, John Odindi and Elfatih Abdel-Rahman

Received: 21 May 2022
Accepted: 21 June 2022
Published: 24 June 2022

Publisher's Note: MDPI stays neutral with regard to jurisdictional claims in published maps and institutional affiliations.

1. Introduction

As an important physical variable driving the energy exchange between the surface and the atmosphere, the surface temperature (LST) is one of the key parameters for studying the energy balance of the surface at global or regional scales. Currently, LST is widely used to assess surface moisture and drought levels [1–4], calculate urban heat island intensity [5–7] and simulate surface energy exchange [8–11]. In urban areas, the spatial and temporal heterogeneity of urban surface temperature is obvious due to the extremely complex surface, the strong differences in three-dimensional spatial geometry and the variety of surface components and types. Therefore, studies of the urban thermal environment and other urban-related research fields usually require LST data with a higher spatiotemporal resolution.

The LST data obtained by thermal infrared remote-sensing technology generally have the problem of conflicting spatial and temporal resolutions. For example, the Moderate Resolution Imaging Spectroradiometer (MODIS) LST product has a high temporal resolution, but the spatial resolution is only 1000 m; Sentinel-3 operates through a binary orbit with a temporal resolution of fewer than 0.9 days, but the spatial resolution of the LST product is also 1000 m; the land surface temperature product retrieved from Landsat TIRS has a spatial resolution of 100 m, but the revisit period is as long as 16 days. Together with the influence of clouds, the available valid Landsat LST data are even further diminished.

High-temporal-resolution data are difficult to generate for refined surface temperature studies at an urban scale due to their low spatial resolution, while the high spatial resolution LST data are unable to study the variation pattern of LST in time due to their low temporal resolution. To solve the contradiction of spatial and temporal resolutions of remote-sensing thermal data, scholars have proposed a large number of technical methods from various perspectives, such as image processing and statistical regression, to obtain land surface temperature data with high spatial and temporal resolutions.

The statistical regression method has gained wide application in LST downscaling studies due to its low computational complexity and satisfactory downscaling results. The application of this method has become relatively mature in suburban and mountainous areas with simple land covers at a large spatial scale [12,13]. However, there are two problems that cannot be ignored when applying the statistical method to urban areas with complex land surface types. Firstly, the larger the spatial resolution span of the downscaling, the lower the accuracy tends to be. From the available thermal infrared remote-sensing data, there are lots of LST products with a higher temporal resolution at about 1000-m spatial resolution. When they are downscaled to the 100-m level or even the 10-m level, the spatial resolution of the downscaled LST differs from the original resolution by a factor of 10 or even 100, and the downscaled accuracy decreases as the spatial resolution increases. The main reason for this problem is that the assumption of a "constant spatial scale relationship" between LST and the driving factor does not hold when the resolution difference is large. Secondly, the traditional two-dimensional remote-sensing spectral indices and surface parameters are not sufficient to accurately describe the spatial pattern of an urban surface. Currently, commonly used remote-sensing indices for downscaling models, such as the normalized difference vegetation index (NDVI), normalized difference moisture index (NDMI), normalized difference water index (NDWI) and normalized difference building index (NDBI) [14] use surface parameters including the DEM, slope, slope direction, latitude, longitude and surface cover type [15,16], as well as multispectral data [17] describing the vegetation cover, moisture status and topographic relief of the land surface from the two-dimensional perspective. In contrast, cities are dominated by buildings and impervious pavements, but the influence of the three-dimensional morphological structure on local land surface temperature is less considered. In fact, a large number of studies have demonstrated that urban spatial morphological parameters such as the sky view factor (SVF), frontal area index (FAI) and building density (BD) are closely related to LST [18–21], meaning they need to be considered in downscaling models.

To address the above two problems, this study aimed to develop a step-by-step LST downscaling method by further considering urban spatial morphological parameters to obtain the urban land surface temperature at a spatial resolution of 30 m with high accuracy. The paper takes the central urban area of Nanjing, Jiangsu Province, China as the study area, and selects multi-source remote-sensing data, three-dimensional spatial distribution data of urban buildings to downscale the Sentinel-3 LST with the spatial resolution of 1000 m to the resolutions of 450, 150 and 30 m step-by-step using surface 2D and 3D parameters as driving factors. The downscaling results are evaluated by Landsat TIRS LST at the resolution of 30 m. Then, the influence of urban spatial morphological parameters on land surface temperature downscaling is discussed. The step-by-step LST downscaling method changes the traditional studies that directly downscale LST from a low spatial resolution to a high one, selecting several spatial resolutions for the intermediate downscaling process. The intermediate downscaling process is equivalent to supplementing the model with the land surface information and reducing the difference in spatial resolution, thus ensuring that the statistical regression model varies less with the spatial scale.

2. Research Review

A lot of research has been carried out on land surface temperature downscaling by scholars around the world. The main methods used for LST downscaling can be divided into two categories: image-based spatiotemporal fusion and kernel-driven downscaling methods.

The image fusion method obtains a high spatial and temporal resolution land of surface temperature by constructing a model to generate fused images, based on the combined weight of spectral, temporal and spatial information, by selecting similar images in the spatiotemporal neighborhood to be fused. Unlike the statistical downscaling method, the image fusion method does not directly model the relationship between land surface temperature and influencing parameters at low-spatial-resolution scales. Classical algorithms are as follows. Weng et al. [22] improved the STARFM model to establish the relationship between MODIS and TM radiometric brightness by linear spectral mixing analysis, and proposed a spatiotemporal adaptive fusion algorithm (SADFAT) for land surface temperature downscaling. Wu et al. [23] proposed a spatiotemporal integrated temperature fusion model (STITFM) for estimating high-spatiotemporal-resolution LST from multi-scale polar and geostationary orbiting satellite observations. The image fusion-based approaches assume that the radiative brightness for similar pixels behaves consistently at any spatial resolution, while in practice, the radiative brightness will inevitably vary in space and time. So, this approach generally performs poorly in urban areas with high-spatial-heterogeneity characteristics.

Kernel-driven downscaling methods can be classified into physical models and statistical regression methods according to whether the model is physically meaningful or not. Physical downscaling methods establish the relationship between LST and auxiliary data by using the physical mechanism of a mixture pixel and the thermal radiation principle. In this way, low-spatial-resolution pixels are decomposed to multiple subpixels to obtain the high-spatial-resolution LST. For example, L.J and Moore [24] developed the pixel block intensity modulation (PBIM) method to improve the spatial information in the low-spatial-resolution thermal infrared band by using multispectral data with a high spatial resolution. Nichol [25] proposed the emissivity modulation (EM) model to improve the spatial resolution of thermal radiation by using land surface emissivity and land cover data. Wang et al. [26] downscaled MODIS LST to a 30-m resolution based on the thermal decomposition equation. However, the design of physical models is usually difficult and the models are computationally complex and time-consuming.

The basic principle of a statistical regression method is to assume that the relationship between land surface temperature and driving factors does not change with the spatial scale. A statistical regression model is built using the low-spatial-resolution LST and the drivers, after which the high-spatial-resolution drivers are added to the model to predict the high-spatial-resolution LSTs. Up to now, the statistical regression method is the most widely used method in LST spatial downscaling studies. Based on the number of driving factors, statistical regression methods can be divided into single- and multi-factor models. For example, Distrad models used NDVI as the driver [27], and the TsHARP model used vegetation cover instead of NDVI [28]. In addition, Anthony et al. [29] developed a high-resolution urban thermal fusion (HUTS) technique to downscale Landsat TIRS to 30 m based on NDVI and surface albedo. Lacerda et al. [30] used the TsHARP model to downscale the MODIS LST to a high spatial resolution of 10 m. Vaculik et al. [31] downscaled the GOES-R LST with the resolution of 2000 m to 30 m by establishing a linear relationship between NDVI and LST. J.M. et al. [32] modified the TsHARP algorithm to downscale MODIS LST data covering one Spanish farm. However, single-factor models are only applicable to a region with high vegetation cover; they do not perform well in urban or arid areas, limited by the predictor variables. Multi-factor models with multiple remote-sensing indices and land surface parameters as driving factors were gradually proposed and applied. For example, Liu et al. proposed the G_Distrad model by adding NDBI and NDWI to the traditional Distrad model [14]. Pereira et al. [33] proposed a geographically weighted regression model (GWRK) by using NDVI and multispectral data to downscale the ASTER thermal infrared data for the natural regions and urban areas of Pantanal, Brazil. Considering the complex nonlinear relationships between land surface temperature and various geophysical parameters in urban areas [12,13], the three-layer structural (TLC) model [34], neural network [35], support vector machines [6], random forests [36] and other multivariate nonlinear statistical models have been continuously developed and applied to urban LST downscaling studies.

Random forest models have been widely used in urban LST downscaling studies in recent years because of their low model complexity, fast training speed and ability to effectively avoid overfitting problems. Li et al. [36] compared three machine learning algorithms, random forest (RF), support vector machine (SVM) and artificial neural network (ANN), to the traditional TsHARP method in both urban and suburban areas of Beijing, and found that the LST downscaling accuracy of the machine learning algorithm was higher than that of the TsHARP algorithm. Wang et al. [37] compared the downscaling results from a multiple linear regression model (MRL), TsHARP model and random forest (RF), and found that the RF model is more suitable for heterogeneous surfaces such as urban areas. Ebrahimy et al. [38] used an adaptive random forest regression method to downscale MODIS LST over Iran to 30 m in the GEE platform. Njuk et al. [39] proposed an improved downscaling method for low-resolution thermal data based on minimizing the spatial mean bias of random forest, and the results demonstrated that the method reduces the inherent mean bias in the LST downscaling process and is more suitable for LST down-scaling applications in complex environments. Here, we comprehensively analyzed most land surface temperature downscaling methods and built global models and assumed that the statistical relationships were spatially invariant, however, global models may produce large errors in local area applications. In recent years, many scholars have devoted their work to using local models to capture the spatial non-stationary characteristics of land surface variables, and then established the local relationships between LST and influencing factors to improve the accuracy of LST downscaling [15].

3. Materials and Methods

3.1. Study Area

The central urban region of Nanjing, Jiangsu Province, China was chosen as the study area because it contains a variety of underlying surface types such as buildings, vegetation and water bodies, which helps to carry out land surface temperature downscaling studies of complex ground cover types. Figure 1 presents a true-color image and building distribution map of the study area at a spatial resolution of 10 m.

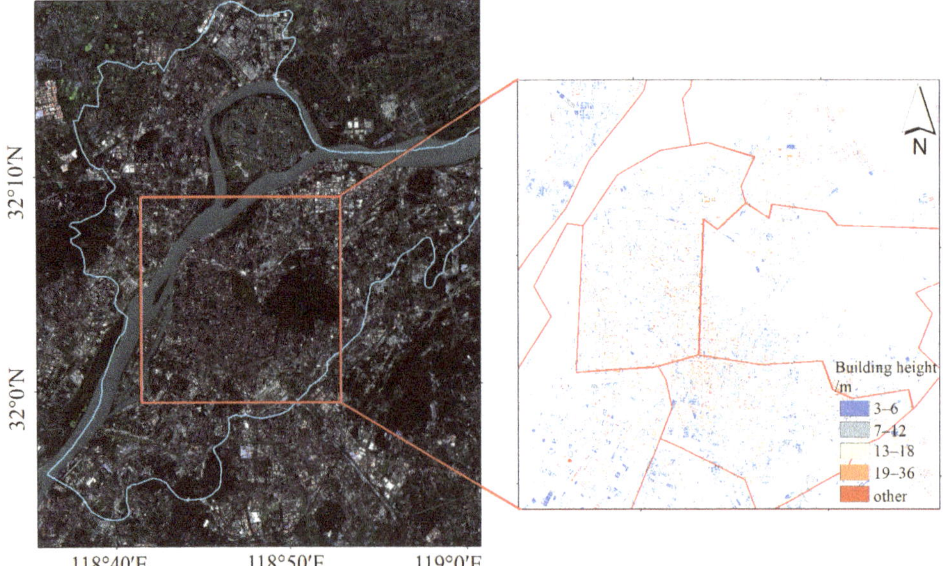

Figure 1. Sentinel-2 true-color composite image and building data of the study area (blue line represents the main urban area of Nanjing; red lines represent the study area boundaries).

The study area includes several urban administrative districts, including the Qixia, Xuanwu, Gulou, Qinhuai and Jianye Districts, with an area of approximately 18×18 km^2. The study area is located in the central region of the lower Yangtze River, with geographic coordinates between 31°14′N and 32°37′N and 118°22′E and 119°14″E. The total built-up area is approximately 823 km^2. Although the study area is near a hilly area, the topography is flat, and there are many low hills and gentle hills. Nanjing has a humid subtropical climate with four distinct seasons, abundant rainfall and significant temperature differences between winter and summer. The average annual precipitation is 1106 mm, and the average annual temperature is 15.4 °C. Nanjing had a population of 10,312,200 at the end of 2019, with a resident population of 8.5 million, including 7.072 million in urban areas, and an urbanization rate of 83.2%. Nanjing is one of the economic-center cities in China, with a regional GDP of $1.6 billion in 2021.

3.2. Data

The Sentinel-3 LST product at a 1000-m spatial resolution for downscaling and Sentinel-2 multispectral image data were downloaded from the ESA website (https://scihub.copernicus.eu/dhus/#/home, accessed on 14 May 2022). The Landsat LST product at a 30-m spatial resolution for validation of downscaling results was downloaded from the USGS website (https://earthexplorer.usgs.gov/, accessed on 30 December 2021). Sentinel-2 visible light and shortwave infrared bands (B2–B4, B8, B11 and B12) were used to calculate normalized remote-sensing spectral indices. Nanjing downtown building and wind data were used to calculate urban spatial morphological parameters. Nanjing wind data from 2016 to 2020 were used to calculate the wind direction frequency, which were downloaded from the China Air Quality Online Monitoring and Analysis Platform (https://www.aqistudy.cn/historydata/, accessed on 14 May 2022). The details of all the data involved in this study are presented as follows.

3.2.1. Landsat LST Product

Landsat LST products were generated by EROS based on a single-channel inversion algorithm, by using the Landsat C2L1 thermal infrared band and other ancillary data [40,41]. Landsat LST products were resampled from 100 to 30 m for release to users by EROS using the nearest-neighbor resampling method. The Landsat LST images used in this study were imaged at 10:37 a.m. on 4 October 2021, with orbital row/column numbers 120/038 [42].

3.2.2. Sentinel-3 LST Product

The Sentinel series is an Earth observation satellite mission of the European Copernicus program. Sentinel-3 carries a variety of payloads, such as OLCI (sea and land colorimeter) and SLSTR (sea and land surface temperature radiometer), which are mainly used for high-precision measurements of the sea surface topography, sea and surface temperatures, ocean water color and soil properties [43]. Both 3A and 3B satellites in orbit have revisit periods of less than one day for areas within 30° latitude of the equator. SLSTR has six solar reflection bands (S1–S6) and four thermal infrared bands (S7–S9, F1, F2) with spatial resolutions of 500 and 1000 m, respectively. The Sentinel-3 LST products are produced by a split-window algorithm using three bands of S7–S9 and other auxiliary data, and the products are accurate to 1 K. The LST product of Sentinel-3B SLSTR was selected for this study, with an imaging time of 10:04 am on 4 October 2021. The Sentinel LST was resampled to a spatial resolution of 900 m by using the bilinear interpolation method for downscaling in this study to match the reference LST with a spatial resolution of 30 m.

3.2.3. Sentinel-2 Multispectral Data

Sentinel-2 is a multispectral high-resolution imaging satellite with two satellites in orbit, 2A and 2B, with a revisit period of five days [44]. Each satellite carries a multispectral imager (MSI), which can cover 13 spectral bands with ground resolutions of 10, 20 and 60 m and an amplitude of 290 km. The blue, green, red, and near-infrared bands needed for this study are the B2, B3, B4 and B8 bands of the Sentinel-2 satellite, each with a resolution of 10 m. B11 and B12 are shortwave infrared bands with a resolution of 20 m, resampled to 10 m to match the visible bands.

3.2.4. Urban Building Data

The building data used in this study were provided by Urban Data Corps, obtained in around 2012. Urban Data Corps is ranked in the top 10 in the big-data field according to the 2017 China Big Data Development Report published by the National Information Center of China. Urban Data Corps can provide a variety of high-precision data for urban research.

The building vector data contain the polygon of the building distribution, building floor data and building height data in a shapefile format with the WGS-84 coordinate system. Comparing the urban building distribution data with satellite images in 2012, we found that the building location and outline highly overlap with the satellite images, and the building floor number is also very consistent with the field survey results, which indicates the high accuracy of the building distribution data. The field survey found that the ground cover types in most of the study areas, such as Gulou District and Qinhuai District, did not change significantly between 2012 and 2021, except for Qixia District. In this study, the building vector data in the shapefile format were firstly converted to raster data in the TIF format with a spatial resolution of 10 m. After that, the urban spatial morphological parameters were calculated based on the building raster data and other auxiliary data using corresponding formulas.

3.3. Calculation of the Downscaling Driving Factor

3.3.1. Calculation of the Remote-Sensing Spectral Index

The L2A-level surface reflectance data from Sentinel-2B were selected for this study to calculate remotely sensed spectral indices that are closely related to surface temperature, including the modified normalized difference water index (MNDWI), normalized difference building index (NDBI), normalized difference built-up and soil index (NDBSI) [28], normalized difference moisture index (NDMI), normalized difference vegetation index (NDVI) and soil adjusted vegetation index (SAVI). The calculation process was performed using SNAP 8.0, a professional piece of software for data-processing in the Sentinel series. The calculation formula is shown in Table 1.

Table 1. Remote-sensing spectral indices required for downscaling and calculation formulas.

Var.	Description	Equations
MNDWI	Improve the normalized difference water body index to highlight water body information	$MNDWI = \frac{\rho 3 - \rho 12}{\rho 3 + \rho 12}$
NDBI	Normalize the difference building index to highlight building information	$NDBI = \frac{\rho 12 - \rho 8}{\rho 12 + \rho 8}$
NDBSI	Indicate the degree of dryness of the ground surface [45]	$IBI = \frac{\frac{2 \times \rho 11}{(\rho 11 + \rho 8)} - \left(\frac{\rho 8}{(\rho 8 + \rho 4)} + \frac{\rho 3}{\rho 3 + \rho 11} \right)}{\frac{2 \times \rho 11}{(\rho 11 + \rho 8)} + \left(\frac{\rho 8}{(\rho 8 + \rho 4)} + \frac{\rho 3}{\rho 3 + \rho 11} \right)}$ $SI = \frac{(\rho 11 + \rho 4) - (\rho 8 + \rho 2)}{(\rho 11 + \rho 4) + (\rho 8 + \rho 2)}$ $NDBSI = \frac{IBI + SI}{2}$
NDMI	Indicate the vegetation moisture	$NDMI = \frac{\rho 8 - \rho 11}{\rho 8 + \rho 11}$
NDVI	Highlight vegetation information	$NDVI = \frac{\rho 8 - \rho 4}{\rho 8 + \rho 4}$
SAVI	Reduce the sensitivity of vegetation indices to changes in reflectance of different soils	$SAVI = \frac{\rho 8 - \rho 4}{\rho 8 + \rho 4 + L} \times (1 + L)$ $L = 0.5$

Notes: $\rho 1 - \rho 12$ denote the surface reflectance of Sentinel-2 bands 1–12.

3.3.2. Calculation of Urban Spatial Morphological Parameters

The building vector data of Nanjing were converted to raster data with a 10-m resolution to calculate urban spatial morphological parameters including the building density (BD), frontal area density (FAD), floor area ratio (FAR), mean height (MH) and sky view factor (SVF). The SVF was calculated by the raster algorithm. The influence of buildings within a radius of 100 m was considered when calculating the SVF of each pixel. FAD was calculated using the building raster data and the wind direction frequency data by a self-developed raster algorithm with a plot area of 100×100 m^2. Other spatial morphological parameters, such as BD, FAR and MH, were calculated using 10-m building raster data and the corresponding equations in Table 2 with a plot area of 100×100 m^2.

Table 2. Urban spatial morphological parameters required for LST downscaling.

Var.	Description	Equations
BD	Building Density	$$BD = \frac{\sum_{i=1}^{n} A_i}{A_T}$$ Ai indicates the *i*th building area and AT indicates the calculated plot area
FAD	Frontal Area Density	$$\lambda_{f(z)} = \sum_{i=1}^{n} \frac{A(\theta)_{proj(z)}}{A_T} \times P_{\theta,i}$$ $\lambda_{f(z)}$ indicates the weighted frontal area density (FAD); $A(\theta)_{proj(z)}$ indicates the projected area at a certain height z in the wind direction θ, $P_{\theta,i}$ indicates the frequency of the wind direction θ, $i = 1, \ldots, 16$
FAR	Floor Area Ratio	$$FAR = \frac{\sum_{i=1}^{n} F_i \times A_i}{A_T}$$ F_i indicates the *i*th building floor number and A_i indicates the *i*th building area
MH	Mean Height	$$MH = \frac{\sum_{i=1}^{n} H_i}{n}$$ H_i indicates the *i*th building height and n indicates the number of buildings
SVF	Sky View Factor	$$\Psi_{sky} = 1 - \sum_{i=1}^{360/\alpha} sin^2\beta \times (\alpha/360)$$ $$\beta = tan^{-1}(H/X)$$ ψ_{sky} indicates SVF, β indicates the building height angle, H indicates the building height, X indicates the calculated radius and is set to 100 m in this study

Note: Except for SVF, the plot area AT calculated by all other parameters takes the value of 100×100 m.

3.4. Downscaling LST Method Based on Random Forest

The core idea of surface temperature downscaling is the invariance of the relationship between LST and driving factors at different spatial resolutions so that the statistical relationship between LST and the regression kernel at a low resolution can be applied to a high spatial resolution to complete the downscaling process. The specific formulas are as follows:

$$T_c' = f(var_c) \tag{1}$$

$$\Delta T = T_c - T_c' \tag{2}$$

$$T_f' = f\left(var_f\right) + \Delta T \tag{3}$$

where var_c denotes the low-resolution explanatory variable, var_f denotes the high-resolution explanatory variable, Tc denotes the low-resolution LST, T_c' denotes the predicted low-resolution LST, ΔT denotes the simulation residual and T_f' denotes the predicted high-resolution LST.

This study used the random forest algorithm to construct the LST downscaling model. Random forest is an integrated decision tree-based learning algorithm proposed by Breiman in 2001 as a supervised learning algorithm [32]. The algorithm uses the bootstrap resampling method to randomly select samples from the training sample set. The extracted bootstrap samples are trained separately for each decision tree, and an algorithm for randomly selecting a subset of features is introduced in the process of splitting the nodes of the decision tree. The prediction results of each decision tree are counted and voted on to obtain the final prediction results of the input data. The random forest algorithm has stronger noise immunity than other algorithms because of the introduction of randomly selected training samples and randomly selected feature subsets that make the correlation among each decision tree smaller. The random forest algorithm is better at handling non-linear problems than traditional statistical regression algorithms. As long as the number of decision trees is sufficient, the random forest algorithm can effectively avoid the overfitting problem, and the training speed is faster. The random forest algorithm can examine the interaction between features during the training process and output the feature importance, which is a reference for analyzing the degree of influence of features. In this study, the experimental dataset was divided into training and validation datasets according to the ratio of 8:2. The main parameters that need to be set manually to build a random forest downscaling model using the Pycharm platform include the number of decision trees (n_estimators) and the maximum number of features (max_features). n_estimators refers to the number of decision trees built in the random forest, which was set to 700 after testing in this study. max_feature refers to the maximum number of features selected when building each decision tree, which was set to log(n_estimators) in this study. Other parameters were set to default values.

3.5. Step-By-Step Random Forest Downscaling Method (SSRFD)

When the spatial resolution spans a larger range, the downscaling results cannot accurately characterize the spatial distribution of LST, which tends to underestimate the surface temperature in building areas and overestimate it for water bodies and vegetated areas. This study proposed a step-by-step downscaling LST method based on the random forest model (SSRFD), which achieves a significant increase in the spatial resolution of LST without excessive loss of accuracy through multiple, small-scale spatial resolution downscaling processes. During the SSRFD model's work, each intermediate downscaling adds additional and finer surface information to the model. In this way, models are trained to accurately express the relationship between land surface temperature and driving factors.

In this study, the direct random forest downscaling (DRFD) method was performed to directly downscale 900-m Sentinel-3 LST to a 30-m resolution, and then two downscaling methods were conducted using the proposed SSRFD method. The first method, named SSRFD, downscaled the 900-m Sentinel-3 LST to 30 m after 450 m and 150 m sequentially, where the SSRFD was driven by the six remote-sensing indices mentioned above. The second method, named SSRFD-M, downscaled 900-m Sentinel-3 LST to 30 m by the same process through SSRFD, where five additional urban spatial morphological parameters were added as the SSRFD driver factors. After that, LST downscaling results of DRFD, SSRFD and SSRFD-M were compared at a 30-m spatial resolution, which were all evaluated by using the 30-m Landsat-8 LST as the reference. Moreover, the influence of urban spatial morphological parameters on LST was analyzed based on SSRFD-M at a 30-m spatial resolution.

3.6. Accuracy Evaluation Methods

Pearson's correlation coefficient (r), the mean absolute error (MAE) and root mean square error (RMSE) were used to comprehensively evaluate the downscaling results. Meanwhile, the maximum/minimum, mean (Mean) and standard deviation (SD) were calculated to evaluate the spatial variability characteristics of LST images before and after downscaling. The SD can reflect the spatial variability of thermal features.

4. Results

4.1. Comparison of the Results Obtained with SSRFD and DRFD

To reduce data differences caused by different sensors and LST inversion algorithms and increase the comparability and verifiability between Landsat-8 and Sentinel-3 LST products, a simple linear correction was applied to Sentinel-3 LST before the downscaling work. After performing the linearity correction, the maximum, minimum and mean values of Sentinel-3 LST changed from 38.80, 27.57 and 35.18 °C to 41.25, 30.84 and 37.90 °C, respectively, which were closer to Landsat LST in the range of values. The RMSE of the two LST products changed from 3.22 to 1.59 °C, indicating that the systematic differences between Sentinel-3 LST and Landsat LST were reduced to some extent.

Comparative plots of the downscaling results are given in Figure 2, where the 900-m Sentinel-3 LST was downscaled to 30 m using the DRFD method and SSRFD method. Comparing Figure 2b,c with Figure 2a, both results captured finer spatial discrepancy characteristics and texture features, and the resulting LST distribution is basically consistent with Landsat LST. However, according to Figure 2b, there are large areas of high-temperature regions in the study area. The results obtained with DRFD as a whole are significantly overestimated. For example, the regions along the northwestern coast of the Yangtze River, Xuanwu Lake and Zijinshan Mountain show obvious temperature overestimation errors. The difference characteristics between the high-temperature region and the sub-high-temperature region are less clearly expressed than Landsat LST in Figure 2a. In contrast to Figure 2b, the results obtained with SSRFD (Figure 2c) captured the spatial distribution differences and textural characteristics more accurately. The distribution characteristics of both the building high-temperature zone and the water and vegetation low-temperature zone are in good agreement with Figure 2a. Overall, the results obtained with DRFD show an overall underestimation of the high-temperature region and an overestimation of the low-temperature region, which cannot accurately depict the spatial distribution pattern of LST in the study area.

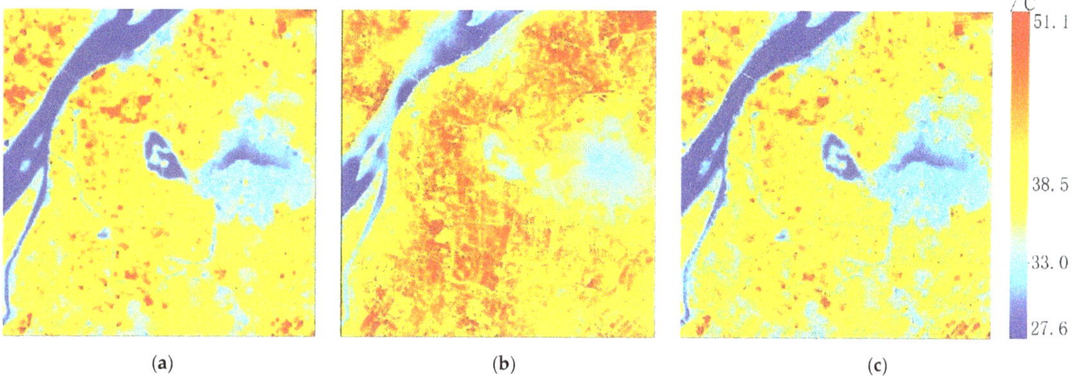

Figure 2. Spatial distribution of LST at 30-m spatial resolution. (**a**) Landsat reference LST. (**b**) Downscaled LST results of DRFD. (**c**) Downscaled LST results of SSRFD.

The statistics (Table 3) show that the results from DRFD, with an SD of 2.64, are 0.82 lower than Landsat LST, but their mean value is 1.41 °C higher than Landsat LST. This is consistent with the performance of the DRFD results in Figure 2b, which further illustrates the overall high surface temperature predicted by DRFD. In comparison, the maximum, mean and SD of the results from SSRFD are 50.76, 36.78 and 3.73 °C, respectively, which only differ from the corresponding index of Landsat LST by approximately 0.3 °C. In summary, the downscaling results of SSRFD are more accurate, which is also demonstrated in Figure 3.

Table 3. Statistical values of LST downscaled from DRFD, SSRFD methods and Landsat reference LST data at a 30-m resolution.

Statistical Variables	Reference LST/°C	DRFD LST/°C	SSRFD LST/°C
Maximum	51.06	42.80	50.76
Minimum	13.94	28.95	24.36
Mean	36.50	37.91	36.78
Standard deviation	3.46	2.64	3.73

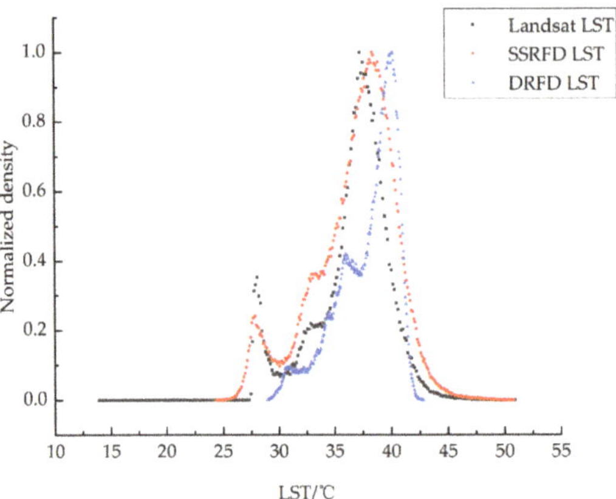

Figure 3. Histogram of downscaled LST and Landsat LST at 30-m resolution (black cubes refer to Landsat LST, red circles refer to downscaled LST obtained by the SSRFD method, blue triangles refer to downscaled LST obtained by the DRFD method).

According to Figure 3, The histogram curves of SSRFD results fit better with that of the Landsat LST, as they both have clear "peak" values between 27.5–29 and 37–39 °C, which indicates that the SSRFD results are reasonable overall. The DRFD results differ significantly from the Landsat LST in terms of histogram shape characteristics, data distribution interval and data value range, e.g., the "peaks" of the results from DRFD are distributed between 40 and 41 °C. Overall, Figure 3 shows that the dense temperature interval of the image element distribution of DRFD is higher than that of SSRFD LST and Landsat LST.

Furthermore, the corresponding scatterplots of the two downscaling results from SSRFD and DRFD with Landsat LST are given in Figure 4a,b, respectively. According to Figure 4, the correlation r value between the DRFD results and Landsat LST is 0.6, while the SSRFD results improve this to 0.81. The MAE and RMSE values of the DRFD results are 2 and 2.6 °C, respectively, while the SSRFD results decrease to 1 and 1.66 °C, respectively. This indicates that using the SSRFD method can obtain a higher-accuracy LST than the DRFD method when the spatial resolution spans a large range.

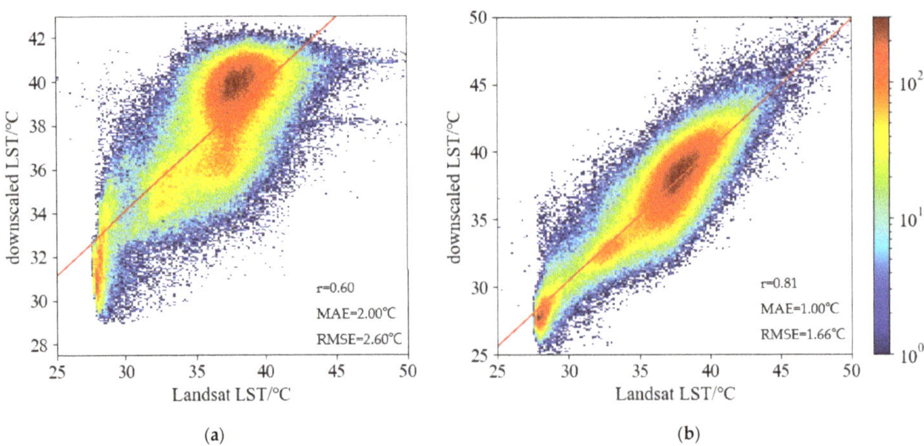

Figure 4. Scatterplots of the correlation analysis between downscaled LST and Landsat reference LST at a 30-m resolution. (**a**) Downscaled LST of Sentinel-3 from DRFD method. (**b**) Downscaled LST of Sentinel-3 from SSRFD method.

4.2. Influence of Urban Spatial Morphological Parameters on Downscaling LST

4.2.1. Analysis of the Overall Downscaling Results in the Study Area

To evaluate the influence of urban spatial morphological parameters on LST downscaling, the five spatial morphological parameters calculated in Table 2 were introduced into the driving factors of SSRFD to downscale Sentinel-3 LST from a spatial resolution of 900 to 30 m. The downscaling results were also validated by Landsat LST.

Figure 5a,b shows the correlation plots of Landsat LST with the results from SSRFD and SSRFD-M, respectively, at a 30-m spatial resolution, where r improves from 0.81 to 0.85 and RMSE decreases from 1.66 to 1.44 °C after adding the spatial morphological parameters. The statistical histograms of Landsat LST and downscaling results are given in Figure 6. Compared to the SSRFD result, SSRFD-M is more matched with Landsat LST in distribution shape, especially between 35 and 40 °C where buildings and concrete pavements are mainly distributed. Otherwise, the features with temperatures between 28 and 35 °C are mainly vegetation and water bodies. The curves of the two downscaling results in this interval are higher than Landsat LST, indicating that there may be some LST overestimation in downscaling results for vegetation and water body areas. Combined with the analysis of Figure 5, we conclude that the SSRFD-M model performs better than SSRFD, especially for dense building areas.

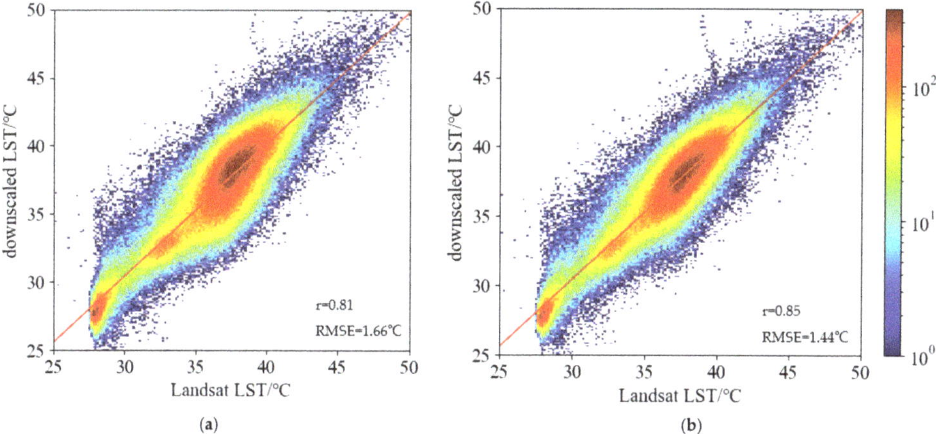

Figure 5. Scatterplots of the correlation analysis between downscaled LST and Landsat LST on the overall region at a 30-m resolution. (**a**) SSRFD result. (**b**) SSRFD-M result.

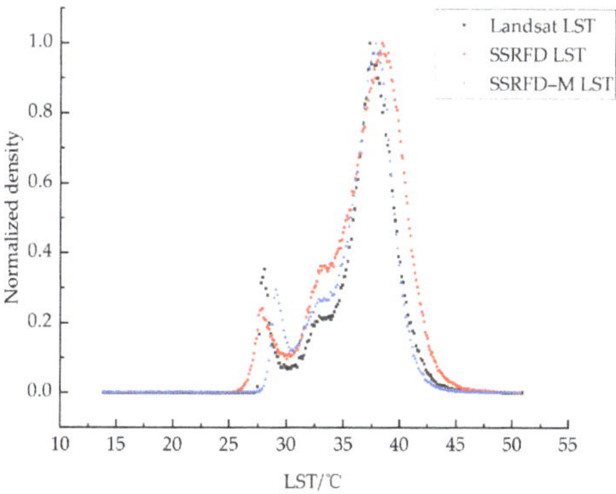

Figure 6. Histogram of downscaled LST and Landsat LST at a 30-m resolution (black cubes refer to Landsat LST, red circles refer to the SSRFD results and blue triangles refer to the SSRFD-M results)

4.2.2. Analysis of Regional Downscaling Results in the Study Area

In Section 4.2.1, we found that the spatial morphological parameters have some favorable effects on the LST downscaling, especially for building areas. However, we remained unaware of how the urban spatial morphological parameters affect the LST downscaling results for different locations within the study. In this section, we intend to discuss the downscaling results for five subregions at a 30-m resolution to further analyze the role of urban spatial morphological parameters in the downscaling process. The distribution of correlations between the downscaling results and Landsat LST for five subdistricts, Qixia, Gulou, Xuanwu, Qinhuai and Jianye, are given in Figure 7. The influence of urban spatial morphological features on LST downscaling can be further verified because of the complexity of the urban ground cover types considered.

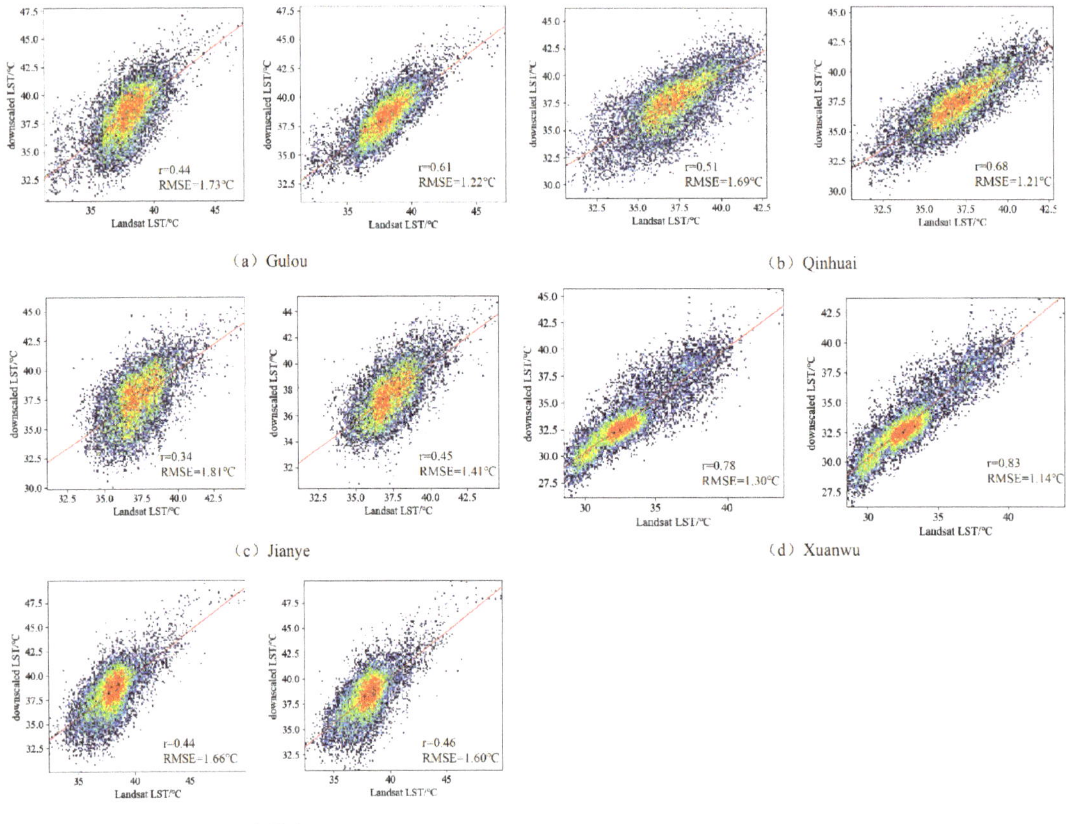

(a) Gulou

(b) Qinhuai

(c) Jianye

(d) Xuanwu

(e) Qixia

Figure 7. Scatterplot of correlation analysis between downscaled LST and Landsat LST at a spatial resolution of 30 m (downscaled LSTs obtained from SSRFD and SSRFD-M correspond to the left and right panels in **a–e**).

Figure 7a–c shows that the r value of the SSRFD-M downscaling results for the Gulou, Qinhuai and Jianye Districts improves from 0.44, 0.51 and 0.34 to 0.61, 0.68 and 0.45, respectively, while the RMSE value decreases from 1.73, 1.69 and 1.81 °C to 1.22, 1.21 and 1.41 °C, respectively. According to the statistical analysis of different areas, all three areas are located in a dense area of middle-rise buildings [33], where buildings and impervious surfaces dominate and the vegetation distribution is relatively low and sparse. Therefore, the LST distribution is closely related to the spatial morphological characteristics of buildings. These figures all exhibit that the downscaling results of SSRFD underestimated the LST for some regions between 35 and 40 °C. According to Figure 7d, r and RMSE values changed from 0.78 and 1.30 °C to 0.83 and 1.14 °C before and after considering spatial morphological parameters in Xuanwu District, respectively, with a slightly smaller improvement in accuracy relative to the Gulou and Qinhuai Districts. Statistically, among the Xuanwu District covered by the study area, non-built-up areas such as Zhong Shan Scenic Area and Xuanwu Lake account for approximately 50% of the district. Therefore, the inclusion of spatial morphological parameters has less influence on the downscaling results of these areas. If the downscaling results of building areas in Xuanwu District are counted separately, the RMSE of SSRFD-M decreases from 1.96 to 1.17 °C compared to the SSRFD results, which proves that SSRFD-M can effectively improve the downscaling effect in dense building areas. Figure 7e indicates that the SSRFD-M results for Qixia only improved/decreased r and RMSE values by 0.02/0.06 °C, respectively, compared to SSRFD. The reason for this is mainly the difference in years between Sentinel-3 LST data and building data. The type of land surface cover in some regions of Qixia has changed significantly. For example, the eastern side of Ningluo Expressway and the northern side of Qixia Avenue have changed from natural surfaces to building and road types. The building data used cannot accurately express the spatial morphological characteristics of these areas and cannot effectively improve the accuracy of LST downscaling. In addition, we found that a spatial resolution of 30 m may not be sufficient to show the surface temperature distribution pattern inside complex building areas.

We further selected five building-dense areas near the western side of Xuanwu Lake, Xinjiekou, Nanjing Museum, Nanjing Forestry University and the Olympic Sports Center, for downscaling results comparison, as shown by A, B, C, D and E in Figure 8a, respectively. A comparative analysis of the downscaling results with and without including spatial morphological parameters was carried out, and the results are shown in Figure 8b. When comparing the local downscaling results of SSRFD with SSRFD-M, it can be found that the regional LST of the vegetation-covered regions in the built-up area changed after considering the spatial morphological parameters. According to the SSRFD results of A1–E1, vegetation-covered areas near buildings, such as streets planted with trees and flowerbeds, showed a clear low-temperature zone (approximately 30–33 °C), which was 5–8 °C lower than the surface temperature of nearby building areas (approximately 35–40 °C). In contrast, the temperature in the corresponding regions illustrated by A2–E2 was only approximately 3–5 °C lower than that of the nearby buildings. No obvious low-temperature regions appeared in A2–E2, which were more consistent with the Landsat LST. Therefore, this study infers that the underestimation of LST generated by the SSRFD using only remote-sensing spectral indices was partially eliminated in SSRFD-M.

4.3. Parameter Importance Analysis of LST Downscaling

The importance of each driver at 90-, 450-, 150- and 30-m resolutions calculated by the random forest model is shown in Figure 9. According to Figure 9a, the NDBI has the highest importance at a 900-m resolution without spatial morphological parameters, which indicates that the NDBI is significantly correlated with LST in urban areas. With the increase in spatial resolution, the importance of the NDBI tends to decrease, and it drops to the lowest value (25%) at a 30-m resolution. Vegetation moisture is an important factor affecting LST in urban areas at any spatial resolution because the importance of the NDMI is maintained at 28–30% as the spatial resolution changes. The importance of the MNDWI increases from 11 to 15%, and the NDVI increases from 8% to 11%, which indicates that the contribution of water bodies and vegetation to LST increases as the resolution increases. The reason could be that some smaller lakes, green areas and narrow rivers are identified at a high spatial resolution. Figure 9b shows that the overall importance of the remote-sensing index did not change significantly in the SSRFD. Among the spatial morphological parameters, the importance of BD, FAD, FAR and MH decreased with increasing spatial resolution, while the SVF increased from approximately 3% to approximately 6%. The reason for this is that the regional spatial morphological parameters, such as BD, FAD and FAR, calculated by a single size (100 × 100 m), cannot represent finer architectural information. Therefore, the regional spatial morphological parameters are less relevant to LST as the resolution increases. As shown in Figure 9a,b, compared to other remote-sensing indices (all variations were in a range of 1–3%), there was a significant decrease in the importance of the NDBI at lower resolutions of 900 and 450 m, from 30 and 29% in Figure 9a to 22 and 23% in Figure 9b, respectively. The spatial morphological parameters at a low resolution to some extent compensated for the deficiency of the NDBI in the description of the spatial morphological features of buildings.

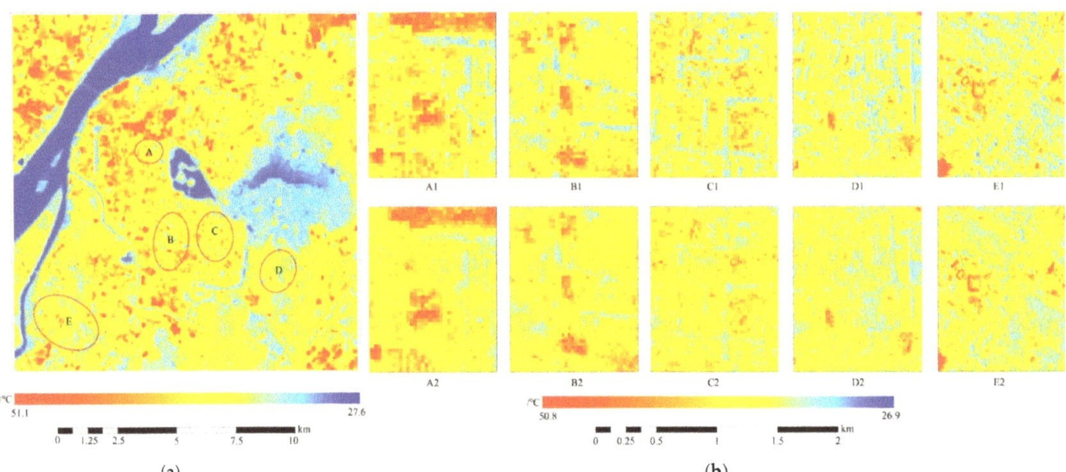

Figure 8. Spatial distribution of the downscaled LST at the resolution of 30 m in five localities of the study area (in **b**, A1–E1 refer to the downscaled LST from SSRFD, and A2–E2 refer to the downscaled LST from SSRFD-M). (**a**) Landsat LST. (**b**) the downscaled LST for representative 5 local area.

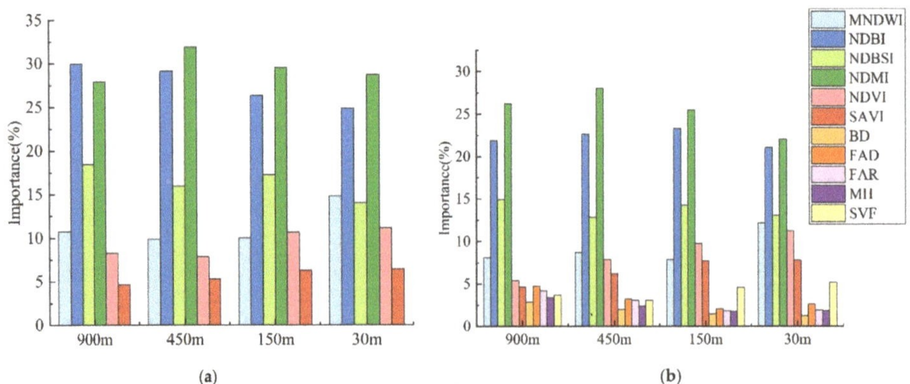

Figure 9. Importance of each driver at spatial resolutions of 900, 450, 150 and 30 m. (**a**) No spatial morphological parameters added. (**b**) Spatial morphological parameters added.

5. Discussion

With respect to the scale effect of the land surface temperature downscaling model, Pu et al. [46,47] concluded that the "constant scale relationship" between LST and driving factors does not hold under certain conditions. Comparing Figure 2a with Figure 2b for the main urban area of Nanjing, when the spatial resolution spans 30 times, the direct downscaling results cannot accurately represent the spatial distribution of land surface temperature, and the root mean square error of the traditional DRFD model is 2.6 °C (Figure 4a). This is similar to LST downscaling results in different regions such as that covered by Njuki et al. [39] in Kenya, Valdes et al. [13] in an arid Antarctic river valley, Zhu et al. [15] in Beijing and Ebrahimy et al. [38] in Iran. When the Sentinel-3 LST was downscaled to 450, 300, 150 and 30 m using the DRFD method, respectively, the research process revealed a phenomenon where r gradually decreases while RMSE increases as the spatial resolution span increases step-by-step, which is consistent with Zhu et al. (2020) [14] and Cao (2020) [48]. Using the step-by-step downscaling model (SSRFD), the root mean square error of the LST downscaling for the whole study area decreases to 1.66 °C, and the accuracy improves by about 1 °C (Figure 4b). Tang et al. [16] conducted a second downscaling procedure using the LST spatial features extracted from the initial downscaling results and obtained a higher accuracy of downscaling result, similar to the results of this study. The step-by-step downscaling method reduces the spatial resolution difference before and after downscaling by adding an intermediate LST downscaling process between low (e.g., 1 km) and high resolutions (e.g., 30 m), from which we can approximate that the statistical downscaling model does not change with smaller scales. The SSRFD can obtain a higher accuracy for LST downscaling and is more suitable for downscaling studies in urban areas with complex surface coverage and high-LST spatial heterogeneity. However, 900 m was used as the initial resolution in the study, and the step-by-step downscaling of resolution was performed subjectively by integer multiples of 2, 3 and 5. To obtain better LST downscaling results, further studies may be needed to determine the optimal spatial resolution change during stepwise downscaling.

In recent years, due to the continuous development of spatial data-acquisition technology, urban 3D spatial distribution data are becoming more and more refined and can be better used to calculate various urban spatial morphological parameters at different scales. There are more and more studies selecting urban morphological parameters to analyze the urban thermal environment. For example, Middel et al. [19] validated the effects of urban morphology and landscape type on local microclimate zones in the semi-arid region of Phoenix, Arizona; Qaid et al. [20] further explored the effect of SVF on the thermal environment of streets with different orientations; Wong et al., Li et al. and Nichol et al. studied the characteristics of urban spatial morphology in Kowloon Peninsula, Hong Kong and confirmed that the urban spatial morphology profoundly affects the urban microclimate [49]. Based on the results of these studies, it appears that the spatiotemporal variability of the urban thermal environment is closely related to the spatial morphological parameters, and the influential role of these three-dimensional parameters needs to be considered in-depth in urban LST downscaling studies. In this study, based on the traditional two-dimensional surface parameters of the downscaling model, five spatial morphological parameters, SVF, FAD, FAR, BD and MH, were added to downscale the Sentinel-3 LST to a high spatial resolution of 30 m. It was found that the urban spatial morphological parameters did affect the spatial distribution of LST, especially for the built-up areas.

The downscaling errors in the five building-dense areas decreased by 0.51, 0.48, 0.4, 0.16 and 0.06 °C (Figure 7). Analysis of the importance of the driving factors showed that the importance of the urban spatial morphological parameters was lower than that of the 2D remote-sensing spectral index (Figure 9). The reason may be that the factors for BD, FAD, FAR and MH were calculated through a moving window of 100×100 m, making it difficult to accurately describe the 3D spatial features at a high spatial resolution of 30 m. Compared to the other four spatial morphological parameters, SVF can be calculated pixel by pixel, and its importance gradually becomes larger as the spatial resolution increases, which generally indicates that the role of urban spatial morphological parameters cannot be neglected for LST downscaling at higher spatial resolutions than 30 m. Li et al. [17] and Liu et al. [50] also mentioned the necessity of considering urban spatial morphological parameters in urban LST downscaling.

In terms of downscaling accuracy validation, due to the lack of real land surface temperature with a higher spatial resolution, we just downscaled LST to the resolution of 30 m. If there are LSTs at a high spatial resolution, such as the data collected by thermal infrared sensors loaded on unmanned aerial vehicles or field measurement temperature probes, LST downscaling for a higher spatial resolution could be performed to further discuss the influence of morphological parameters.

Finally, a global random forest downscaling model was established in the study. Considering the characteristics of different urban regions and establishing sub-regional local models to improve LST downscaling accuracy need to be further discussed. For example, Stewart et al. [51] proposed to establish local urban climate zones by considering building and land cover types, an approach that can be used to downscale LST for various climate zones. More accurate downscaling results may be obtained by further analysis of the relationship between LST and urban spatial morphological parameters based on local urban climate zones.

6. Conclusions

This study carried out downscaling LST in urban areas using the SSRFD and DRFD methods, and the downscaling effects of the two methods were compared and analyzed. Urban spatial morphological parameters were introduced in the driver to verify their role in high-spatial-resolution downscaling LST. From the above investigations, the following conclusions were drawn:

(1) The 900-m LST was downscaled step-by-step on the order of 450, 150 and 30 m. Compared to the results obtained with DRFD, the r value between the SSRFD results and Landsat LST was improved by 0.21, and the RMSE value was reduced by 0.94 °C. The SSRFD results more accurately captured the spatial distribution characteristics of the surface temperature, including the high-temperature zone of buildings and the low-temperature zone of water and vegetation. The underestimation/overestimation phenomenon of DRFD resulting in large errors in the high/low temperature zone was avoided or attenuated when using the SSRFD method.

(2) The results obtained with SSRFD-M were partially significantly improved in the Gulou, Qinhuai and Jianye built-up areas compared to SSRFD, in which r and RMSE values improved/decreased by approximately 0.15 and 0.46 °C, respectively. The phenomenon of low-temperature zones in vegetation-covered areas when only remote-sensing spectral indices were used was improved. The SSRFD-M results to some extent compensated for the deficiency of remote-sensing spectral indices used for urban LST downscaling.

In this study, 900 m was used as the initial resolution, and downscaling was carried out in integer multiples of 2, 3 and 5. Further research is needed to determine the optimal downscaling resolution change multiples to obtain better downscaling results. In addition, this paper established a random forest downscaling LST model for the study area as a whole, while urban climate zones could be divided according to building and land cover types. More accurate downscaling results may be obtained by further analysis of the relationship between LST and urban spatial morphological parameters based on urban local climate zones.

Author Contributions: Conceptualization, G.Z. and X.L.; methodology, G.Z. and S.Z.; software, X.L.; validation, Y.X.; data curation, X.L.; writing—original draft preparation, X.L.; writing—review and editing, G.Z. and Y.X.; project administration, G.Z.; funding acquisition, S.Z. and Y.X. All authors have read and agreed to the published version of the manuscript.

Funding: This research was supported by the National Natural Science Foundation of China (42171101, 41871028).

Data Availability Statement: Not applicable.

Acknowledgments: The authors would like to acknowledge the United States Geological Survey for the provision of the Landsat-8 LST product, and the European Space Agency for the provision of the Sentinel-3 LST product and Sentinel-2 remote-sensing data.

Conflicts of Interest: The authors declare no conflict of interest.

References

1. Jaafar, H.; Mourad, R.; Schull, M. A Global 30-m ET Model (HSEB) Using Harmonized Landsat and Sentinel-2, MODIS and VIIRS: Comparison to ECOSTRESS ET and LST. *Remote Sens. Environ.* **2022**, *274*, 112995. [CrossRef]
2. Paolini, G.; Escorihuela, M.J.; Bellvert, J.; Merlin, O. Disaggregation of SMAP Soil Moisture at 20 m Resolution: Validation and Sub-Field Scale Analysis. *Remote Sens.* **2022**, *14*, 167. [CrossRef]
3. Joshi, R.C.; Ryu, D.; Sheridan, G.J.; Lane, P.N.J. Modeling Vegetation Water Stress over the Forest from Space: Temperature Vegetation Water Stress Index (TVWSI). *Remote Sens.* **2021**, *13*, 4635. [CrossRef]
4. Chu, H.-J.; Wijayanti, R.F.; Jaelani, L.M.; Tsai, H.-P. Time Varying Spatial Downscaling of Satellite-Based Drought Index. *Remote Sens.* **2021**, *13*, 3693. [CrossRef]
5. Baqa, M.F.; Lu, L.; Chen, F.; Nawaz-ul-Huda, S.; Pan, L.; Tariq, A.; Qureshi, S.; Li, B.; Li, Q. Characterizing Spatiotemporal Variations in the Urban Thermal Environment Related to Land Cover Changes in Karachi, Pakistan, from 2000 to 2020. *Remote Sens.* **2022**, *14*, 2164. [CrossRef]
6. Sismanidis, P.; Bechtel, B.; Perry, M.; Ghent, D. The Seasonality of Surface Urban Heat Islands across Climates. *Remote Sens.* **2022**, *14*, 2318. [CrossRef]
7. Niu, L.; Zhang, Z.; Peng, Z.; Jiang, Y.; Liu, M.; Zhou, X.; Tang, R. Research on China's surface urban heat island drivers and its spatial heterogeneity. *China Environ. Sci.* **2021**, *13*, 4428.
8. Romaguera, M.; Vaughan, R.G.; Ettema, J.; Izquierdo-Verdiguier, E.; Hecker, C.A.; van der Meer, F.D. Detecting Geothermal Anomalies and Evaluating LST Geothermal Component by Combining Thermal Remote Sensing Time Series and Land Surface Model Data. *Remote Sens. Environ.* **2018**, *204*, 534–552. [CrossRef]
9. Hrisko, J.; Ramamurthy, P.; Yu, Y.; Yu, P.; Melecio-Vázquez, D. Urban Air Temperature Model Using GOES-16 LST and a Diurnal Regressive Neural Network Algorithm. *Remote Sens. Environ.* **2020**, *237*, 111495. [CrossRef]
10. Westermann, S.; Langer, M.; Boike, J. Spatial and Temporal Variations of Summer Surface Temperatures of High-Arctic Tundra on Svalbard—Implications for MODIS LST Based Permafrost Monitoring. *Remote Sens. Environ.* **2011**, *115*, 908–922. [CrossRef]
11. Mahour, M.; Tolpekin, V.; Stein, A.; Sharifi, A. A Comparison of Two Downscaling Procedures to Increase the Spatial Resolution of Mapping Actual Evapotranspiration. *ISPRS J. Photogramm. Remote Sens.* **2017**, *126*, 56–67. [CrossRef]
12. Bartkowiak, P.; Castelli, M.; Notarnicola, C. Downscaling Land Surface Temperature from MODIS Dataset with Random Forest Approach over Alpine Vegetated Areas. *Remote Sens.* **2019**, *11*, 1319. [CrossRef]
13. Lezama Valdes, L.-M.; Katurji, M.; Meyer, H. A Machine Learning Based Downscaling Approach to Produce High Spatio-Temporal Resolution Land Surface Temperature of the Antarctic Dry Valleys from MODIS Data. *Remote Sens.* **2021**, *13*, 4673. [CrossRef]
14. Zhu, J.; Zhu, S.; Yu, F.; Zhang, G.; Xu, Y. A downscaling method for ERA5 reanalysis land surface temperature over urban and mountain areas. *Natl. Remote Sens. Bull.* **2021**, *25*, 1778–1791.
15. Zhu, X.; Song, X.; Leng, P.; Hu, R. Spatial downscaling of land surface temperature with the multi-scale geographically weighted regression. *Natl. Remote Sens. Bull.* **2021**, *25*, 1749–1766.
16. Tang, K.; Zhu, H.; Ni, P. Spatial Downscaling of Land Surface Temperature over Heterogeneous Regions Using Random Forest Regression Considering Spatial Features. *Remote Sens.* **2021**, *13*, 3645. [CrossRef]
17. Li, N.; Wu, H.; Luan, Q. Land Surface Temperature Downscaling in Urban Area: A Case Study of Beijing. *Natl. Remote Sens. Bull.* **2021**, *25*, 1808–1820.
18. Golany, G.S. Urban Design Morphology and Thermal Performance. *Atmos. Environ.* **1996**, *30*, 455–465. [CrossRef]
19. Middel, A.; Häb, K.; Brazel, A.J.; Martin, C.A.; Guhathakurta, S. Impact of Urban Form and Design on Mid-Afternoon Microclimate in Phoenix Local Climate Zones. *Landsc. Urban Plan.* **2014**, *122*, 16–28. [CrossRef]
20. Qaid, A.; Lamit, H.B.; Ossen, D.R.; Rasidi, M.H. Effect of the Position of the Visible Sky in Determining the Sky View Factor on Micrometeorological and Human Thermal Comfort Conditions in Urban Street Canyons. *Theor. Appl. Climatol.* **2018**, *131*, 1083–1100. [CrossRef]
21. Xiong, Y.; Peng, F.; Zou, B. Spatiotemporal Influences of Land Use/Cover Changes on the Heat Island Effect in Rapid Urbanization Area. *Front. Earth Sci.* **2019**, *13*, 614–627. [CrossRef]
22. Weng, Q.; Fu, P.; Gao, F. Generating Daily Land Surface Temperature at Landsat Resolution by Fusing Landsat and MODIS Data. *Remote Sens. Environ.* **2014**, *145*, 55–67. [CrossRef]
23. Wu, P.; Shen, H.; Zhang, L.; Göttsche, F.-M. Integrated Fusion of Multi-Scale Polar-Orbiting and Geostationary Satellite Observations for the Mapping of High Spatial and Temporal Resolution Land Surface Temperature. *Remote Sens. Environ.* **2015**, *156*, 169–181. [CrossRef]
24. Guo, L.J.; Moore, J.M. Pixel Block Intensity Modulation: Adding Spatial Detail to TM Band 6 Thermal Imagery. *International J. Remote Sens.* **1998**, *19*, 2477–2491. [CrossRef]

25. Nichol, J. An Emissivity Modulation Method for Spatial Enhancement of Thermal Satellite Images in Urban Heat Island Analysis. *Photogramm. Eng. Remote Sens.* **2009**, *75*, 547–556. [CrossRef]
26. Wang, J.; Schmitz, O.; Lu, M.; Karssenberg, D. Thermal Unmixing Based Downscaling for Fine Resolution Diurnal Land Surface Temperature Analysis. *ISPRS J. Photogramm. Remote Sens.* **2020**, *161*, 76–89. [CrossRef]
27. Kustas, W.P.; Norman, J.M.; Anderson, M.C.; French, A.N. Estimating Subpixel Surface Temperatures and Energy Fluxes from the Vegetation Index-Radiometric Temperature Relationship. *Remote Sens. Environ.* **2003**, *85*, 429–440. [CrossRef]
28. Agam, N.; Kustas, W.P.; Anderson, M.C.; Li, F.; Neale, C.M.U. A Vegetation Index Based Technique for Spatial Sharpening of Thermal Imagery. *Remote Sens. Environ.* **2007**, *107*, 545–558. [CrossRef]
29. Dominguez, A.; Kleissl, J.; Luvall, J.C.; Rickman, D.L. High-Resolution Urban Thermal Sharpener (HUTS). *Remote Sens. Environ.* **2011**, *115*, 1772–1780. [CrossRef]
30. Lacerda, L.N.; Cohen, Y.; Snider, J.; Huryna, H.; Liakos, V.; Vellidis, G. Field Scale Assessment of the TsHARP Technique for Thermal Sharpening of MODIS Satellite Images Using VENμS and Sentinel-2-Derived NDVI. *Remote Sens.* **2021**, *13*, 1155. [CrossRef]
31. Vaculik, A.F.; Rachid Bah, A.; Norouzi, H.; Beale, C.; Valentine, M.; Ginchereau, J.; Blake, R. Downscaling of Satellite Land Surface Temperature Data Over Urban Environments. In Proceedings of the IEEE International Geoscience and Remote Sensing Symposium, Yokohama, Japan, 28 July–2 August 2019; pp. 7475–7477.
32. Sánchez, J.M.; Galve, J.M.; González-Piqueras, J.; López-Urrea, R.; Niclòs, R.; Calera, A. Monitoring 10-m LST from the Combination MODIS/Sentinel-2, Validation in a High Contrast Semi-Arid Agroecosystem. *Remote Sens.* **2020**, *12*, 1453. [CrossRef]
33. Pereira, O.J.R.; Melfi, A.J.; Montes, C.R.; Lucas, Y. Downscaling of ASTER Thermal Images Based on Geographically Weighted Regression Kriging. *Remote Sens.* **2018**, *10*, 633. [CrossRef]
34. Guo, F.; Hu, D.; Schlink, U. A New Nonlinear Method for Downscaling Land Surface Temperature by Integrating Guided and Gaussian Filtering. *Remote Sens. Environ.* **2022**, *271*, 112915. [CrossRef]
35. Guijun, Y.; Ruiliang, P.; Wenjiang, H.; Jihua, W.; Chunjiang, Z. A Novel Method to Estimate Subpixel Temperature by Fusing Solar-Reflective and Thermal-Infrared Remote-Sensing Data with an Artificial Neural Network. *IEEE Trans. Geosci. Remote Sens.* **2010**, *48*, 2170–2178. [CrossRef]
36. Li, W.; Ni, L.; Li, Z.-L.; Duan, S.-B.; Wu, H. Evaluation of Machine Learning Algorithms in Spatial Downscaling of MODIS Land Surface Temperature. *IEEE J. Sel. Top. Appl. Earth Obs. Remote Sens.* **2019**, *12*, 2299–2307. [CrossRef]
37. Wang, R.; Gao, W.; Peng, W. Downscale MODIS Land Surface Temperature Based on Three Different Models to Analyze Surface Urban Heat Island: A Case Study of Hangzhou. *Remote Sens.* **2020**, *12*, 2134. [CrossRef]
38. Ebrahimy, H.; Aghighi, H.; Azadbakht, M.; Amani, M.; Mahdavi, S.; Matkan, A.A. Downscaling MODIS Land Surface Temperature Product Using an Adaptive Random Forest Regression Method and Google Earth Engine for a 19-Years Spatiotemporal Trend Analysis Over Iran. *IEEE J. Sel. Top. Appl. Earth Obs. Remote Sens.* **2021**, *14*, 2103–2112. [CrossRef]
39. Njuki, S.M.; Mannaerts, C.M.; Su, Z. An Improved Approach for Downscaling Coarse-Resolution Thermal Data by Minimizing the Spatial Averaging Biases in Random Forest. *Remote Sens.* **2020**, *12*, 3507. [CrossRef]
40. Malakar, N.K.; Hulley, G.C.; Hook, S.J.; Laraby, K.; Cook, M.; Schott, J.R. An Operational Land Surface Temperature Product for Landsat Thermal Data: Methodology and Validation. *IEEE Trans. Geosci. Remote Sens.* **2018**, *56*, 5717–5735. [CrossRef]
41. Cook, M.; Schott, J.R.; Mandel, J.; Raqueno, N. Development of an Operational Calibration Methodology for the Landsat Thermal Data Archive and Initial Testing of the Atmospheric Compensation Component of a Land Surface Temperature (LST) Product from the Archive. *Remote Sens.* **2014**, *6*, 11244–11266. [CrossRef]
42. LSDS-1619_Landsat-8-9-C2-L2-ScienceProductGuide-v4.pdf. Available online: https://d9-wret.s3.us-west-2.amazonaws.com/assets/palladium/production/s3fspublic/media/files/LSDS-1619_Landsat-8-9-C2-L2-ScienceProductGuide-v4.pdf (accessed on 30 December 2021).
43. SLSTR_Level-2_LST_ATBD.pdf. Available online: https://sentinel.esa.int/documents/247904/0/SLSTR_Level_2_LST_ATBD.pdf/8a4322ef-c7e0-4abc-9cac-8f5fd69e1fd7 (accessed on 14 May 2022).
44. S2-PDGS-TAS-DI-PSD-V14.9.pdf. Available online: https://sentinel.esa.int/documents/247904/4756619/S2-PDGS-TAS-DI-PSD-V14.9.pdf/3d3b6c9c-4334-dcc4-3aa7-f7c0deffbaf7 (accessed on 14 May 2022).
45. Xu, H. A remote sensing urban ecological index and its application. *Acta Ecol. Sin.* **2013**, *33*, 7853–7862.
46. Pu, R. Assessing Scaling Effect in Downscaling Land Surface Temperature in a Heterogenous Urban Environment. *Int. J. Appl. Earth Obs. Geoinf.* **2021**, *96*, 102256. [CrossRef]
47. Pu, R.; Bonafoni, S. Reducing Scaling Effect on Downscaled Land Surface Temperature Maps in Heterogenous Urban Environments. *Remote Sens.* **2021**, *13*, 5044. [CrossRef]
48. Cao, C.; Yang, Y. Downscaling Multi-resolution Land surface Temperature Research. *Mod. Surv. Mapp.* **2018**, *41*, 3–8.
49. Wong, M.S.; Nichol, J.; Ng, E. A Study of the "Wall Effect" Caused by Proliferation of High-Rise Buildings Using GIS Techniques. *Landsc. Urban Plan.* **2011**, *102*, 245–253. [CrossRef]
50. Liu, Y.; Xu, Y.; Zhang, F.; Shu, W. Influence of Beijing Spatial Morphology on the Distribution of Urban Heat Island. *Acta Geogr. Sin.* **2021**, *76*, 1662–1679.
51. Stewart, I.D.; Oke, T.R. Local Climate Zones for Urban Temperature Studies. *Bull. Am. Meteorol. Soc.* **2012**, *93*, 1879–1900. [CrossRef]

remote sensing

Article

How to Measure the Urban Park Cooling Island? A Perspective of Absolute and Relative Indicators Using Remote Sensing and Buffer Analysis

Wenhao Zhu [1], Jiabin Sun [1], Chaobin Yang [1,2,3,*], Min Liu [2], Xinliang Xu [4] and Caoxiang Ji [5]

[1] School of Civil and Architectural Engineering, Shandong University of Technology, Zibo 255000, China; 20507020762@stumail.sdut.edu.cn (W.Z.); 20507020766@stumail.sdut.edu.cn (J.S.)

[2] Shanghai Key Lab for Urban Ecological Processes and Eco-Restoration, East China Normal University, Shanghai 200241, China; mliu@re.ecnu.edu.cn

[3] Research Center for Eco-Environmental Sciences, Chinese Academy of Sciences, Beijing 100085, China

[4] State Key Laboratory of Resources and Environmental Information System, Institute of Geographic Sciences and Natural Resources Research, Chinese Academy of Sciences, Beijing 100101, China; xuxl@lreis.ac.cn

[5] Shenyang Meteorological Bureau, Shenyang 110168, China; verio@sina.com

* Correspondence: yangchaobin@sdut.edu.cn

Citation: Zhu, W.; Sun, J.; Yang, C.; Liu, M.; Xu, X.; Ji, C. How to Measure the Urban Park Cooling Island? A Perspective of Absolute and Relative Indicators Using Remote Sensing and Buffer Analysis. *Remote Sens.* **2021**, *13*, 3154. https://doi.org/10.3390/rs13163154

Academic Editors: Yuyu Zhou, Elhadi Adam, John Odindi and Yuji Murayama

Received: 23 June 2021
Accepted: 3 August 2021
Published: 9 August 2021

Abstract: Urban parks have been proven to cool the surrounding environment, and can thus mitigate the urban heat island to an extent by forming a park cooling island. However, a comprehensive understanding of the mechanism of park cooling islands is still required. Therefore, we studied 32 urban parks in Jinan, China and proposed absolute and relative indicators to depict the detailed features of the park cooling island. High-spatial-resolution GF-2 images were used to obtain the land cover of parks, and Landsat 8 TIR images were used to examine the thermal environment by applying buffer analysis. Linear statistical models were developed to explore the relationships between park characteristics and the park cooling island. The results showed that the average land surface temperature (LST) of urban parks was approximately 3.6 °C lower than that of the study area, with the largest temperature difference of 7.84 °C occurring during summer daytime, while the average park cooling area was approximately 120.68 ha. The park cooling island could be classified into four categories—regular, declined, increased, and others—based on the changing features of the surrounding LSTs. Park area (PA), park perimeter (PP), water area proportion (WAP), and park shape index (PSI) were significantly negatively correlated with the park LST. We also found that WAP, PP, and greenness (characterized by the normalized difference vegetation index (NDVI)) were three important factors that determined the park cooling island. However, the relationship between PA and the park cooling island was complex, as the results indicated that only parks larger than a threshold size (20 ha in our study) would provide a larger cooling effect with the increase in park size. In this case, increasing the NDVI of the parks by planting more vegetation would be a more sustainable and effective solution to form a stronger park cooling island.

Keywords: park cooling island; driving factors; land surface temperate; buffer analysis; Jinan

1. Introduction

The world is currently experiencing rapid urbanization and industrialization, which has led to dramatic changes in the land use, land cover, and local climatic conditions, thereby exacerbating urban heat island (UHI) effects [1–4]. The negative effects from UHIs have significantly increased the energy consumption for cooling, physical discomfort, and even death [5–7]. It is projected that 60% of the world's population will live in cities by 2030, and making cities inclusive, safe, resilient, and sustainable is one of the 17 goals proposed by the United Nations to transform our world [8]. In recent years, studies on alleviating the UHI effect have become more popular than its spatial–temporal monitoring [9–12]. Urban parks, which usually have a combination of blue and green spaces, are among the most important determinants of mitigating UHI effects [13–16]. Therefore, enhancing the understanding of the cooling effect provided by parks and the drivers of its variations will help in developing policies to mitigate UHI and designing future parks.

Urban parks have been proven to considerably mitigate the UHI because of shading and evaporation effects [17–19]. The land surface temperature (LST) of shaded areas is 19 °C lower than that of unshaded areas [17]. Additionally, urban parks cool their surroundings through energy exchange [20]. Several previous studies showed that the average LST of internal urban parks was 1–2 °C, and sometimes 4–8 °C lower than the surroundings, thereby generating a "cooling island" [19–23]. A stronger "cooling island" affects a larger area and provides more outdoor thermal comfort for residents. Consequently, assessing the park cooling island is the most fundamental and important aspect of our study.

One basic question that needs to be solved in relevant studies is the definition of "park cooling island", or the comprehensive depiction of the cooling effect. Park cooling intensity (PCI) is a commonly used indicator to characterize "cooling islands" [18,23]. In Cao's study, the PCI was defined as the LST difference between the inside and outside of parks, and the LST outside the park was manually set as the average LST of a 500-m buffer [24]. However, the 500-m buffer may not be applicable to other parks in different study areas. Additionally, other scholars introduced the park cooling distance (PCD) to measure the park cooling island, which is defined as the spatial extent up to which the park could have a significant influence [18,22]. PCD is usually identified using buffer analysis [21]. For instance, the PCD was defined as the distance from the park boundary to the buffer whose LST difference was less than 0.1 °C [25]. The concept of "first turning point" of the LST curve proposed in Peng's study was used to define the largest cooling distance (or PCD), which was the distance between the park boundary and the first turning point [22]. These two indicators can be treated as absolute features of the park cooling island, such as the absolute temperature difference and the absolute cooling area. However, the surrounding thermal environment of a park changes continuously, and the same PCI might occur at different distances. Additionally, the parks that have the same values of park cooling area (PCA; or the same cooling effect) might have different sizes, and the differences between such parks cannot be measured by simply using absolute indicators. Therefore, relative indicators of the park cooling island—such as the changes in surrounding LST per meter or the cooling effect per unit area of the park—should be considered in assessing the park cooling island from the perspective of spatial accumulation [22].

Park characteristics are vital for urban planning and urban climate studies. Numerous studies have explored the relationship between urban characteristics and urban park cooling islands, and the results showed that the park cooling effect could be affected by the park's size, shape, type, greenness, and other factors [25–27]. However, the park cooling island was mainly characterized by a single indicator of PCI, and the effect of urban park features on the relative indicators of the cooling island is still uncertain. Additionally, the effect of the land use/land cover, especially regarding the configuration of park components, was not fully assessed. A review paper also showed that most studies on cooling effects are subject to a single park, and only few have explored the cooling effects of parks on the surrounding environment [28]. These limitations might be attributed to the high cost of such studies. Fortunately, obtaining detailed characteristics of urban parks is facilitated by the development of high-spatial-resolution remote sensing or LiDAR [29,30]. For instance, even an individual tree can be identified using high-resolution images. The thermal infrared remote sensing image can be used to retrieve the LST for each pixel [31,32]; this helps in analyzing the surrounding thermal environment of a park, which is more convenient than in situ air temperature measurements [33–35]. The explanation of the LST and air temperature differences remains rooted, and these are only likely to be predicted by fully coupled surface–atmosphere models [4]. Skin temperature was poorly correlated with both air temperature and apparent temperature [36]. However, several studies showed that the relationship between traditional air temperature and LST is statistically significant [37,38].

Therefore, using remote sensing images to conduct park cooling island studies can obtain robust results.

Considering the insufficiencies mentioned above, and to provide implications for the studies of urban park cooling islands, we selected the case of Jinan city, China to: (1) map the detailed characteristics of urban parks, including their composition and configuration features; (2) analyze the features of the park cooling island by comprehensively proposing and using absolute and relative indicators; and (3) identify the most important urban park characteristics that determine the variations of the urban park cooling island.

2. Data and Methodology

2.1. Study Area and Data Sources

Jinan, the capital of Shandong province, which is in the eastern part of China, serves as the political, economic, cultural, educational, and financial center of the region, and is located to the north of Mount Tai. Jinan is famous for its 72 springs, and is known as the "City of Springs". Jinan has a warm, temperate, continental monsoon climate zone, with four distinct seasons and sufficient sunshine. According to statistics from the Jinan government, the annual average air temperature of Jinan is 14.2 °C. January is the coldest month, with an average air temperature of 0.2 °C, and the highest temperature was observed in July, with an average air temperature of 28.3 °C. The annual average precipitation is 548.7 mm. By the end of 2019, there were 8.91 million permanent residents in the city. The green coverage rate of the built-up area is 40.7%, and the per capita park green area is only 13.1 m^2 based on the statistics from Bureau of Forestry and Landscaping of Jinan. Consequently, building and managing urban parks to maximize their ecosystem services—such as climate regulation and thermal environment improvement—has become an urgent problem. The core of Jinan city was selected as the study area, with a size of 295.49 km^2 (Figure 1). There are several mountains in the south of Jinan, and to avoid their effects on the local thermal environment and park cooling islands, we selected 32 urban parks that were not adjacent to them (Figure 1).

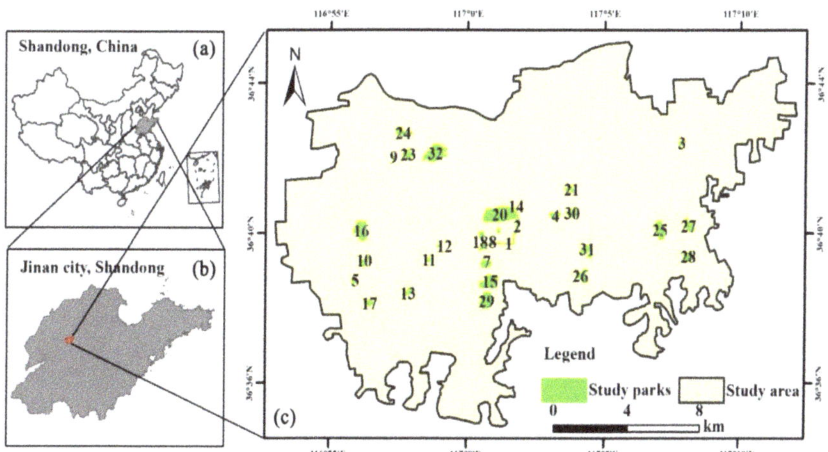

Figure 1. Location of study area: (**a**) Shandong province, China; (**b**) Jinan city in Shandong; (**c**) study parks in Jinan.

High-spatial-resolution GF-2 images (1 × 1 m) acquired on 4 October 2018, from the Resource and Environmental Science and Data Center, Chinese Academy of Sciences (https://www.resdc.cn/), were used to manually obtain the detailed vector information of urban parks. Three cloud-free Landsat 8 images acquired at about 10:47 a.m. local time for the study area (path 122, row 35, acquired on 17 June 2017-LC81220352017168LGN00, 20 June 2018-LC81220352018171LGN00, and 7 June 2019-LC81220352019158LGN00) were utilized to retrieve the LST, and were obtained from the geospatial data cloud (http://www.gscloud.cn/). Using only one Landsat image may yield an uncertain LST map, as it is subject to the time of satellite overpass. Consequently, the LSTs retrieved from Landsat 8 images acquired over multiple years were averaged to obtain robust results (mean LST) regarding the climate. Data pre-processing (radiometric correction and atmospheric correction) was conducted on the GF-2 and Landsat 8 OLI images prior to the interpretation and LST retrievals.

2.2. Characterizing Urban Parks

The GF-2 images with high spatial resolution facilitated the procurement of detailed information inside a park in this study. Based on the local environment and our prior knowledge, six land cover types (water, trees, shrubs and grass, impervious surfaces, buildings, and bare soil) were extracted by visual interpretation based on image features (e.g., color, size, texture, spatial relationship, etc.). An example of the resulting urban park vector is shown in Figure 2.

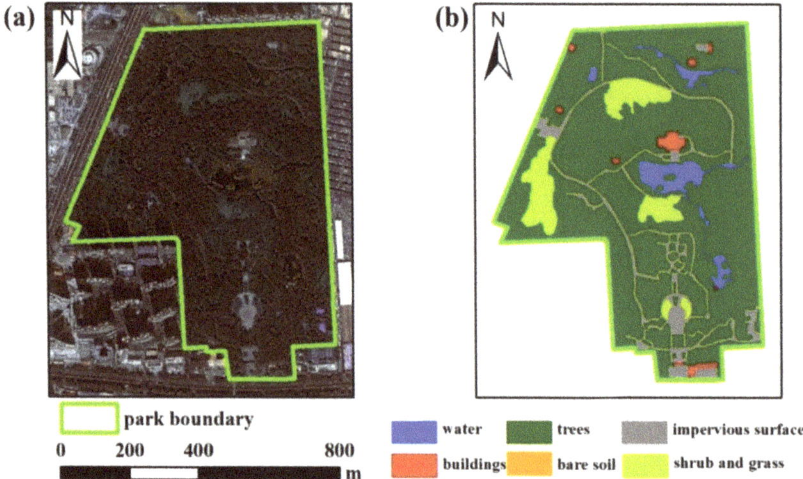

Figure 2. (**a**) GF-2 image (true color) of an urban park. (**b**) Land-cover map of an urban park.

In this research, six widely used landscape metrics and the normalized difference vegetation index (NDVI) (Table 1) were employed to measure the urban park features based on previous studies [21,22], and were classified into two categories: park composition, and park configuration. The composition metrics include the percentage of landscape (PLAND), park area (PA), park perimeter (PP), and the average NDVI of the park (PNDVI). PLAND comprises the water area proportion (WAP), tree area proportion (TAP), shrub and grass area proportion (SGAP), impervious surface area proportion (ISAP), building area proportion (BAP), bare soil area proportion (BSAP), and tree and water area proportion (TWAP). The configurations comprise the park patch density (PPD), park edge density (PED), and park shape index (PSI). These metrics were selected based on previous studies and the following principles: (1) easy calculation and understanding; (2) important in both theory and practice; and (3) comprehensive depiction of urban parks [21,39].

Table 1. Landscape metrics selected in this study.

Metrics and Abbreviation	Calculation	Description
Composition		
Percentage of landscape, PLAND	PLAND = Ai/PA; Ai = area of land cover i;	Measures the area proportion of each type of land cover in the park
Park area, PA	PA = area of park	Measures the area of the park
Park perimeter, PP	PP = perimeter of park	Measures the perimeter of the park
the average NDVI of park, PNDVI	PNDVI = NDVI of park	Measures the NDVI of the park
Configuration		
Park patch density, PPD	PPD = n/A × 10,000, n = the number of patches in a park	Measures the patch density of the park
Park edge density, PED	$PED = \sum_{i=1}^{m} Ei/A \times 10,000$, Ei = perimeter of patch i.	Measures the shape complexity of the park
Park shape index, PSI	PSI = 0.25 × PP/$\sqrt{PA}$	Measures the ratio of park perimeter to area

2.3. LST Retrieval

The LST was retrieved from Landsat 8 TIRS images. First, the digital number (DN) of band 10 (10.60–11.19 um) was converted into the radiation intensity value using Equation (1) [40]:

$$Radiance = M_L \times Q_{cal} + A_L \tag{1}$$

where *Radiance* is the spectral radiance, M_L = the radiance multiplicative scaling factor for band 10, A_L = the radiance additive scaling factor for band 10, and Q_{cal} = the level 1 pixel value in DN. All scaling factors can be obtained from the header file.

Subsequently, TIRS data can also be converted from radiance to brightness temperature, which is the effective temperature viewed by the satellite under an assumed emissivity of unity. The conversion formula is as follows:

$$T = K_2 / \ln\left(\frac{K_1}{Radiance} + 1\right) \tag{2}$$

where T is the top of the atmospheric brightness temperature (Kelvin), and K_2 and K_1 are conversion constants from the metadata.

Finally, LST can be calculated using Equation (3):

$$LST = T / (1 + (\lambda T / \alpha) \ln \varepsilon) \tag{3}$$

where λ is the wavelength of band 10 (=10.9 μm), $\alpha = 1.43 \times 10^{-2}$ mK, and ε is the land surface emissivity, which is a crucial variable in the LST retrieval [41]. The calculated LST value (Kelvin) is then converted to °C [31].

In this study, land surfaces' emissivity was estimated using the NDVI threshold method. NDVI can be calculated using the following equation:

$$NDVI = \frac{OLI5 - OLI4}{OLI5 + OLI4} \tag{4}$$

where *OLI5* and *OLI4* represent the atmospherically corrected surface reflectance of bands 5 and 4 for Landsat 8. The land surface emissivity can then be obtained from different NDVI values [42]. If NDVI was greater than 0.5, the land surface emissivity was 0.99, whereas if it was smaller than 0.2, the land surface emissivity was 0.97. In other cases:

$$\varepsilon = 0.986 + 0.004 \times F_v \tag{5}$$

where F_v is the vegetation proportion, which can be obtained according to Equation (6):

$$F_v = \left(\frac{NDVI - NDVI_{min}}{NDVI_{max} - NDVI_{min}}\right)^2 \tag{6}$$

where $NDVI_{min} = 0.2$ and $NDVI_{max} = 0.5$.

2.4. Characterizing Park Cooling Island

Urban parks are known to cool their surroundings; however, the cooling effect could decrease with distance from the boundary of the park, and disappear after a certain distance [43]. In previous studies, the urban park cooling island included the following aspects: park cooling intensity (PCI), park cooling distance (PCD), temperature gradient of park cooling gradient (TPCI), park cooling area (PCA), and park cooling efficiency (PCE) [24,25,27,43,44]. In this study, the urban park cooling island was divided into four aspects, as presented in Table 2.

Table 2. Four aspects of the park cooling island in this study.

Park Cooling Island	Abbreviation	Description
Absolute perspective		
Park cooling intensity	PCI	PCI = DLST − Park LST, where DLST is the LST of the first turning point. PCI measure the temperature different between park LST and surroundings.
Park cooling area	PCA	The largest area in which the urban park could have influence on the thermal environment.
Relative perspective		
Park cooling efficiency	PCE	PCE = (PCI × PCA)/PA, which measures the cooling effect produced by the unit area of a park.
Park cooling gradient	PCG	PCG = PCI/PCD, where PCD is the park cooling distance; PCG measures the rate of temperature increase with distance.

As shown in Figure 3a, the surrounding LST increased with increasing distance from the park boundary. Subsequently, the LST curve reached the first turning point, after which the LST changed slightly and even decreased. To characterize the park cooling island, multiple buffer rings (11 in total) with intervals of 90 m at a distance of 990 m from the park boundary were created. Considering that the spatial resolution of the resampled Landsat TIRS is 30 m, the buffer interval was set to 90 m to yield robust results for the LST of each buffer. We assumed that the thermal environment of the latter buffer was not influenced by the urban park if the LST difference between two adjacent buffers was lower than 0.1 °C [21]. Moreover, the distance from the park boundary to the former buffer was defined as the PCD (Figure 3b). The average LST of the buffer with the PCD is treated as the LST of the first turning point, and the area of the buffer is considered to be the PCA. Therefore, the temperature difference between the park and the buffer is the PCI. Consequently, the PCE and PCG can be calculated using the PCA and PCI values.

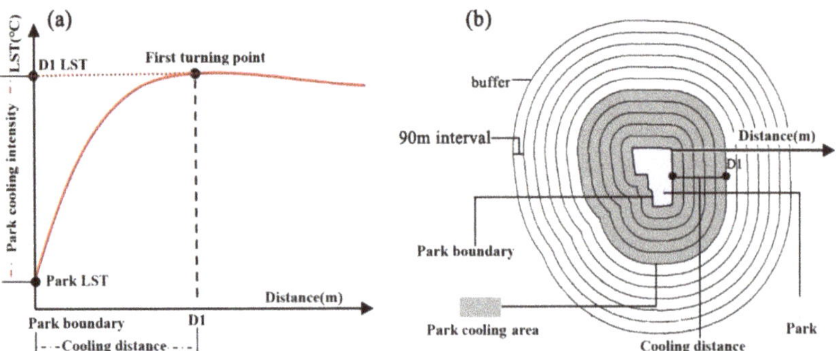

Figure 3. (**a**) Illustration of the LST variations. (**b**) Illustration of the buffer established in this study.

2.5. Statistical Analysis

First, Pearson's correlation analysis was applied to assess the relationship between the park LST, four aspects of the park cooling island, and its characteristics, in order to examine the landscape metric, showing a significant relationship with the park cooling island.

Subsequently, the ordinary least squares linear regression model was used to explore the influence of a single urban park feature on the park cooling island. The four aspects of the park cooling island were the dependent variables, and park features comprised the independent variables. Finally, the different effects of these landscape metrics on the park cooling island were tested using a multiple linear regression model. All statistical analyses were conducted using SPSS software.

3. Results

3.1. Mapping the Characteristics of Urban Parks

A total of 32 urban parks were included in this study. The proportion of land cover types varied among these parks, as shown in Figure 4a. The size of these parks ranged from 2.89 to 107.32 ha, with an average area of 20.81 ha. It can be seen that trees are the most important components of almost every park. Nearly 75% of the parks had a water area. Shrub and grass, together with bare soil, covered a small area in these parks. The configuration landscape metrics and NDVI for each park are shown in Figure 4b. For example, the PSI of these parks ranged from 0.97 to 2.02, with a mean of 1.34, indicating that the shape of parks varied greatly. More detailed statistics of the urban park features are presented in Table 3.

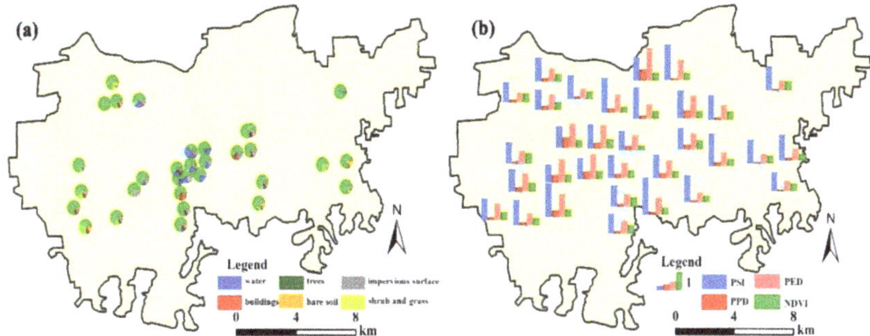

Figure 4. (a) The proportion of land cover and (b) landscape metrics for each park, respectively.

Table 3. The statistics of urban park characteristics.

Statistics.	PA (ha)	PP (km)	WAP	TAP	SGAP	ISAP	BAP	BSAP	TWAP	NDVI	PPD	PED	PSI
Max	107.32	4.45	0.48	0.96	0.31	0.30	0.22	0.12	1	0.57	0.58	1.73	2.02
Min	2.89	0.79	0	0.37	0	0	0	0	0.45	0.26	0.06	0.51	0.97
Ave	20.81	2.12	0.07	0.71	0.05	0.10	0.05	0.01	0.79	0.43	0.21	0.89	1.34

3.2. Characteristics of the Urban Park Cooling Island

The spatial patterns of the average LST are shown in Figure 5a. High LSTs were distributed in the central and eastern parts of the city, where dense buildings and industrial districts exist. Most of the low-LST regions were located in the southern part of the city, with several mountains providing good vegetation coverage as well as a cooling effect. Additionally, the Yellow River in the north generated a "blue cold belt". The highest, lowest, and mean LST in Jinan city were 48.84, 22.55, and 38.46 °C, respectively. Typically, the LST of urban parks in Jinan city is lower than that of the surrounding areas. The average LST of all 32 parks was 34.86 °C; this was 3.60 °C lower than the average LST of the entire study area, indicating that they provide a cooling effect. Daming Lake Park is a famous scenic area in Jinan city, attracting thousands of residents and tourists each day to enjoy the water and classic garden (Figure 5b1). This area was selected to further illustrate the distribution of the thermal environment. The most important land cover types were water and trees, with area ratios of 48.16 and 36.97%, respectively. This generated a very low-temperature zone inside the park compared to the surrounding areas. As the distance increased from the park boundary (green circle), the color gradually changed from yellow to red, indicating an increase in the LST.

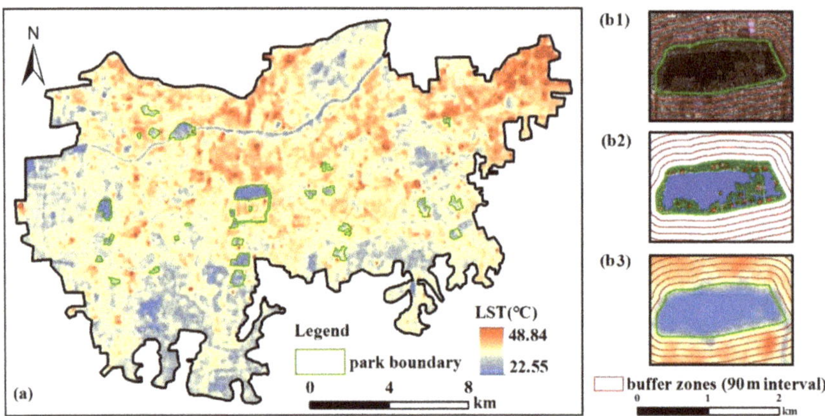

Figure 5. (**a**) The distribution of the average LST, (**b1**) the GF-2 image of Daming Lake Park, (**b2**) the land cover of Daming Lake Park, and (**b3**) the LST of Daming Lake Park and its surroundings.

To further compare the different LST and PCI scores among these parks, the average internal LST, corresponding PCI (LST difference between the first turning point and park LST), and largest PCI (defined as the LST difference between the highest LST of all the buffers and park LST) are shown in Figure 6. It can be seen that the average LST in parks varied greatly. For example, park 20 had the lowest internal LST (31.72 °C) and the strongest PCI (7.84 °C), while park 31 had the highest LST (37.42 °C). The largest LST difference among these parks was 5.7 °C. However, the PCI of park 31 (1.46 °C) was larger than that of park 13, whose PCI (0.85 °C) was the weakest. Therefore, the high internal LST of the park did not imply a low PCI, and the PCI was also influenced by its surrounding features. Additionally, the average PCA, PCE, and PCG of these parks were 120.68 ha, 0.71, and 35.45, respectively. The largest PCA, PCE, and PCG were 319.20 ha (park 20), 1.31 (park 20), and 135.28 (park 8), respectively.

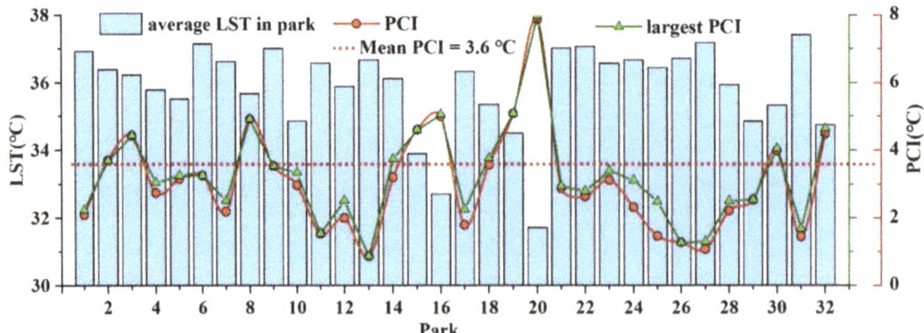

Figure 6. The internal LST of 32 urban parks and their PCIs.

In this study, a distance of 990 m with 11 buffers (90 m intervals) was established to explore the four aspects of the park cooling island based on the changes in the LST of each buffer. The urban parks were classified into four categories according to the characteristics of the surrounding LST curves: regular parks, declined parks, increased parks, and other parks. As shown in Figure 7a, most parks were regular ones, whose surrounding LST had the following features: a gradual increase in LST with distance from the park boundary, and subsequently reaching the first turning point, after which the LST varied slightly. Seven parks belonged to the second category: declined parks whose surrounding LSTs increased sharply at the beginning and had an obvious decreasing trend after the first turning point. The largest LST difference between the first turning point and the later buffers could reach up to 0.99 °C (park 19), and this value was 0.31 °C for regular parks. For increased parks (Figure 7c), the surrounding LSTs increased from the park boundary to the end of the buffer width and, notably, no decrease in the LSTs was observed. There were also two parks whose surrounding LSTs had no changing features, and these were treated as other parks. In particular, the LST increased sharply from 360 to 630 m for park 8. Moreover, there were several commercial districts with very high LSTs in the buffers, leading to this unusual phenomenon.

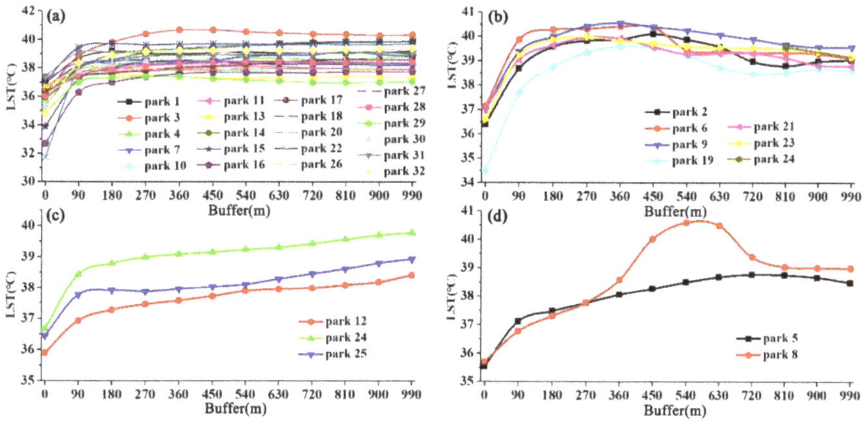

Figure 7. Variation of surrounding LSTs from the urban park boundary: (**a**) regular parks, (**b**) decreased parks, (**c**) increased parks, and (**d**) other parks.

3.3. The Effects of Landscape Metrics on the Park Cooling Island

3.3.1. The Relationship between Urban Park Features and LST

Pearson's correlation analysis was applied to explore the relationships between urban park features and their internal and surrounding LSTs. Only those results of the coefficients that are significant are presented in Table 4. Notably, the internal LST of the parks had negative relationships with PA, PP, and WAP, indicating that large areas and more water bodies could generate a low LST for the parks. The PSI was positively correlated with the park LST. Regarding the relationship between park features and surrounding LSTs, the results showed that the PA had significant positive correlations with the LSTs of all of the buffers. The coefficients gradually decreased from 90 m (R = 0.73) to 450 m (R = 0.56); they subsequently became stronger from 450 to 990 m (R = 0.68). The relationships between PP and surrounding LSTs were not significant from 270 to 540 m, and their positive correlations were weaker compared to those between PA and LSTs. Similarly to PA, WAP had a positive relationship with the LST of all buffers. BAP had negative relationships with the LSTs of buffers between 180 and 450 m, and their correlations were not significant among other buffers.

Based on the results, PA and WAP were seemingly the two most important factors that could affect the park and buffer LSTs. However, the effects of PA and WAP on the park LST were negative, but positive for buffer LSTs. Notably, the park was removed from all buffers, and the LST of buffers did not include park LST.

Table 4. The correlation coefficients for urban park features and LST.

Characteristics.	Park LST	90 m	180 m	270 m	360 m	450 m	540 m	630 m	720 m	810 m	900 m	990 m
PA	−0.73 **	0.73 **	0.62 **	0.59 **	0.58 **	0.56 **	0.58 **	0.60 **	0.64 **	0.67 **	0.67 **	0.68 **
PP	−0.49 **	0.48 **	0.37 *	–	–	–	–	0.35 *	0.39 *	0.42 *	0.42 *	0.44 *
WAP	−0.47 **	0.61 **	0.62 **	0.61 **	0.63 **	0.67 **	0.66 **	0.65 **	0.62 **	0.59 **	0.59 **	0.59 **
BAP	–	–	−0.37 *	−0.37 *	−0.38 *	−0.36 *	–	–	–	–	–	–
PSI	−0.39 **	–	–	–	–	–	–	–	–	–	–	–

** Significance at the 0.01 level; * significance at the 0.05 level; – not significant.

3.3.2. The Relationship between the Park Features and Park Cooling Island

Based on the results of Section 3.3.1, it can be inferred that large PA and WAP would generate low park LST and high buffer LST. Consequently, this may lead to a strong PCI. Therefore, Pearson's correlation analysis was initially employed to study the relationships between urban park features and four aspects of park cooling islands. The results of the coefficients are presented in Table 5. PA had positive effects on the PCI (R = 0.58), PCA (R = 0.73), and PCG (R = −0.45), but had negative effects on PCE. The relationship between PP, PCI, and PCG was not significant, and a large PP could lead to a low PCE. In this study, the effect of TAP on PCI was not significant, and differed from other studies wherein trees could improve the park cooling effect. Consistent with expectations, the WAP was positively correlated with the PCI, PCA, and PCG, indicating that more water would produce a better cooling effect.

Table 5. The correlation coefficients for urban park features and the park cooling island.

Characteristics.	PCI	PCA	PCG	PCE
PA	0.58 **	0.73 **	0.46 **	−0.45 *
PP	0.26	0.45 *	0.32	−0.61 **
TAP	−0.06	−0.43 *	−0.09	0.07
WAP	0.79 **	0.41 *	0.55 **	0.26
BAP	−0.41	−0.14	−0.33	−0.37 *
NDVI	0.46 **	0.23	0.39 **	0.06

** Significance at the 0.01 level; * significance at the 0.05 level.

To further assess the effects of the two most important factors (PA and WAP) on the park cooling island, simple linear regression models were built for the PA, WAP, and PCI, PCA, PCG, and PCE, respectively. The original result of the linear regression model built for PA and PCI using all of the park samples was not satisfactory, as the adjusted R^2 was smaller than 0.30. After exploring the scatter diagram between PCI and PA, a PA threshold (20 ha) was found to divide the parks into two categories. If the PA was smaller than 20 ha (blue points in Figure 8a), the relationship between the PA and PCI was not significant. In this case, the PA had little influence on the PCI. However, if the PA was larger than 20 ha, it had a very strong positive relationship with the PCI. The PCI would increase by 0.7 °C if the PA was 10 ha larger. The threshold of 20 ha for PA was also practical for studying its relationship with the PCA and PCG. Similarly, parks with areas larger than 20 ha had significant positive effects on the PCA and PCG. Regarding the effects of PA on PCE, the latter decreased sharply when PA was smaller than 20 ha, but this rate of decrease was much lower if the PA was larger than 20 ha.

The relationships between WAP and the park cooling island are shown in Figure 9. Typically, water has a strong cooling effect owing to its large specific heat capacity. Consequently, more water in the park would generate an obvious cooling island. In this study, the PCI would increase by 0.89 °C if there were a 10% increase in the WAP. If the PA was larger than 20 ha, the WAP was positively correlated with the PCA. However, the effects of WAP on PCA were not significant if the PA was less than 20 ha. Compared to the PCI and PCA, the relationships between WAP and PCG were much weaker. Our results also showed that no obvious statistical relationships existed between the WAP and PCE. The

PCE measures the cooling ability produced per unit area of the park and, thus, WAP was not the dominant factor.

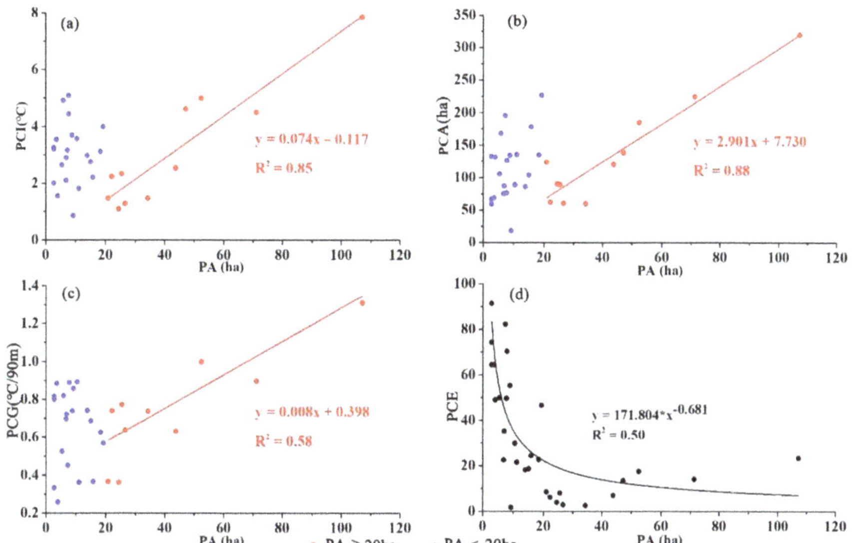

Figure 8. Linear regression models of the PA and (**a**) PCI, (**b**) PCA, (**c**) PCG, and (**d**) PCE.

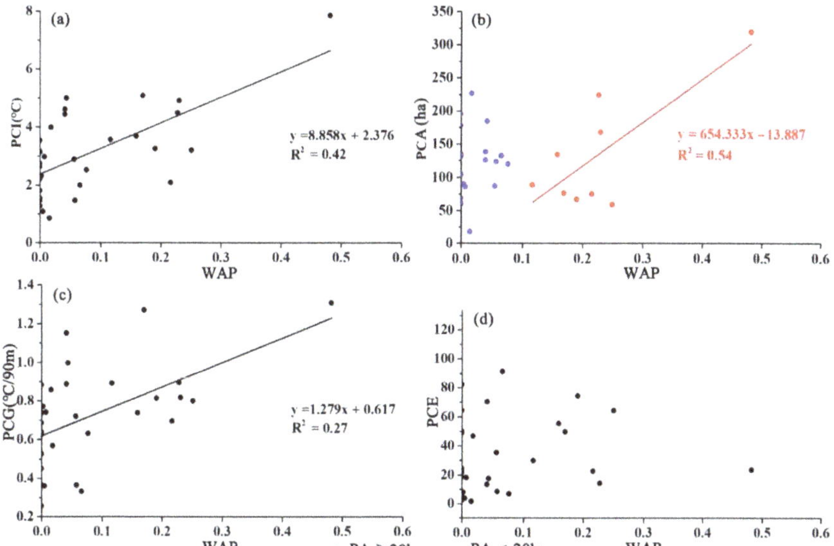

Figure 9. Linear regression models of the WAP and (**a**) PCI, (**b**) PCA, (**c**) PCG, and (**d**) PCE.

4. Discussion

4.1. Characterizing the Urban Park Cooling Island

This study demonstrated a park cooling island in Jinan, China. The primary question that needs to be resolved involves understanding the entirety of the park cooling island. In this study, the characteristics of the park cooling island were divided into four aspects—PCI, PCA, PCG, and PCE—based on previous studies and our research [21,22]. PCI and PCA are absolute indicators that reflect the cooling effect of the park as a whole; PCI measures the temperature differences between the park and its surroundings, and the PCA is the largest area whose thermal environment can influence the park. In previous studies, the PCE was used to measure the cooling island [21,27]. However, this is similar to the largest cooling distance in our study, which is a one-dimensional distance. A large and small park may have the same PCE, but the area influenced by the two parks could vary greatly. Consequently, PCA was used instead of PCE in this study. PCG and PCE are different indicators of a park cooling island that measure the relative cooling effect or cooling ability per unit of distance or area. For example, a park with a large area might have a large PCI or PCA but a small PCG or PCE. In our opinion, the four aspects could characterize the park cooling island in a relatively comprehensive manner.

Upon clarifying the ontology of the urban park cooling island, the next question is the method of measuring the park cooling island. In this study, popular Landsat 8 remote sensing images were used to retrieve the LST to characterize the thermal environment. Buffer analysis has proven to be applicable for studying thermal environments [45,46]. However, the different choices of width and interval of buffers could affect and add more uncertainty to the results. Many previous studies used 10–20 buffers, each 30 or 60 m wide [22,27]. As the original spatial resolution of Landsat 8 TIRS was 100 m and resampled to 30 m, 11 buffers with intervals of 90 m were established to identify the cooling distance of the park and determine the PCI and PCA in our study. The total width of these buffers was 990 m, which is larger than that in many previous studies [21,22]. This was because some parks in our cases had a maximum cooling distance of 630 m, and a buffer width of 300 or 450 m could not facilitate the identification of the PCA. Therefore, the methodology used in this study to measure the park cooling islands could provide insights or references for other relevant studies in the future.

4.2. The Effects of Landscape Metrics on the Park LST and Park Cooling Island

During the daytime in summer, our results showed that the park size and WAP had significant negative relationships with the park LSTs. The results were consistent with those of other studies that utilized remote sensing to assess the park cooling island [14,23]. The PSI measures the shape complexity of a park. Our results showed that the coefficient between PSI and park LST was −0.39, indicating that parks with irregular shapes could have lower LSTs, which was in accordance with earlier studies. Other landscapes—such as the PED, PPD, and BAP—had little effect on the park LST. However, other studies showed that the PPD was positively correlated with the park LST, which was not found in this study [21]. This might be attributable to the differences in the studied parks and different climate backgrounds.

The characteristics of the park cooling island are as follows: the average, maximum, and minimum PCIs of the 32 urban parks were 3.60, 7.84, and 0.85 °C, respectively, which is consistent with previous studies [47,48]. Peng's study showed that the average LST difference between the inside and surrounding of parks was 3.06 °C [22]. Additionally, the average PCA in our study was 120.68 ha, which was much larger than that in Peng's study (81.3 ha). This might be ascribed to the difference in the buffer interval (30 m in Peng's study and 90 m in ours). As we proposed a new definition of PCE owing to the limited current literature, it was difficult to compare the PCE to previous studies.

Among the many landscape metrics and features of a park, the dominant factors that determine the park cooling island were further evaluated. Multivariate linear regression models were constructed for this purpose to explore the relative contributions of the variables to the park cooling island. In these models, landscape metrics were independent variables, while PCI, PCA, PCG, and PCE were the dependent variables. The proportion of water was the most important factor in the PCI (Tables 5 and 6), which is consistent with previous studies [11,49–51]. Water helps to reduce LST owing to its large specific heat capacity and the enhanced heat exchange occurring between water and the surrounding environment [52]. Waterbodies are important components of urban parks. However, owing to the accessibility of water resources, increasing the water area in the park to enhance the PCI is unsustainable. The size of the park is another important determining factor in the PCI (Table 5 and Figure 8) [24]. PA had a positive relationship with the PCI, and similar results were also obtained in previous studies [24]. In particular, a large park would produce a strong cooling effect. However, the relationship between the PA and PCI is not a simple linear relationship. Our results showed that almost no correlation existed between the PA and PCI for small parks, with only larger parks leading to stronger PCI for parks larger than a threshold of 20 ha. Similar correlations were also observed between the green space and LST in previous studies [53,54]. The NDVI was the third most important urban feature significantly affecting PCI in our study. The NDVI represents the greenness of the park, and a high NDVI indicates more vegetation [55,56]. Vegetation can reduce the LST through shadows and evapotranspiration, which has been proven by many studies [30,57–59]. Consequently, if the PA was smaller than the threshold, increasing the NDVI of parks by adding more vegetation would provide a stronger cooling effect [54]. However, the tree area proportion was not included in the regression model presented in Table 6. One possible reason for this was that in our case, almost all of the parks had relatively high TAP values, and the differences between them were not obvious, thereby not making it a determining factor. This does not imply that trees are not important in urban park planning [28,60]. The PA and WAP also had positive effects on the PCA. Moreover, the effects of the independent variables on the PCG and PCE were not significant, except that the PP had an obvious negative effect on the PCE (Table 6).

Table 6. Regression results of landscape metrics, NDVI, and PCI, PCA, PCG, and PCE. R^2 represents the proportions of the park cooling island variations that can be explained by the model. Colors represent the relative contribution of each metric to variations in the park cooling island. Red and blue represent positive and negative effects, respectively.

Characteristics.	PCI	PCA	PCG	PCE
PA	0.61	0.74	0.02	−0.18
PP	−0.20	−0.43	0.28	−0.65
SGAP	−0.098	−0.10	0.01	−0.16
WAP	0.97	0.56	0.48	0.59
BAP	0.12	0.30	−0.11	−0.12
ISAP	0.17	0.40	−0.29	0.32
BSAP	0.07	0.28	−0.26	0.11
PSI	0.30	0.02	−0.21	−0.02
PPD	−0.24	−0.32	0.10	−0.15
NDVI	0.53	0.29	0.23	0.10
R^2	0.78	0.63	0.56	0.68
Adjusted R^2	0.68	0.45	0.35	0.53
−1		0		1

Based on our findings, the WAP, PA, and NDVI (greenness) of the park are the three dominant factors that determine the PCI. They deserve more attention in future urban park designs from the perspective of regulating the thermal environment.

4.3. Limitations and Future Research Directions

Our study had several limitations. First, the detailed features of the urban parks were extracted from high-resolution GF-2 images (1 m). However, the inside and surrounding LSTs of parks were of relatively low resolution because of the performance of Landsat 8 TIRS. Simultaneously, the LST data only represent the thermal environment when the satellite transits late in the morning. Temperature is an environmental variable that changes daily. Therefore, finer spatial and temporal resolution LST data acquired by unmanned aerial vehicles equipped with high-performance thermal cameras are expected in future studies. Using both LST and air temperature in the perspective of surface energy balance may generate more comprehensive and robust results to obtain a better understanding of the thermal regime of urban parks [20,61,62]. Second, the number of parks studied was limited, and studying more parks or changing the studied city may yield different results. Third, due to the limited availability of data, the effects of socio-economic factors on park cooling islands were not explored. Future studies can obtain more robust results by applying big data such as the point of interest (POI).

5. Conclusions

To enhance the understanding of the detailed characteristics of the urban park cooling island and the driving mechanisms of its variations, high-spatial-resolution remote sensing images (GF-2) were used to extract the detailed land cover types of urban parks, and landscape metrics were employed to depict the park features. Buffer analysis was conducted to analyze the four aspects of the park cooling island: PCI, PCA, PCG, and PCE. Statistical models were constructed to explore the effects of driving factors on the park cooling island. We found that the average LST of urban parks was approximately 3.60 °C lower than that of the surrounding environment, indicating that parks can provide significant cooling effects. Based on the changing features of surrounding LSTs, the park cooling island could be classified into four categories: regular, increased, decreased, and other parks. Additionally, our results showed that the PA, WAP, and NDVI were the three most important factors that positively affected the park cooling intensity. Theoretically, increasing the waterbody area is the most effective way to enhance the cooling effects of urban parks. However, this might not be a sustainable solution because of the limited water resources, and adding more vegetation to increase the NDVI could prove more feasible. Unfortunately, there was no significant relationship between the PA and PCI if the area was smaller than a threshold of 20 ha in our study. The findings of our study provide valuable insights for the planning and design of urban parks.

Author Contributions: Conceptualization, C.Y.; methodology, W.Z.; software, W.Z.; validation, J.S.; formal analysis, J.S.; investigation, C.Y.; resources, C.J.; data curation, X.X.; writing—original draft preparation, W.Z.; writing—review and editing, C.Y.; visualization, W.Z.; supervision, X.X.; project administration, C.Y.; funding acquisition, M.L. All authors have read and agreed to the published version of the manuscript.

Funding: This research was funded by Shandong Provincial Natural Science Foundation, China, grant number ZR2019BD036, and the Open Fund of Shanghai Key Lab for Urban Ecological Processes and Eco-Restoration, grant number SHUES2021A10.

Institutional Review Board Statement: Not applicable.

Informed Consent Statement: Not applicable.

Conflicts of Interest: The authors declare no conflict of interest.

References

1. Arnfield, A.J. Two decades of urban climate research: A review of turbulence, exchanges of energy and water, and the urban heat island. *Int. J. Climatol.* **2003**, *23*, 1–26. [CrossRef]
2. Rizwan, A.M.; Dennis, L.Y.C.; Liu, C. A review on the generation, determination and mitigation of Urban Heat Island. *J. Environ. Sci.* **2008**, *20*, 120–128. [CrossRef]
3. Deilami, K.; Kamruzzaman, M.; Liu, Y. Urban heat island effect: A systematic review of spatio-temporal factors, data, methods, and mitigation measures. *Int. J. Appl. Earth Obs. Geoinf.* **2018**, *67*, 30–42. [CrossRef]
4. Voogt, J.A.; Oke, T.R. Thermal remote sensing of urban climates. *Remote Sens. Environ.* **2003**, *86*, 370–384. [CrossRef]
5. Rizvi, S.H.; Alam, K.; Iqbal, M.J. Spatio -temporal variations in urban heat island and its interaction with heat wave. *J. Atmos. Solar Terr. Phys.* **2019**, *185*, 50–57. [CrossRef]
6. Wong, P.P.-Y.; Lai, P.-C.; Low, C.-T.; Chen, S.; Hart, M. The impact of environmental and human factors on urban heat and microclimate variability. *Build. Environ.* **2016**, *95*, 199–208. [CrossRef]
7. Santamouris, M.; Cartalis, C.; Synnefa, A.; Kolokotsa, D. On the impact of urban heat island and global warming on the power demand and electricity consumption of buildings—A review. *Energy Build.* **2015**, *98*, 119–124. [CrossRef]
8. Bebbington, J.; Unerman, J. Achieving the United Nations sustainable development goals. *Account. Audit. Account. J.* **2018**, *31*, 2–24. [CrossRef]

9. Lehnert, M.; Brabec, M.; Jurek, M.; Tokar, V.; Geletič, J. The role of blue and green infrastructure in thermal sensation in public urban areas: A case study of summer days in four Czech cities. *Sustain. Cities Soc.* **2021**, *66*, 102683. [CrossRef]
10. Cheung, P.K.; Fung, C.K.; Jim, C. Seasonal and meteorological effects on the cooling magnitude of trees in subtropical climate. *Build. Environ.* **2020**, *177*, 106911. [CrossRef]
11. Lin, Y.; Wang, Z.; Jim, C.Y.; Li, J.; Deng, J.; Liu, J. Water as an urban heat sink: Blue infrastructure alleviates urban heat island effect in mega-city agglomeration. *J. Clean. Prod.* **2020**, *262*, 121411. [CrossRef]
12. Lai, D.; Liu, W.; Gan, T.; Liu, K.; Chen, Q. A review of mitigating strategies to improve the thermal environment and thermal comfort in urban outdoor spaces. *Sci. Total Environ.* **2019**, *661*, 337–353. [CrossRef]
13. Akbari, H.; Kolokotsa, D. Three decades of urban heat islands and mitigation technologies research. *Energy Build.* **2016**, *133*, 834–842. [CrossRef]
14. Aram, F.; Solgi, E.; Baghaee, S.; García, E.H.; Mosavi, A.; Band, S.S. How parks provide thermal comfort perception in the metropolitan cores; a case study in Madrid Mediterranean climatic zone. *Clim. Risk Manag.* **2020**, *30*, 100245. [CrossRef]
15. Du, H.; Cai, Y.; Zhou, F.; Jiang, H.; Jiang, W.; Xu, Y. Urban blue-green space planning based on thermal environment simulation: A case study of Shanghai, China. *Ecol. Indic.* **2019**, *106*, 105501. [CrossRef]
16. Chan, S.; Chau, C.; Leung, T. On the study of thermal comfort and perceptions of environmental features in urban parks: A structural equation modeling approach. *Build. Environ.* **2017**, *122*, 171–183. [CrossRef]
17. Vidrih, B.; Medved, S. Multiparametric model of urban park cooling island. *Urban For. Urban Green.* **2013**, *12*, 220–229. [CrossRef]
18. Chang, C.-R.; Li, M.-H. Effects of urban parks on the local urban thermal environment. *Urban For. Urban Green.* **2014**, *13*, 672–681. [CrossRef]
19. Xu, M.; Hong, B.; Mi, J.; Yan, S. Outdoor thermal comfort in an urban park during winter in cold regions of China. *Sustain. Cities Soc.* **2018**, *43*, 208–220. [CrossRef]
20. Spronken-Smith, R.A.; Oke, T.R. The thermal regime of urban parks in two cities with different summer climates. *Int. J. Remote Sens.* **1998**, *19*, 2085–2104. [CrossRef]
21. Yang, C.; He, X.; Yu, L.; Yang, J.; Yan, F.; Bu, K.; Chang, L.; Zhang, S. The Cooling Effect of Urban Parks and Its Monthly Variations in a Snow Climate City. *Remote Sens.* **2017**, *9*, 1066. [CrossRef]
22. Peng, J.; Dan, Y.; Qiao, R.; Liu, Y.; Dong, J.; Wu, J. How to quantify the cooling effect of urban parks? Linking maximum and accumulation perspectives. *Remote Sens. Environ.* **2021**, *252*, 112135. [CrossRef]
23. Chen, X.; Su, Y.; Li, D.; Huang, G.; Chen, W.; Chen, S. Study on the cooling effects of urban parks on surrounding environments using Landsat TM data: A case study in Guangzhou, southern China. *Int. J. Remote Sens.* **2012**, *33*, 5889–5914. [CrossRef]
24. Cao, X.; Onishi, A.; Chen, J.; Imura, H. Quantifying the cool island intensity of urban parks using ASTER and IKONOS data. *Landsc. Urban Plan.* **2010**, *96*, 224–231. [CrossRef]
25. Toparlar, Y.; Blocken, B.; Maiheu, B.; Heijst, G.V. The effect of an urban park on the microclimate in its vicinity: A case study for Antwerp, Belgium. *Int. J. Climatol.* **2018**, *38*, e303–e322. [CrossRef]
26. Sodoudi, S.; Zhang, H.; Chi, X.; Müller, F.; Li, H. The influence of spatial configuration of green areas on microclimate and thermal comfort. *Urban For. Urban Green.* **2018**, *34*, 85–96. [CrossRef]
27. Feyisa, G.L.; Dons, K.; Meilby, H. Efficiency of parks in mitigating urban heat island effect: An example from Addis Ababa. *Landsc. Urban Plan.* **2014**, *123*, 87–95. [CrossRef]
28. Bowler, D.E.; Buyung-Ali, L.; Knight, T.M.; Pullin, A.S. Urban greening to cool towns and cities: A systematic review of the empirical evidence. *Landsc. Urban Plan.* **2010**, *97*, 147–155. [CrossRef]
29. Chen, J.; Jin, S.; Du, P. Roles of horizontal and vertical tree canopy structure in mitigating daytime and nighttime urban heat island effects. *Int. J. Appl. Earth Obs. Geoinf.* **2020**, *89*, 102060. [CrossRef]
30. Bartesaghi-Koc, C.; Osmond, P.; Peters, A. Mapping and classifying green infrastructure typologies for climate-related studies based on remote sensing data. *Urban For. Urban Green.* **2019**, *37*, 154–167. [CrossRef]
31. Ranagalage, M.; Murayama, Y.; Dissanayake, D.; Simwanda, M. The Impacts of Landscape Changes on Annual Mean Land Surface Temperature in the Tropical Mountain City of Sri Lanka: A Case Study of Nuwara Eliya (1996–2017). *Sustainability* **2019**, *11*, 5517. [CrossRef]
32. Dissanayake, D. Land Use Change and Its Impacts on Land Surface Temperature in Galle City, Sri Lanka. *Climate* **2020**, *8*, 65. [CrossRef]
33. Bahi, H.; Mastouri, H.; Radoine, H. Review of methods for retrieving urban heat islands. *Mater. Today Proc.* **2020**, *27*, 3004–3009. [CrossRef]
34. Weng, Q. Thermal infrared remote sensing for urban climate and environmental studies: Methods, applications, and trends. *ISPRS J. Photogramm. Remote Sens.* **2009**, *64*, 335–344. [CrossRef]
35. Mohajerani, A.; Bakaric, J.; Jeffrey-Bailey, T. The urban heat island effect, its causes, and mitigation, with reference to the thermal properties of asphalt concrete. *J. Environ. Manag.* **2017**, *197*, 522–538. [CrossRef] [PubMed]
36. Ho, H.C.; Knudby, A.; Xu, Y.; Hodul, M.; Aminipouri, M. A comparison of urban heat islands mapped using skin temperature, air temperature, and apparent temperature (Humidex), for the greater Vancouver area. *Sci. Total Environ.* **2016**, *544*, 929–938. [CrossRef] [PubMed]
37. Yang, C.; Yan, F.; Zhang, S. Comparison of land surface and air temperatures for quantifying summer and winter urban heat island in a snow climate city. *J. Environ. Manag.* **2020**, *265*, 110563. [CrossRef] [PubMed]

38. Schwarz, N.; Schlink, U.; Franck, U.; Großmann, K. Relationship of land surface and air temperatures and its implications for quantifying urban heat island indicators—An application for the city of Leipzig (Germany). *Ecol. Indic.* **2012**, *18*, 693–704. [CrossRef]

39. Xiao, X.D.; Dong, L.; Yan, H.; Yang, N.; Xiong, Y. The influence of the spatial characteristics of urban green space on the urban heat island effect in Suzhou Industrial Park. *Sustain. Cities Soc.* **2018**, *40*, 428–439. [CrossRef]

40. USGS. Landsat 8 Hand Book. 2016. Available online: https://www.usgs.gov/media/files/landsat-8-data-users-handbook (accessed on 22 June 2021).

41. Sobrino, J.A.; Jimenez-Munoz, J.C.; Soria, G.; Romaguera, M.; Guanter, L.; Moreno, J.; Plaza, A.; Martinez, P. Land Surface Emissivity Retrieval From Different VNIR and TIR Sensors. *IEEE Trans. Geosci. Remote Sens.* **2008**, *46*, 316–327. [CrossRef]

42. Sobrino, J.A.; Jiménez-Muñoz, J.C.; Paolini, L. Land surface temperature retrieval from LANDSAT TM 5. *Remote Sens. Environ.* **2004**, *90*, 434–440. [CrossRef]

43. Lin, W.; Yu, T.; Chang, X.; Wu, W.; Zhang, Y. Calculating cooling extents of green parks using remote sensing: Method and test. *Landsc. Urban Plan.* **2015**, *134*, 66–75. [CrossRef]

44. Dai, Z.; Guldmann, J.-M.; Hu, Y. Spatial regression models of park and land-use impacts on the urban heat island in central Beijing. *Sci. Total Environ.* **2018**, *626*, 1136–1147. [CrossRef] [PubMed]

45. Zhao, Z.-Q.; He, B.-J.; Li, L.-G.; Wang, H.-B.; Darko, A. Profile and concentric zonal analysis of relationships between land use/land cover and land surface temperature: Case study of Shenyang, China. *Energy Build.* **2017**, *155*, 282–295. [CrossRef]

46. Dong, J.; Lin, M.; Zuo, J.; Lin, T.; Liu, J.; Sun, C.; Luo, J. Quantitative study on the cooling effect of green roofs in a high-density urban Area—A case study of Xiamen, China. *J. Clean. Prod.* **2020**, *255*, 120152. [CrossRef]

47. Yu, Z.; Yang, G.; Zuo, S.; Jørgensen, G.; Koga, M.; Vejre, H. Critical review on the cooling effect of urban blue-green space: A threshold-size perspective. *Urban For. Urban Green.* **2020**, *49*, 126630. [CrossRef]

48. Kim, G.; Coseo, P. Urban Park Systems to Support Sustainability: The Role of Urban Park Systems in Hot Arid Urban Climates. *Forests* **2018**, *9*, 439. [CrossRef]

49. Hathway, E.A.; Sharples, S. The interaction of rivers and urban form in mitigating the Urban Heat Island effect: A UK case study. *Build. Environ.* **2012**, *58*, 14–22. [CrossRef]

50. Peng, J.; Liu, Q.; Xu, Z.; Lyu, D.; Du, Y.; Qiao, R.; Wu, J. How to effectively mitigate urban heat island effect? A perspective of waterbody patch size threshold. *Landsc. Urban Plan.* **2020**, *202*, 103873. [CrossRef]

51. Yang, G.; Yu, Z.; Jørgensen, G.; Vejre, H. How can urban blue-green space be planned for climate adaption in high-latitude cities? A seasonal perspective. *Sustain. Cities Soc.* **2020**, *53*, 101932. [CrossRef]

52. Ampatzidis, P.; Kershaw, T. A review of the impact of blue space on the urban microclimate. *Sci. Total Environ.* **2020**, *730*, 139068. [CrossRef] [PubMed]

53. Yu, Z.; Guo, X.; Jørgensen, G.; Vejre, H. How can urban green spaces be planned for climate adaptation in subtropical cities? *Ecol. Indic.* **2017**, *82*, 152–162. [CrossRef]

54. Shah, A.; Garg, A.; Mishra, V. Quantifying the local cooling effects of urban green spaces: Evidence from Bengaluru, India. *Landsc. Urban Plan.* **2021**, *209*, 104043. [CrossRef]

55. Alexander, C. Normalised difference spectral indices and urban land cover as indicators of land surface temperature (LST). *Int. J. Appl. Earth Obs. Geoinf.* **2020**, *86*, 102013. [CrossRef]

56. Yuan, F.; Bauer, M.E. Comparison of impervious surface area and normalized difference vegetation index as indicators of surface urban heat island effects in Landsat imagery. *Remote Sens. Environ.* **2007**, *106*, 375–386. [CrossRef]

57. Bartesaghi-Koc, C.; Osmond, P.; Peters, A. Quantifying the seasonal cooling capacity of 'green infrastructure types' (GITs): An approach to assess and mitigate surface urban heat island in Sydney, Australia. *Landsc. Urban Plan.* **2020**, *203*, 103893. [CrossRef]

58. Massetti, L.; Petralli, M.; Napoli, M.; Brandani, G.; Orlandini, S.; Pearlmutter, D. Effects of deciduous shade trees on surface temperature and pedestrian thermal stress during summer and autumn. *Int. J. Biometeorol.* **2019**, *63*, 467–479. [CrossRef] [PubMed]

59. Ren, Z.; Zhao, H.; Fu, Y.; Xiao, L.; Dong, Y. Effects of urban street trees on human thermal comfort and physiological indices: A case study in Changchun city, China. *J. For. Res.* **2021**. [CrossRef]

60. Pamukcu-Albers, P.; Ugolini, F.; La Rosa, D.; Grădinaru, S.R.; Azevedo, J.C.; Wu, J. Building green infrastructure to enhance urban resilience to climate change and pandemics. *Landsc. Ecol.* **2021**. [CrossRef]

61. Spronken-Smith, R.A.; Oke, T.R.; Lowry, W.P. Advection and the surface energy balance across an irrigated urban park. *Int. J. Climatol.* **2000**, *20*, 1033–1047. [CrossRef]

62. Spronken-Smith, R.A.; Oke, T.R. Scale Modelling of Nocturnal Cooling in Urban Parks. *Bound. Layer Meteorol.* **1999**, *93*, 287–312. [CrossRef]

remote sensing

Article

Sentinel-Based Adaptation of the Local Climate Zones Framework to a South African Context

Tshilidzi Manyanya [1,*], Janne Teerlinck [1], Ben Somers [1], Bruno Verbist [1] and Nthaduleni Nethengwe [2]

[1] Division of Forest Nature and Landscape, Department Earth & Environmental Sciences, Katholieke Universiteit Leuven, BE-3001 Leuven, Belgium; janne.teerlinck@kuleuven.be (J.T.); ben.somers@kuleuven.be (B.S.); bruno.verbist@kuleuven.be (B.V.)

[2] Geography and Geo-Information Sciences, Faculty of Science, Engineering and Agriculture, University of Venda, Thohoyandou 0950, South Africa; nthaduleni.nethengwe@univen.ac.za

* Correspondence: tshilidzicloudia.manyanya@kuleuven.be

Abstract: The LCZ framework has become a widely applied approach to study urban climate. The standard LCZ typology is highly specific when applied to western urban areas but generic in some African cities. We tested the generic nature of the standard typology by taking a two-part approach. First, we applied a single-source WUDAPT-based training input across three urban areas that represent a gradient in South African urbanization (Cape Town, Thohoyandou and East London). Second, we applied a local customized training that accounts for the unique characteristics of the specific area. The LCZ classification was completed using a random forest classifier on a subset of single (SI) and multitemporal (MT) Sentinel 2 imagery. The results show an increase in overall classification accuracy between 17 and 30% for the locally calibrated over the generic standard LCZ framework. The spring season is the best classified of the single-date imagery with the accuracies 7% higher than the least classified season. The multi-date classification accuracy is 13% higher than spring but only 9% higher when a neighborhood function (NF) is applied. For acceptable performance of the LCZ classifier in an African context, the training must be local and customized to the uniqueness of that specific area.

Keywords: local climate zones; random forests; neighborhood function; multitemporal classification

Citation: Manyanya, T.; Teerlinck, J.; Somers, B.; Verbist, B.; Nethengwe, N. Sentinel-Based Adaptation of the Local Climate Zones Framework to a South African Context. *Remote Sens.* **2022**, *14*, 3594. https://doi.org/10.3390/rs14153594

Academic Editors: Yuyu Zhou, Elhadi Adam, John Odindi and Elfatih Abdel-Rahman

Received: 25 June 2022
Accepted: 21 July 2022
Published: 27 July 2022

Publisher's Note: MDPI stays neutral with regard to jurisdictional claims in published maps and institutional affiliations.

1. Introduction

Since before the dawn of civilization, the global population has been increasing both in isolated as well as connected communities [1,2]. This continuous rise in population has resulted in civilization and furthermore has created the barrier between urban and rural regions [3,4]. While urbanization comes with prospects of technologically advanced livelihoods for the inhabitants and easier access to amenities, it brings with it some undesirable side effects [5]. One of these side effects is the urban heat island (UHI) phenomenon. The urban heat island is a term coined by [6] to refer to a phenomenon where urban regions experience warmer surface and atmospheric climatic conditions as compared to their surrounding rural areas. The earliest documentation of this phenomenon was in a 1820 study on the London climate [7]. While the UHI is created by urban infrastructural developments, planning and design as well as population growth, it is however projected to be more intensified by climate change through extreme heat waves [8,9]. With anthropogenically-driven climate change becoming an even bigger threat to livelihoods, there is a need to establish living conditions that do not exacerbate but adapt to the changing climate. Urban planners propose that the increase in urban population demands innovation toward sustainable cities while some propose low energy buildings [10,11]. These are cities that have systems in place to curb and adapt to the effects of UHI while accommodating as much as 68% of the global population as projected for 2050 [12,13]. Even with this much awareness being raised in the urban planning and climate change circles, studies on UHI are still limited and only localized in Asia, Europe and North America with very little literature available on Africa, Mid and South America as well as Oceania [14].

What these limited studies highlight is that the biggest challenge in the study of UHI has been and is still that of relating the observed surface and air temperature patterns with the features on the ground [14,15]. This has been addressed primarily by the development and application of local climate zones to study both surface and atmospheric UHI [16–18]. Local climate zones (LCZ) are regions of homogenous surface features that experience uniform climatic conditions [19]. The local climate zones typology currently accepted as the standard framework for classification was designed by [15]. The framework was developed specifically to study UHI on the hypothesis that surface and atmospheric patterns in UHI can be attributed to the spatial and structural characteristics of surface features.

Even with the development of local climate zones to address the spatial distribution challenge, studies are still not evenly distributed geographically. Between 1970 and 2020, 57% of the publications have been in Asia, 23% have been in North America, 14% have been in Europe and only about 3% have been in Africa [14]. The majority of the African studies on local climate zones were part of global studies and not particularly focused on local African cities. Local climate zones studies in Africa have been limited as compared to the other continents, which creates a big gap in the literature in this area of study [20]. Urban climatology studies in South Africa have focused on a diversity of urban climate variables ranging from temperature-focused studies, Koppen's climatic zones and rainfall and drainage but not local climate zones [21,22]. This current study intended to not only play a part in filling the gap in the lack of LCZ classifications in South Africa but also contributing to the knowledge increase in African studies in general.

For the purposes of universal application and generalizability, the standard LCZ was designed to be culturally neutral [16]. Ref. [23] explored the relationship between culture and urban form and found them strongly correlated. Urban form is defined by [24] as the description of the city's physical characteristics. This covers everything from urban design, type of building material, arrangement of infrastructure and type of ground material among others. It was observed in Southwestern Saudi Arabia that cultural laws also influenced urban form [25]. Seventy-seven metropolitan cities in Asia, US, Europe, Latin America and Australia were sampled as part of a study to assess urban form across different continents. This study by [26] found that apart from differences of density and height, there are urban form features that are common across all the cities and yet there are some urban form features that are only unique to some cities. This suggests that a classification framework typology must be flexible enough to allow training based on the standard classes, combinations of the standard classes as well as the addition of classes that are unique to that local environment. The flexibility of the standard framework to manipulation is even more important in African urban regions due to how diverse they are and how different they are from each other and from western urban regions [27]. This current study explored the extent to which the standard LCZ framework can be manipulated and modified to accommodate the uniqueness of an urban area. All this was completed with the aim of developing a classification protocol that can be possibly applicable to similar urban regions that do not strictly resemble the cities the standard framework was based on.

While the standard typology remains the same, its implementation over the past decade since its conceptualization had adopted different methodologies. These approaches have included in situ measurements, GIS-based mapping as well as remote sensing-based approaches [28]. Within each approach, there are different sub-methods which can result in differences in accuracy even within the same general approach. A study on local climate zones on the Zimbabwean capital Harare adopted a remote sensing approach in attempt to compare the machine learning technique of support vector machines with the World Urban Database and Access Portal Tools (WUDAPT) generator approach. The WUDAPT is a global initiative of online tools to create local climate zone maps for a given city using a standard methodology [29]. This study by [30] found out that the WUDAPT approach yielded higher accuracies as compared to the SVM approach. Ref. [28] proposes that the GIS-based and remote sensing-based classifications produce different accuracies and level of detail as a function of scale [31]. The GIS-based method is more suitable for micro-scale classification, while remote sensing is suitable for larger-scale classification [32]. The WUDAPT approach in general yields higher accuracies as compared to other remote sensing as well as GIS-based methods. Ref. [28] attributed the accuracy of the WUDAPT methodology to its generic nature. According to [29], the satellite imagery used by the online WUDAPT classifier has pixel size of 100–120 m. This is the same size as the minimum size of a local climate zone training input, which makes the accuracy higher.

The WUDAPT method was developed for global urban morphology data collection for global climate models [33]. This suggests that it was not designed for implementation at the local level. For lower mesoscale and local scale climate models, more detail is required in urban morphology. This has resulted in the use of higher resolution data sources such as Landsat-8 and Sentinel-2 among others. These products have been used in classifications using mostly machine learning and deep learning techniques [20,30]. The use of coarser imagery of 100–200 m resolution uses the surface reflectance values for each pixel at a size that is comparable to the minimum size of a local climate zone. However, when finer resolution data of less than 60 m are used, the pixel size becomes much smaller than the size of the LCZ training input. When the pixel size is smaller than the object, the variation in pixels belonging to the same class becomes larger [34]. This then necessitates a neighborhood function kernel to aggregate the pixels to the level of the local climate zone. At the end, the result is an aggregated value instead of the original classified value.

The first objective of this study was to characterize the spatial designs and layout of three cities, namely, Cape Town, East London and Thohoyandou that cover a gradient in urbanization within the South African context. This looks at the historical influences that shape the morphology and urban form of each of these urban areas. The second objective was to determine the extent to which the standard LCZ framework covers the nature and morphology of South African cities. This is by applying the standard LCZ framework as it is. For this objective to be carried out, a combination of a field survey together with digital imagery was used to develop a training input of spectrally distinct classes found within these urban regions, which was the third objective of the study. For this objective, all these urban regions are assumed to be similar, and only one training is developed using Cape Town and then applied across the rest of the urban regions. Lastly, a more specific training is developed for a more customized LCZ classification protocol with each city having its own training sample.

2. Materials and Methods

2.1. Study Area

The study was conducted in Cape Town, Thohoyandou and East London urban areas of the Western Cape, Limpopo and Eastern Cape provinces of South Africa, respectively (Figure 1). These are urban areas of different size, urban form and land-use systems, representing a gradient across South African urbanization. Their geographical location and dispersion through South Africa put them in different climatological systems. Cape Town and East London are coastal cities, while Thohoyandou is a remote inland small town (Figure 1).

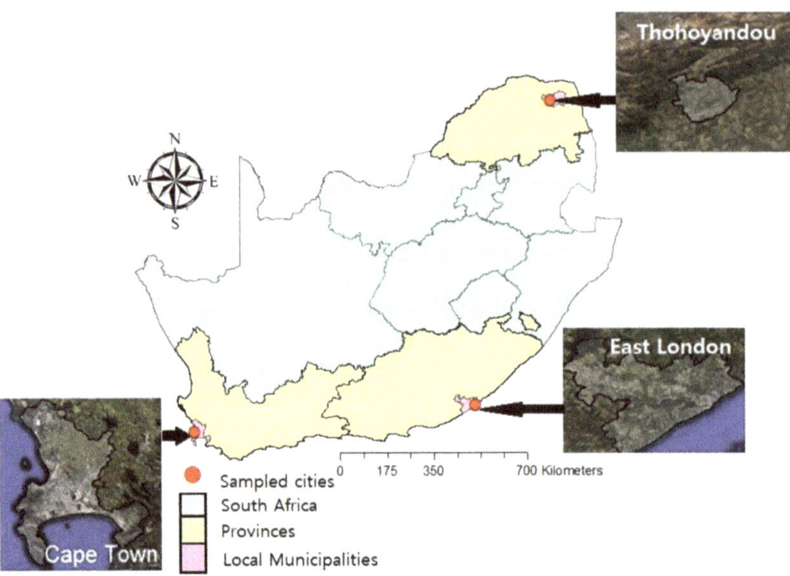

Figure 1. Geographic location of the Thohoyandou, East London and Cape Town relative to their respective municipalities, provinces and South Africa in general.

Cape Town, located at the southernmost tip of Africa, experiences a Mediterranean climate characterized by cold wet winters and hot dry summers [35]. East London experiences a maritime climate with cool winters and mild summers that is moist throughout the year because of proximity to the ocean [36]. Thohoyandou experiences cold dry winters and hot wet summers [37]. Cape Town as a city is a monument of the historical occupations and influences of the diverse people groups who contributed to its development. This has greatly influenced urban form resembling Portuguese, Indian, Dutch and British architecture [38]. The original development of Cape Town into somewhat of an urban area began with the Portuguese explorers in the 14th century in an area that belonged to the Khoikhoi people. This was followed by the Dutch period in the 17th century and then the British period in the 19th century. Finally, the South African period began in the early 20th century and extends to date. The 20th century ushered in rapid urban expansion in Cape Town from multiple epicenters, resulting in an overall design that is comprised of multiple administration and suburban residential areas resembling Harris and Ullman's (1945) multiple nuclei model [39]. Cape Town currently sits at 400 km^2 with a population density of 17,500 per km^2.

East London similar to Cape Town was also developed originally as a harbor town. However, East London does not have as long a history of cultural diversity from different developers as cape town. The city has remained a harbor city of mostly British influence in style, but the expansion outwards into the indigenous communities has brought indigenous cultural influences into the East London urban form [40]. It is currently sitting at an area of 168.9 km^2 and a population density of 2745 per km^2.

Thohoyandou, which was developed as a capital for the Venda Bantustan in the latter half of the 20th century, does not benefit from the diversity of historical western influences in its urban form and design that East London and Cape Town encompasses [41]. The development of Thohoyandou was originally for a shopping center and government administration offices [42]. Among the three urban areas, Thohoyandou is the most integrated with features associated with the rural environment. The area of Thohoyandou is 42.62 km^2 with a population density of 2051 per km^2.

2.2. Materials and Methods

2.2.1. Sentinel-2 Multispectral Imagery

The Sentinel 2 Top-of-Atmosphere (ToA) (L1C) product was obtained from the Copernicus Open Access Hub (COAH) of the European Space Agency (ESA). Sen2Cor288 algorithm was used in SNAP to correct for atmospheric interference and thus convert the L1C product to Sentinel L2A, which is Bottom-of-Atmosphere (BoA) reflectance [43]. Sen2Cor creates BoA reflectance images, terrain and cirrus corrected reflectance, aerosol optical thickness, water vapor, scene classification maps and quality indicators for cloud and snow probabilities [44]. The central month image of each season with a cloud coverage filter of less than 0.5% was selected as the most suitable to reflect peak season dynamics (Table 1).

Table 1. Date and seasons for the Sentinel-2 imagery used throughout the classification process.

	Sentinel-2			
	Summer	**Autumn**	**Winter**	**Spring**
Cape Town	25 February 2018	16 May 2018	24 August 2018	2 November 2018
East London	7 January 2018	28 March 2018	6 July 2018	14 September 2018
Thohoyandou	13 December 2018	27 April 2018	26 July 2018	24 September 2018

2.2.2. Definition of LCZ Classification

The standard [15] LCZ classification framework was selected for this study to be applied across all three urban areas. A remote sensing-based approach was chosen over a GIS-based approach. A GIS approach to classification would require manual digitization of the entire image, which is laborious and time consuming. The processing of remotely sensed data also requires the manipulation and interpretation of digital data [45]. This tends to be a mathematically complex process due to the heterogeneity of materials and geometry of the features [46]. However, the advantage of the remote sensing-based route is that it can be automated. The irregularities of the geometry of the local climate zones becomes a challenge to both pixel and object-based classifications [45,47,48]. A pixel-based classification which was adopted for this study overcomes this geometric non-uniformity challenge by assigning every pixel into a single class based on the reflectance value [49].

This application of Stewart and Oke's LCZ classification framework was used in two approaches that differ in the creation of training data. Approach 1 followed strictly the LCZ region of interest (RoI) creation as outlined by WUDAPT to create a standard training set based on Cape Town to be applied on all three cities. The application of this remote training on Thohoyandou and East London indirectly assesses whether the influence of origin and culture on urban form affects the LCZ classification. This informs whether there is a need to have a customized training for LCZ classification in South African cities for all cities or locally for each urban region. Cape Town was chosen because it is the only urban area of the three that experiences all four seasons and contains all 17 LCZ standard classes [50]. Theoretically, this means that the impacts of phenology would be more apparent in Cape Town than in East London and Thohoyandou, which also would make it ideal for identifying the best season for single image classification across all three urban regions. In defining these LCZ classes, a separability analysis was performed. A spectral separability is an assessment of the performance of the Sentinel-2A multispectral instrument bands to differentiate between the classes of the typology. For the purposes of this study, histograms, scatter and box plots were used to perform this separability according to guidelines from [51]. The second approach to the classification still took off from the traditional [15] typology but explored combinations and subclasses of the standard typology based on the unique features of each urban region.

i.   Model Training

Two types of model training sets were developed for the classification where the first was standardized and the second was customized to the context of the urban area. The ground reference data were collected initially using digital globe resources due to COVID restrictions and were validated and revised in situ on February and March 2022 in Cape Town, Thohoyandou and East London. Based on a digital globe basemap, circular regions of interest were made using QGIS having a diameter of 100 m following the specifications of the standard typology as outlined by WUDAPT. The field campaign was then used as a manual method to verify the regions of interest (Figure 2). Most (80%) of the regions of interest were used for training and the remaining 20% were used for validation. The field campaign was also used to observe and document the unique elements of each urban region for the discriminations of subclasses to feed into the more specific training for Approach 2.

Figure 2. Some of the sampled locations across Cape Town, East London and Thohoyandou.

a. Approach 1

The WUDAPT LCZ typology guidelines for the development of training data was applied. These guidelines are divided into subcategories depending on the scale of the total area, classification methods and the intended use of the final product [47]. These guidelines have a strict protocol for training with the objective of using the WUDAPT online LCZ generator as well as a more flexible protocol for developing training data to be used outside the WUDAPT generator [16]. The development of these training polygons depends on a combination of general typology elements such as cover, material, geometry and function taken to different levels of detail depending on the scale of the imagery and the purpose of the classification. Cities are then mapped using the scheme of [15], which classifies the urban landscape into 10 urban and seven natural classes. Each class in the typology represents a LCZ described in terms of specific landscape parameters of mean building height, canyon width, aspect ratio, building surface ratio and impervious area. These training areas are used to characterize the reflectance properties of each LCZ, which is then used to develop a model that assigns every other untrained pixel of the image into the LCZ classes within the framework.

A three-step sampling method (Figure 3) was adopted from [52]. This block-based system was developed for LCZ classification at the city block scale primarily as GIS-based. In this study, this method was used to guide the development of training samples following the three steps.

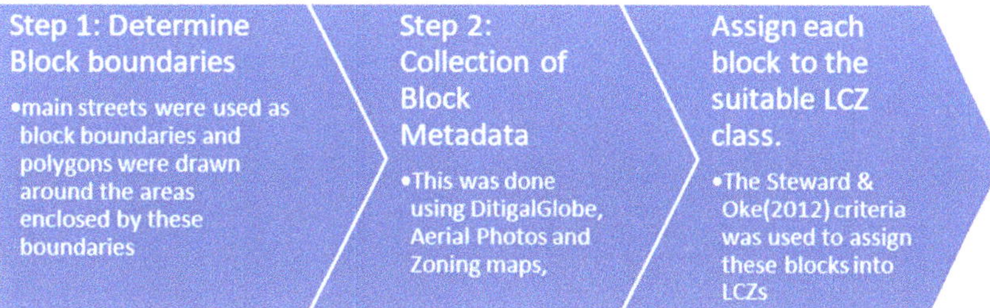

Figure 3. An outline of the city block method steps for accurately creating training polygons and assigning them to different LCZ for the classification [15].

The natural city blocks are easier to assign to LCZ classes because they are homogenous, but the urban classes are much harder even with the physical access to the area. Urban LCZ metadata variables were thus limited to mean building height (H_s), which is the number of stories as collected in the field, mean building height (BH), canyon width (CW), aspect ratio, building surface ratio and impervious area (Figure 4). When the number of stories per building is less than 10, every story is assumed to be 3 m; otherwise, Equation (1) is used for buildings with more than 10 stories [53]. Buildings with one to three stories were considered low-rise, four to nine stories were considered mid-rise, and more than nine stories were classified as high rise (Figure 5). Canyon width is estimated by the average distance between two buildings. Aspect ratio is estimated by the ratio of the building height (BH) to the canyon width (W). Building surface fraction (BSF) and impervious surface fraction (ISF) were estimated using simple calculations in QGIS following the polygonization method (Figure 4) for areas of built and impervious surfaces. These variables were all used to assign each one of the blocks into LCZ classes (Figure 5). Within each block, multiple points were placed at a distance of 200 m from each other. A circular buffer of 50 m was created around each point, ultimately becoming the circular LCZ training polygon. Each of these training polygons was at least 100 m from the next one. A total of 200 training polygons were selected for the model training.

$$\text{Building Height (BH)} = H\_(s) \times 3.5 + 9.6 + 2.6 \times (H\_(s)/25): H\_(s) \qquad (1)$$

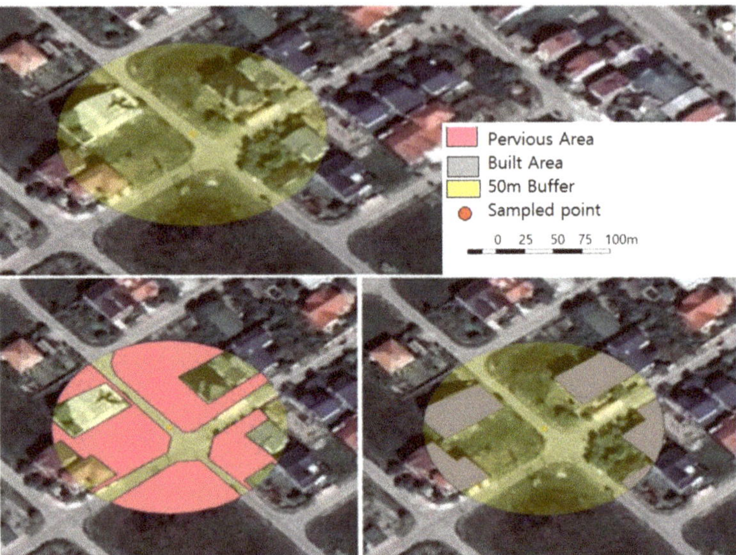

Figure 4. A display of the method for mapping the features within each polygon in order to classify it to the suitable LCZ [53].

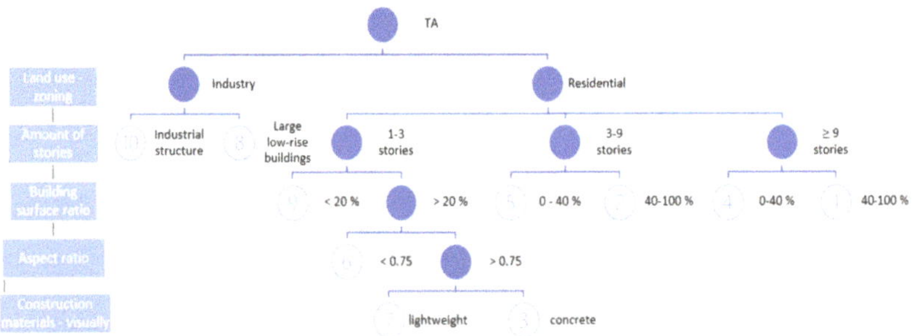

Figure 5. A flow diagram of the final stage of the method for assigning training polygons to their suitable LCZ classes.

b. Local Customized Training Input (Approach 2)

This approach is developed based on the layout and design of each city taking into account features that are common across all three cities and features that are unique to the specific urban region. In Thohoyandou, the urbanized city center and the immediate blocks around the city center have rural features integrated into the urban landscape. The intra-block streets in Thohoyandou are not homogenous in material; some are asphalted while some are gravel. While in a standard framework, the building density and height stand out as the main discriminators for local climate zones, the street canyon material stands out just as significantly in Thohoyandou. This is also noted by [27], who stated that as a unique general feature, remote African urban areas tend to have more bare soils than western urban. While this might not be statistically significant for a highly urbanized and highly westernized city such as Cape Town, its significance in a small town such as Thohoyandou cannot be neglected without investigation. However, spectrally separating bare soil from impervious surfaces in a remote sensing approach at the level of the pixel (10 m) has inherent confusion in spectral signatures [54]. Therefore, Following

Jin's blocking, a GIS approach was developed in order to create a criterion for separating blocks that are completely asphalted from blocks that are gravel within the urban area without the inherent confusion of a remote sensing approach. A digitization process was applied to digital globe imagery to create asphalted and bare soil inter-street blocks (Figure 6).

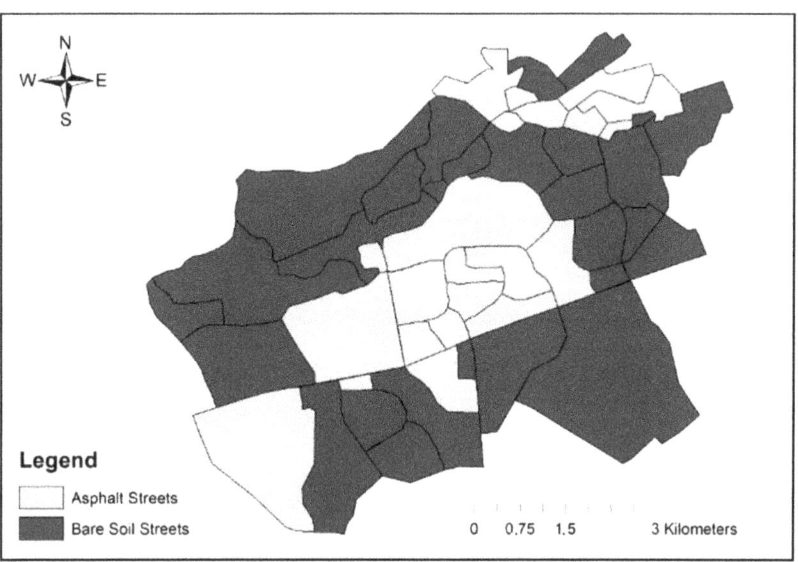

Figure 6. Asphalted vs. bare soil (gravel) streets in Thohoyandou urbanized region.

The output of this digitization was used to separate training inputs that are in gravel blocks from those that are in asphalted blocks. LCZ 3 was the only class observed to be present in both block types. The buildings within LCZ 9 in Thohoyandou are also further apart than they are in Cape Town and East London. The space between houses is thus confused with Shrublands (LCZ 14) due to the dry nature of the shrublands in Thohoyandou. In order to reduce this confusion, the low plants were thus combined with shrubs lands to form a single class. Scattered trees were also combined with dense trees to form a single class. The number of the natural classes was thus minimized so that the built classes can then stand out more spectrally. Impervious surface and bare soil were also combined to form a single class. This is because they are the least represented classes in the area, and combining them makes them a slightly larger class. This thus makes the updated training set for Thohoyandou to become LCZ 3a, 3b, 6, 8, 9, 11 & 12, 13 & 14, 15 & 16, 17.

In East London, the ground truthing process in lightweight zones (LCZ 7) revealed a unique class that is a hybrid between lightweight and compact low-rise (Figure 7). A unique feature of South African light weight squatter camps is that they have no designated stand-numbers. This means they have no yards, and it is common for houses to share walls with their neighbors on all sides except the front. Because of this nature of South African squatter camps, the integration of LCZ 3 and LCZ 7 in these East London zones happens below the minimum size of the local climate (100 m). This is thus treated as a unique class and incorporated as LCZ 7a into the updated East London training, which then becomes LCZ 2, 3, 5, 6, 7, 7a, 8, 9, 10, 11, 13, 14, 16, 17.

Figure 7. East London squatter camp types; LCZ 7 showing a standard lightweight region while LCZ 7a shows a hybrid region of concrete and light weight.

ii. Remote Sensing Classification Protocol

The choice of the LCZ classification method is guided by the nature of the data, the available computational resources and the application purpose [45]. The classification protocol was performed via a coded script in R on R-Studio using mainly the CARET package through a Random Forests (RF) classifier. This was designed to extract the training pixels from the image stack, build predictive models that assign the rest of the image pixels into the most fitting class based on surface reflectance values and to validate the assigned pixels. This was performed on a single image stack as well as a multitemporal image stack.

a. Single Image vs. Multitemporal Classification and Neighborhood Function

The first classification method is the most straightforward application of a LCZ classification and consists of applying the iterative process on a single date image. The seasonality was therefore analyzed in terms of meteorological seasons, namely winter, spring, summer and autumn, each in turn, by one scene at the center of the season [55]. For the multitemporal approach, accuracies of single-image classifications of each season will therefore be compared with those of a classification combining images of all four seasons. This is in order to account for the spectral and spatial changes in the natural vegetation that is caused by seasonal changes. This has the potential to increase the accuracy of the classification. This eliminates confusing between seasonal classes such as bare soil in the dry period, which is covered by low vegetation in the rainy periods.

b. Neighborhood Function

A neighborhood function or contextual classifier can contribute to increased accuracies in the classification of urban areas that are internally highly differentiated or heterogeneous, resulting from historical urbanization patterns that reflect the locality and the culture. In addition, most classification methods, including the original WUDAPT protocol, do not take this spatial variation into account. Moreover, the WUDAPT workflow causes a loss of spectral variability information before the actual classification by resampling the Landsat images during the pre-processing phase to a spatial resolution of 100 m [56–60]. At 10 m resolution, the sensitivity of the neighborhood function was tested by increasing the size of the kernel window for an optimal cell number.

iii. Validation

The ground truth data were randomly split in R into a training (80%) and validation (20%) set. The validation set is used to validate the model using accuracy metrics. The first accuracy metric performed was visual comparison of the output with satellite imagery. The User Accuracy (UA) is the probability that the predicted value is correct; the Producer's Accuracy (PA) is the probability that a value in a certain class was classified correctly. The Overall Accuracy (OA), the Kappa coefficient, is a measure of the agreement between classification and truth values. All were calculated according to the guidelines in [48]. The F1-score was calculated as the harmonic mean of the UA and PA, which is even more useful when the classes are not balanced.

3. Results
3.1. Local and Remote Standard LCZ Training (Approach 1)
3.1.1. Visual Analysis of the Classification Outputs

A visual inspection of the Cape Town output reveals a clear separation of the built from the natural classes (Figure 8). What this pattern also reveals is that the compact classes are rather concentrated next to each other and the open classes are further away. The city center toward the harbor is composed of compact high-rises and compact mid-rises, as expected in a modern city such as Cape Town. The agricultural lands up north are also visible from a visual inspection with minor patches of LCZ 9 through them. What is also worth noting is the confusion that arises between the paved surfaces and built classes. Roads at the city center are wrongfully classified as either compact high-rise (LCZ 1) or heavy industry (LCZ 10).

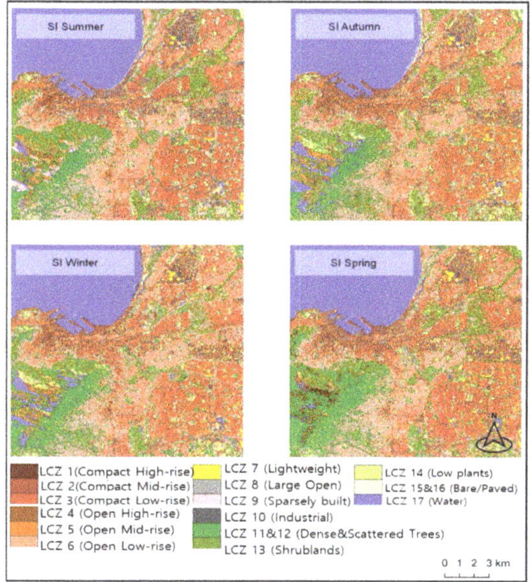

Figure 8. Cape Town single image classifications across all 4 seasons with standard training according to WUDAPT guidelines.

When the remote Cape Town-based standardized training is applied on a Thohoyandou multitemporal image, LCZ 3 is completely absent (Figure 9A), but it reappears immediately when a local standardized training is applied (Figure 9B). Using a local standardized training shows the Nandoni area to be mostly LCZ 9 with some vegetation; this is in line with the pre-study survey and google globe imagery that show the area being a rural village. However, the remote standardized training shows the same area as being mostly composed of LCZ 8 (Large Open), which is mostly found in the city center. Single-image classification using remote training shows LCZ 7 (light low-rise) at the city center, which according to google globe is a misclassification, as Thohoyandou does not have squatter camps, and this remote training also shows forested environment at the Nandoni region. This is different from the output of the local training, which is almost in complete agreement with the local MT classification showing no LCZ 7 at the city center and low plants and sparse trees at the Nandoni dam area. The application of remote vs. local training in East London does not yield significant differences in the appearance of classification outputs.

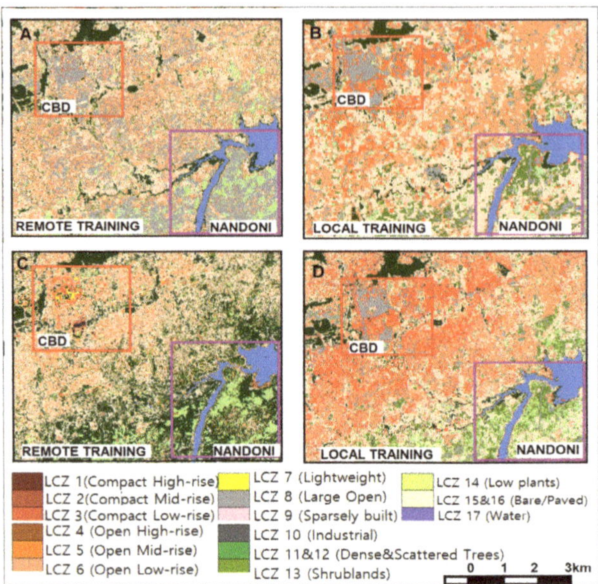

Figure 9. Single image (**C,D**) and multi-temporal (**A,B**) classifications of Thohoyandou using a standardized remote (**A,C**) and local (**B,D**) training.

3.1.2. Single Versus Multi-Seasonal Images and Neighborhood Functions Comparing Performance of Remote and Local Standard Trainings

While the multitemporal Cape Town image is the most accurately classified with an overall accuracy of 44%, the spring image has the highest accuracy of the single image classifications with an accuracy of 42% (Table 2). The difference in Kappa and OA metrics between the spring image and the multitemporal image is relatively small: 1.6% for Kappa and 1.4% for OA. This suggests that the spring image could be an acceptable representation of the Cape Town area when there are no multitemporal data. The spring image OA is a 6% improvement on the summer classification, which is at 36%. However, even this 6% is mostly due to the natural vegetation (OA-nat), which is 15% higher in spring than in the summer. The difference between accuracies of the urban classes (OA-urb) is only 2%. This also proves that the effects of phenology are greatest among natural classes, as expected. The summer image as the least successfully classified has the lowest Kappa at 31.2%, which is a 68.8% disagreement between the training and the output. This indicates that the least represented classes in the training are not very well classified. According to the Kappa statistic, the least classified (summer) and the best classified (multitemporal) all have Kappa values that fall within the same range. This means about 60% disagreement between training and output, suggesting that only 40% of the data can be relied upon to produce the observed results. The difference between the F1 score and the OA is also Indicative of patterns in classification of individual classes, since the F1 is a harmonic mean of the PA and UA. From Table 2, the summer and autumn images F1-scores and OA are almost identical which indicates confusion even in the classes which are rather well classified in other seasons.

Table 2. Accuracy metrics for single image and multitemporal random forest classifications over Cape Town.

Metrics	SI Summer	SI Autumn	SI Winter	SI Spring	Multitemporal
OA	36.20%	40.10%	41.80%	42.70%	44.10%
OA-urb	29.80%	28.70%	31.80%	32.70%	33.80%
OA-nat	47.20%	68.20%	64.30%	63.90%	68.50%
Kappa	31.20%	35.50%	37.20%	38.00%	39.60%
F1–metric	36.50%	40.40%	42.60%	43.20%	45.20%

The confusion matrix of the spring image shows that the built classes have more confusion than the natural classes (Table 3). While LCZ 3 and LCZ 6 are visually well represented across all reasons, the confusion matrix reveals that they have the highest confusion in general. A lot of bare lands in the summer and autumn are classified as built classes. This trend is also seen in the spring season with LCZ 16 confused with LCZ 9 and LCZ 10. However, the spring image is still the best classified single date image and is the best season to test the Cape Town based remote training in Thohoyandou and East London. The other tool that improves the classification is the contextual classifier, and the optimal kernel size must be determined for application across all urban areas. This fragmentation (salt and pepper effect) of classes is most visible from the classification output (Figures 8 and 9).

Table 3. Confusion matrix from the standard LCZ training single date spring image over Cape Town.

Classified	\(Reference Classes\) 1	2	3	4	5	6	7	8	9	10	11	13	14	16	17	User's Accuracy
1	42	15	0	10	8	2	0	63	0	9	0	0	7	0	0	27%
2	27	120	17	61	71	19	41	87	6	44	7	1	9	0	0	24%
3	55	156	506	77	146	326	213	155	55	58	0	0	4	0	0	29%
4	16	46	7	67	96	33	2	46	26	26	6	33	13	2	1	16%
5	43	100	26	82	118	22	22	63	58	13	11	9	54	0	8	19%
6	80	166	75	171	195	352	0	1	315	6	59	59	4	4	0	24%
7	1	46	438	11	5	120	847	103	65	46	0	7	5	0	0	50%
8	53	107	42	48	66	64	107	247	34	116	0	6	13	0	0	27%
9	27	45	82	66	124	180	50	40	462	28	134	229	13	38	0	30%
10	37	34	32	41	57	9	100	218	15	607	0	0	0	99	0	49%
11	3	1	0	25	37	3	1	2	64	0	763	147	88	3	82	63%
13	3	12	27	24	11	46	41	27	106	124	30	702	254	30	0	49%
14	0	6	2	5	9	0	0	22	104	1	0	30	429	0	0	71%
16	0	3	1	9	0	5	65	25	36	17	0	23	130	293	0	48%
17	0	0	0	0	0	0	0	0	0	0	0	0	0	78	929	92%
Producer's Accuracy	11%	14%	40%	10%	13%	30%	57%	22%	34%	55%	76%	56%	42%	54%	91%	Overall Accuracy 43%

The neighborhood function aggregates the pixels within a certain threshold into a single value and results in a smoother output as compared to the raw data classification (Figure 10). This is tested on Cape Town and the optimal kernel size applied to Thohoyandou and East London to compare their results with Cape Town. The accuracy remains constant at 42.7% and Kappa at 43.2% for all kernel sizes below 11 cells and reaches its peak at 13 cells, after which the accuracy starts to decrease (Table 4).

Table 4. Cape Town classification with 13 pixel neighborhood function.

	SI	SI + NH			MT	MT + NH		
Kernel Size	*I*	**11 × 11**	**13 × 13**	**15 × 15**	*I*	**11 × 11**	**13 × 13**	**15 × 15**
OA	42.70%	48.50%	48.70%	48.40%	44.10%	49.10%	49.80%	49.50%
OA-urb/nat	88.80%	91.30%	91.00%	90.50%	87.70%	89.70%	89.40%	89.00%
OA-urb	32.70%	37.10%	37.50%	37.00%	33.80%	38.50%	39.40%	39.40%
OA-nat	63.90%	72.70%	72.70%	72.70%	68.50%	71.70%	71.80%	71.20%
Kappa	38.00%	44.10%	44.40%	44.10%	39.60%	44.90%	45.60%	45.40%
F1–metric	43.20%	49.20%	49.40%	49.10%	45.20%	50.10%	50.90%	50.70%

With a 13 × 13 kernel of 130 m at a resolution of 10 m, the accuracy of the spring single image classification is improved by 6.0%, and the multitemporal classification is improved by 5.7%. However, in general, the multitemporal combined with the neighborhood function still yields higher accuracies than the single image with the neighborhood function. The highest overall accuracy of the multitemporal is 49.8% at a 13 × 13 moving window size, bringing the total improvement over the general single image to 7.2%.

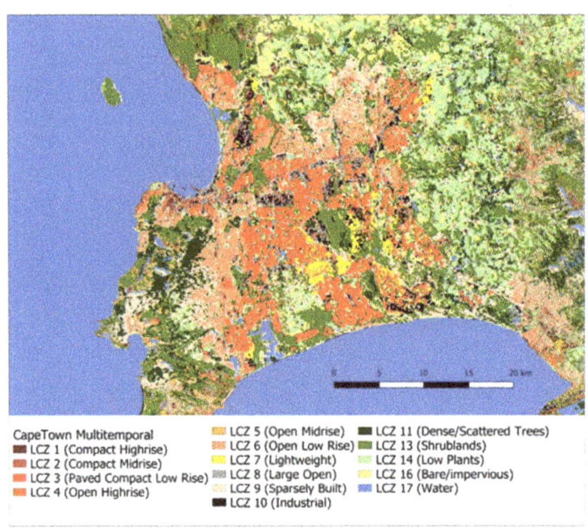

Figure 10. Classification output for Cape Town with a neighborhood function of size 13.

Nevertheless, it is only a slight improvement of 1.1% compared to the single-image classification, which has an overall accuracy of 47.7%. Table 5 also shows that for both the single and multitemporal classifications, the overall accuracy of the combinations of both natural and built classes (OA-nat/urb) metric is highest at a kernel size of 11 × 11 or 110 m, indicating that confusion between natural classes and built classes is lower in these classifications. However, this is only a 0.3% difference from the 13-cell kernel application, which individually has higher OA-nat and OA-urb. For the purposes of this classification, a standard 13 cell NH kernel was adopted as the optimal kernel size for comparison across the three urban regions.

The application of remote training yields the highest results in a Thohoyandou multi-date stack at 53.2%; however, a local standardized training yields a 7% increase in overall accuracy and a 10% increase in Kappa. Urban classes are less successfully classified across the board. However, they are better classified in East London than they are in Thohoyandou. The use of remote vs. local classification in East London does not seem to have as big an impact as it does on Thohoyandou. The multi-date classification using local standardized training is only 2.1% higher than its remote training counterpart. The single-date classification using remote training yields the lowest overall accuracy, Kappa and F1-metric across the board, while the multi-date local training yields the highest values of the same metrics. There seems to be a similarity in the LCZ layout between Cape Town and East London, which is seen through the accuracy metrics, which are almost similar for both remote and local training.

Table 5. Accuracies of single-date and multitemporal classification using remote (local to Cape Town but remote to Thohoyandou and East London) and local (standard training collected in Thohoyandou and East London).

	CPT		Thohoyandou				East London			
Training	Local		Remote		Local		Remote		Local	
Stack	SI	MT	SI	MT	SI	MT	SI	MT	SI	MT
OA	48.70%	49.80%	31.40%	53.20%	54.8186	60.50%	39.10%	41.30%	40.70%	43.40%
OA-urb/nat	91.00%	89.40%	72.90%	81.50%	81.00%	86.20%	91.70%	85.50%	89.00%	84.30%
OA-urb	37.50%	39.40%	16.80%	37.40%	31.00%	34.00%	22.90%	26.10%	34.60%	42.10%
OA-nat	72.70%	71.80%	41.80%	94.00%	75.00%	80.70%	61.90%	72.70%	79.10%	80.60%
Kappa	44.40%	45.60%	22.10%	45.50%	45.7	55.30%	33.30%	35.70%	35.10%	37.60%
F1–metric	49.40%	50.90%	24.00%	53.50%	53.10%	56.30%	33.50%	38.40%	33.90%	40.10%

3.2. Classification Using Local Customized Training Data (Approach 2)

3.2.1. Thohoyandou Classification

a.   Random Splitting of Training Data

Using a randomized selection of training and test sample makes the model more robust, but it does not create an even number of pixels throughout. The challenge even in the development of the original training protocol was that some classes had better representation than others. While this can be avoided by creating an even amount of training samples throughout, there simply is not always enough land area to create regions of interest in some classes. Other classes also call for a higher number of training due to their inter-class variability that must be accounted for in the training sample. The output of the randomly selected training pixels (Figure 11A) shows that classes 14, 15 and 16 do not have enough representation in the sample. This was solved by joining them to other classes that have similar spectral signatures. Class 14 was combined with class 13; class 15 was combined with class 16. The result of this ends up with the lowest number of pixels in the training going from 250 pixels to above 700 (Figure 11B).

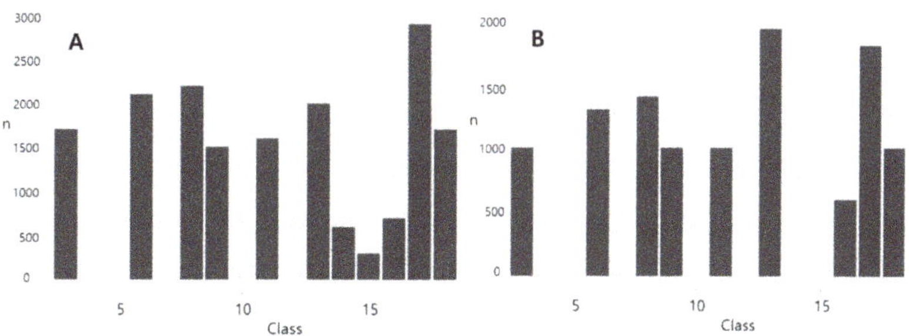

Figure 11. Pixels per class for the randomly selected training input for Approach 2 in Thohoyandou where (**A**) is unmerged and (**B**) is merged.

b.   Single Versus Multi-Seasonal Classification and Comparison with Standard Training

What has been observed in Approach 1 classifications is still maintained in the Approach 2 results. This is the observation that the spring image provides the highest accuracy of all the single image classifications (Table 6). However, natural vegetation is more accurately classified in the winter NH than in any other season, while urban is highest in the spring. The multitemporal image that combines all the seasons is the most accurately classified with an accuracy of 82.68% as compared to 53.2% of Approach 1.

Table 6. Thohoyandou accuracy metrics for SI and MT raw and NH with local customized training.

	Summer		Autumn		Winter		Spring		Multitemporal	
	Raw	**NF**	**Raw**	**NF**	**Raw**	**NF**	**Raw**	**NF**	**Raw**	**NF**
OA	60.64	71.74	54.92	71.36	58.25	71.8	59.15	73.42	73.92	82.68
OA-nat	81.65	87.36	79.1	87.44	85.06	87.83	80.61	85.54	83.65	88.58
OA-urb	43.44	62.11	40.92	55.06	42.01	58.61	47.11	63.49	53.2	69.66
Kappa	53.12	65.19	46.08	64.58	50.4	65.69	51.83	67.5	70.31	75.6
F1 score	0.554	0.65	0.5324	0.627	0.5513	0.661	0.5766	0.694	0.695	0.764

What the confusion matrix of the multitemporal classification reveals is that all the classes are more accurately classified than misclassified except for LCZ 16, which is a compound class of LCZ 15 and original LCZ 16. The highest confusion of LCZ 16 is with LCZ 8, which is due to the paved spaces between large open buildings, and LCZ 13, which is due to the large bare regions in between shrubs (Table 7).

Table 7. Thohoyandou confusion matrix for multitemporal classification.

Classified	Reference										User's Accuracy
	3	**6**	**8**	**9**	**11**	**13**	**16**	**17**	**3a**	**Total**	
3	738	0	29	0	0	0	17	0	33	817	90%
6	47	1284	1	0	0	186	113	0	110	1741	74%
8	85	0	1138	0	0	156	258	0	0	1637	70%
9	0	0	0	900	0	131	0	0	0	1031	87%
11	0	0	0	0	943	0	29	0	0	972	97%
13	42	11	4	0	57	1416	256	0	0	1786	79%
16	0	0	106	0	0	42	27	0	0	175	15%
17	0	0	0	0	0	0	0	1800	0	1800	100%
3a	188	5	22	0	0	0	0	0	957	1172	82%
Total	1100	1300	1300	900	1000	1931	700	1800	1100	11,131	
Producer's Accuracy	67%	99%	88%	100%	94%	73%	4%	100%	87%		Overall Accuracy 82.7%

The analysis of the multitemporal classification model reveals that the higher bands (Band 9—11) of each season are the most important in classifying the pixels (Figure 12). These are the SWIR (shortwave infrared) bands of the Sentinel 2A image stack.

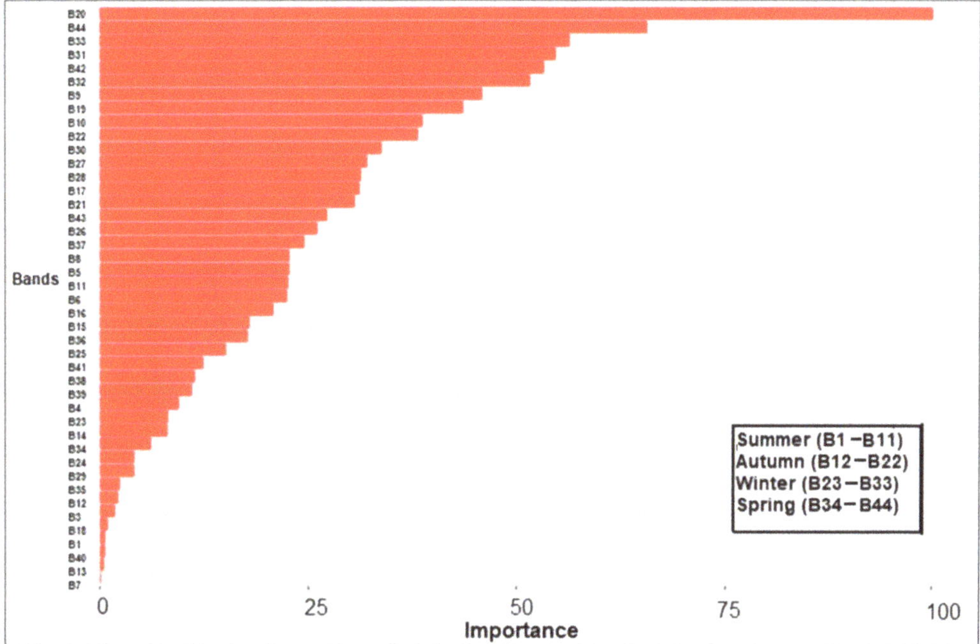

Figure 12. Band priority and importance in the class discrimination process for multitemporal stack in Thohoyandou.

The discrimination between the gravel LCZ 3a and the asphalted LCZ 3 seems to be very good according to a visual inspection of the classification output (Figure 13). The confusion matrix also confirms this with 17% of the LCZ 3 classified as LCZ 3a. However, the highest confusion with LCZ 3a is with LCZ 6. This is due to them sharing similar open spaces and also their proximity with 11.5% of LCZ 3a pixels classified as LCZ 6.

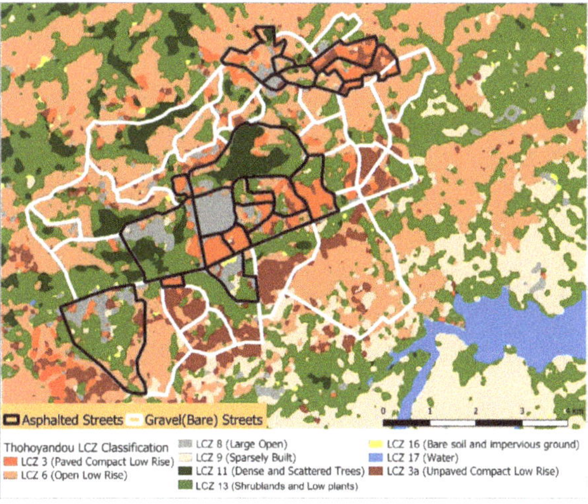

Figure 13. Thohoyandou classification output for multitemporal with neighborhood function with local customized training showing asphalted and gravel streets.

3.2.2. East London Classification

Using a local and customized training input yielded higher overall accuracy and Kappa values. The highest OA value for East London using Approach 1 was 41.3%, which has increased to 58.5% using Approach 2 which is a 17.5% increase. However, contrary to observations from the Approach 1 results, the multitemporal image does not appear to be the best fit for both built and natural classes. The spring image with a 13-cell neighborhood function yields the highest overall Kappa, F1-Score, and overall accuracy (Table 8). This is in agreement with the observation made in Approach 1, where the spring image was the highest of the seasonal single image classifications. What is also similar is that the summer image has the lowest values for all the overall metrics. It is worth noting that the spring image still has the highest number of lowest individual class Kappa values spread out across built classes; this suggests that the higher overall metrics are due to the perfect and near perfect classification of the natural classes. The highest individual LCZ Kappa values are not localized to a single season but spread out through different seasons. A common trend, however, is that the highest individual Kappa values fall within the classifications that have been smoothed out with the 13-cell kernel, while the lowest values are within the raw image classification.

Table 8. East London, individual Kappa and overall accuracy metrics for single image and multitemporal raw and neighborhood function with local customized training.

	Summer		Autumn		Winter		Spring		Multitemporal	
CLASS KAPPA	Raw	NH	Raw	NH	Raw	NH	Raw	NH	Raw	NH
LCZ 2	0.216	0.238	0.422	0.425	0.443	0.457	0.453	0.456	0.436	0.465
LCZ 3	0.331	0.373	0.364	0.586	0.295	0.489	0.236	0.497	0.325	0.454
LCZ 5	0.136	0.009	0.130	−0.008	0.161	0.020	0.152	0.030	0.135	0.065
LCZ 6	0.299	0.419	0.340	0.369	0.295	0.456	0.296	0.456	0.350	0.440
LCZ 7	0.267	0.504	0.391	0.991	0.381	0.873	0.381	0.877	0.455	0.749
LCZ 8	0.555	0.653	0.513	0.612	0.503	0.659	0.502	0.661	0.458	0.516
LCZ 9	0.346	0.410	0.300	0.372	0.264	0.356	0.255	0.356	0.350	0.408
LCZ 10	0.076	0.157	0.149	−0.003	0.154	−0.006	0.157	−0.006	0.147	−0.005
LCZ 11	0.837	0.690	0.856	0.784	0.848	0.765	0.838	0.765	0.853	0.728
LCZ 13	0.417	0.598	0.430	0.665	0.289	0.631	0.275	0.631	0.763	0.936
LCZ 14	0.125	nan	0.145	nan	0.524	1.000	0.512	1.000	−0.070	nan
LCZ 16	1.000	1.000	0.989	1.000	1.000	1.000	1.000	1.000	0.988	1.000
LCZ 17	0.997	1.000	0.964	0.959	0.922	0.873	0.832	0.873	0.972	1.000
LCZ 18	0.435	0.739	0.395	0.434	0.387	0.560	0.335	0.558	0.468	0.651
MAIN KAPPA	0.431	0.522	0.456	0.553	0.462	0.581	0.445	0.582	0.474	0.570
OA	45.951	53.530	48.305	56.528	46.736	58.401	46.723	58.553	50.240	58.540
F1	0.444	0.497	0.465	0.525	0.467	0.565	0.462	0.567	0.481	0.538
OA-urb/nat	76.800	76.500	68.000	75.600	76.700	75.600	76.700	75.600	77.200	76.300

What the spring neighborhood function image (Figure 14) reveals visually is that LCZ 7 and 7a are in close proximity to LCZ 3. This goes from a strict LCZ 3 and moves into LCZ 7a, which is an integration of 7 and 3, and then ultimately moves into a strict LCZ 7.

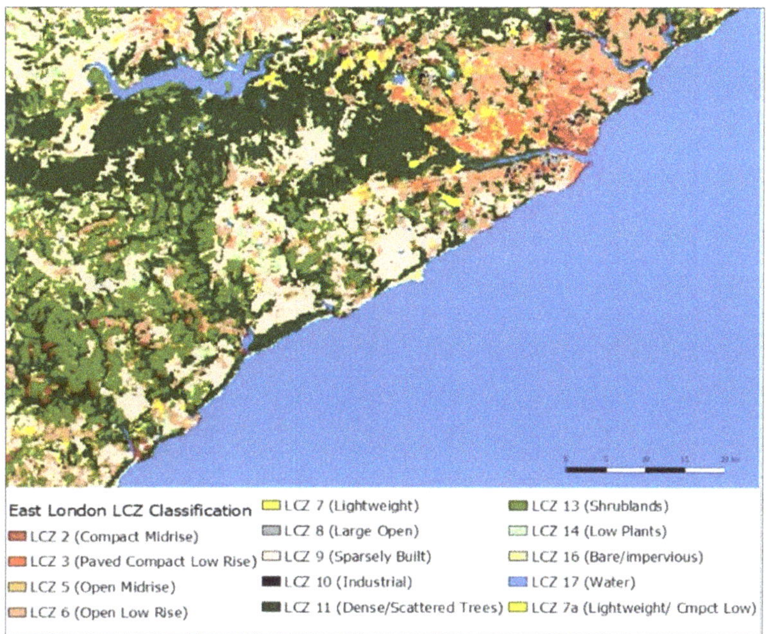

Figure 14. East London classification output for multitemporal with neighborhood function using local customized training.

4. Conclusions

Cape Town is a multi-nuclei urban region of multi-cultural origin, East London is a harbor town, and Thohoyandou is a small town that originated as the administrative capital of the Venda Bantustan. These urban regions represent a gradient within urbanization in South Africa. These different historical backgrounds contribute to the uniqueness of the layout and feature type in each region, which is a phenomenon also noted in the Middle East [25]. These unique features become an element of importance, as they could potentially explain the poor performance of the standard framework when performed using multispectral imagery at the local scale in Africa. Cape Town as an urban area resembles closely the cities of the west; as such, the standard LCZ framework typology would best fit Cape Town with minimal to no adjustment in the guidelines for RoI creation. However, the development of a localized and customized training for East London and Thohoyandou individually creates a classification protocol that considers these unique local features stemming from influences of their unique origin and cultural evolution as they herd toward modernization.

The nature of the training input was the major difference between Approach 1 and Approach 2. Where Approach 1 used a single all-inclusive training input for all three cities, Approach 2 used a local customized training input for each urban region and yields better results. The biggest challenge in this study was the lack of a height layer in the stack as a discriminator for the algorithm. Ref. [61] stated that the presence of a height layer is essential for cities with LCZs belonging to different height classes (low, mid and high-rise). What is immediately noticeable in the accuracies metrics is the big difference between values obtained in homogenous-height Thohoyandou across all LCZs and the heterogenous-height Cape Town and East London using both Approaches 1 and 2 (Tables 6–8). Without a height discriminator in the classification stack, there is inter-class confusion within the compact classes as well as the open ones (Table 3). Ref. [62] addressed the height challenge by using an abridged version of the LCZ classification that considers surface feature density but eliminates height altogether. While this land cover-based framework by [62] was also proven to explain trends in urban heat islands, the local scale suffers from detail loss. This compromise of detail over accuracy renders the output less useful and to some degree even unsuitable for local climate models. The best way to address the height data gap challenge while maintaining detail resolution remains using locally calibrated training as opposed to the [62] land cover approach.

Both approaches revealed that a local customized training sample is a better fit for the random forest LCZ classifier than using a standardized training input for all regions. This is seen through the classifier performance being better in Approach 2 as opposed to Approach 1. The literature would also dictate that seasonality would mostly affect natural LCZ classes because of plant phenology [63–65]. However, as seen in the metric tables (Tables 3, 6 and 7), urban classes are also classified to varying degrees of accuracy at different seasons. While the higher short-ware infrared (SWIR) bands are always the most important in the automated LCZ discrimination protocol, the lower bands range from minimally important to completely negligible. Studies by [66,67] isolate the variations in band priority for different seasons as a function of the physical properties of surface features. This is not limited to biotic but also abiotic features such buildings. The multitemporal classification was the most accurately classified of all classifications. Ref. [68] stated that the effects of seasonality are addressed by taking a multitemporal stack which covers classing during all stages of annual variability. While the actual seasons are classified to varying degrees, the multitemporal local customized training would still be more representative of the LCZ classes than using a single image from any season.

The size of the pixel also determines the accuracy. As such, a contextual classifier (NF) significantly improves the accuracy of the model [69]. While applying a neighborhood function does not change the pixel of the image, it reduces the level of detail in the classification output. This is seen by visually looking at the raw data output (Figure 9) as compared to the NF output (Figure 12). The fragmentation is less apparent when the pixel size is higher than 100 m, as seen when the WUDAPT online generator is used [66]. The disadvantage is that classifying LCZ with a local-scale pixel size (100 m) reduces the level of detail that that would otherwise be found in using higher-resolution imagery, which in this study was 10 m. This is crucial while working with urban climate models. The challenge in using a contextual classifier is in finding a kernel size that balances detail, accuracy and aesthetic for the specific goal for which the classification is intended to be used. While for the purpose of this study, the aim of the NF was to achieve the highest accuracy, the nested algorithm is flexible enough to modify the kernel size should the purpose of the classification be different.

An application of these methods in future studies should consider using more training samples for the less represented classes. In addition, whether the accuracy in Thohoyandou would improve if a height discriminator is part of the protocol is worth exploring further. Height is definitely recommended as an important addition to the classification stack for East London, Cape Town or any other city with mid- and high-rise classes. The findings of this particular study as well as the methodological protocols would be recommended for adoption by any future study that aims at studying UHIs in the African context, particularly investigating the spatial correlation between the patterns that are observed in the UHIs with the underlying LCZ classification.

Author Contributions: Conceptualization, T.M. and J.T., B.S., B.V.; Data curation, T.M. and J.T.; Formal analysis, T.M. and J.T.; Funding acquisition, B.S., B.V. and N.N.; Methodology, T.M., J.T., B.S., B.V.; Software, T.M. and J.T.; Visualization, T.M. and J.T.; Writing—original draft, T.M.; Writing—review and editing, T.M., B.S., B.V. All authors have read and agreed to the published version of the manuscript.

Funding: This work was funded by VLIR-UOS, the Flemish University Council for University Development Cooperation through the ReSider project and the Flemish government funded SAF-ADAPT project. Funding number: 000000166183.

Data Availability Statement: Data are available on request from the authors.

Acknowledgments: I would like to acknowledge the help of Anna Van Eyck in collecting the validation data in Thohoyandou and Cape Town in February-March 2022.

Conflicts of Interest: The authors declare no conflict of interest. The funders had no role in the design of the study; in the collection, analyses, or interpretation of data; in the writing of the manuscript; or in the decision to publish the results.

References

1. Adams, R.M. The origin of cities. *Sci. Am.* **1960**, *203*, 153–172. [CrossRef]
2. Robinson, E.; Zahid, H.J.; Codding, B.F.; Haas, R.; Kelly, R.L. Spatiotemporal dynamics of prehistoric human population growth: Radiocarbon 'dates as data' and population ecology models. *J. Archaeol. Sci.* **2019**, *101*, 63–71.
3. Pierson, W.A. Spatial Assessment of Urban Growth in Cities of the Decapolis; and the Implications for Modern Cities. Ph.D. Thesis, University of Arkansas, Fayetteville, AR, USA, 2021.
4. Schaedel, R.P.; Hardoy, J.E.; Scott-Kinzer, N. (Eds.) *Urbanization in the Americas from Its Beginning to the Present*; Walter de Gruyter: Berlin, Germany, 2011.
5. Chen, S.; Chen, B.; Feng, K.; Liu, Z.; Fromer, N.; Tan, X.; Alsaedi, A.; Hayat, T.; Weisz, H.; Schellnhuber, H.J.; et al. Physical and virtual carbon metabolism of global cities. *Nat. Commun.* **2020**, *11*, 182. [CrossRef] [PubMed]
6. Balchin, W.G.V.; Pye, N. A micro-climatological investigation of bath and the surrounding district. *Q. J. R. Meteorol. Soc.* **1947**, *73*, 297–323. [CrossRef]
7. Stewart, I.D. Why should urban heat island researchers study history? *Urban. Clim.* **2019**, *30*, 100484. [CrossRef]
8. Lhotka, O.; Kyselý, J.; Farda, A. Climate change scenarios of heat waves in Central Europe and their uncertainties. *Theor. Appl. Climatol.* **2018**, *131*, 1043–1054. [CrossRef]
9. Profiroiu, C.M.; Bodislav, D.A.; Burlacu, S.; Rădulescu, C.V. Challenges of sustainable urban development in the context of population Growth. *Eur. J. Sustain. Dev.* **2020**, *9*, 51. [CrossRef]
10. Al-Thani, H.; Koç, M.; Isaifan, R.J. A review on the direct effect of particulate atmospheric pollution on materials and its mitigation for sustainable cities and societies. *Environ. Sci. Pollut. Res.* **2018**, *25*, 27839–27857. [CrossRef] [PubMed]
11. Longo, S.; Montana, F.; Sanseverino, E.R. A review on optimization and cost-optimal methodologies in low-energy buildings design and environmental considerations. *Sustain. Cities Soc.* **2019**, *45*, 87–104. [CrossRef]
12. Griffith, D.A.; Can, A. Spatial statistical/econometric versions of simple urban population density models. In *Practical Handbook of Spatial Statistics*; CRC Press: Boca Raton, FL, USA, 2020; pp. 231–249.
13. Zhou, X.; Okaze, T.; Ren, C.; Cai, M.; Ishida, Y.; Watanabe, H.; Mochida, A. Evaluation of urban heat islands using local climate zones and the influence of sea-land breeze. *Sustain. Cities Soc.* **2020**, *55*, 102060. [CrossRef]
14. Zhou, D.; Xiao, J.; Bonafoni, S.; Berger, C.; Deilami, K.; Zhou, Y.; Frolking, S.; Yao, R.; Qiao, Z.; Sobrino, J.A. Satellite remote sensing of surface urban heat islands: Progress, challenges, and perspectives. *Remote Sens.* **2019**, *11*, 48. [CrossRef]
15. Stewart, I.D.; Oke, T.R. Local climate zones for urban temperature studies. *Bull. Am. Meteorol. Soc.* **2012**, *93*, 1879–1900. [CrossRef]
16. Bechtel, B.; Alexander, P.J.; Böhner, J.; Ching, J.; Conrad, O.; Feddema, J.; Mills, G.; See, L.; Stewart, I. Mapping local climate zones for a worldwide database of the form and function of cities. *ISPRS Int. J. Geo-Inf.* **2015**, *4*, 199–219. [CrossRef]
17. Mushore, T.D.; Mutanga, O.; Odindi, J. Determining the Influence of Long Term Urban Growth on Surface Urban Heat Islands Using Local Climate Zones and Intensity Analysis Techniques. *Remote Sens.* **2022**, *14*, 2060. [CrossRef]
18. Yang, J.; Wang, Y.; Xiu, C.; Xiao, X.; Xia, J.; Jin, C. Optimizing local climate zones to mitigate urban heat island effect in human settlements. *J. Clean. Prod.* **2020**, *275*, 123767. [CrossRef]
19. Zheng, Y.; Ren, C.; Xu, Y.; Wang, R.; Ho, J.; Lau, K.; Ng, E. GIS-based mapping of Local Climate Zone in the high-density city of Hong Kong. *Urban. Clim.* **2018**, *24*, 419–448. [CrossRef]
20. Brousse, O.; Georganos, S.; Demuzere, M.; Vanhuysse, S.; Wouters, H.; Wolff, E.; Linard, C.; Nicole, P.M.; Dujardin, S. Using local climate zones in sub-Saharan Africa to tackle urban health issues. *Urban. Clim.* **2019**, *27*, 227–242. [CrossRef]
21. Engelbrecht, C.J.; Engelbrecht, F.A. Shifts in Köppen-Geiger climate zones over southern Africa in relation to key global temperature goals. *Theor. Appl. Climatol.* **2016**, *123*, 247–261. [CrossRef]
22. Wichmann, J. Heat effects of ambient apparent temperature on all-cause mortality in Cape Town, Durban and Johannesburg, South Africa: 2006–2010. *Sci. Total Environ.* **2017**, *587*, 266–272. [CrossRef]
23. Kiet, A. Arab culture and urban form. *Focus* **2011**, *8*, 10. [CrossRef]
24. Živković, J. Urban form and function. *Clim. Action* **2020**, 862–871. [CrossRef]
25. Saleh, M.A.E. The impact of Islamic and customary laws on urban form development in southwestern Saudi Arabia. *Habitat Int.* **1998**, *22*, 537–556. [CrossRef]
26. Huang, J.; Lu, X.X.; Sellers, J.M. A global comparative analysis of urban form: Applying spatial metrics and remote sensing. *Landsc. Urban Plan.* **2007**, *82*, 184–197. [CrossRef]
27. Akinyemi, F.O.; Ikanyeng, M.; Muro, J. Land cover change effects on land surface temperature trends in an African urbanizing dryland region. *City Environ. Interact.* **2019**, *4*, 100029. [CrossRef]
28. Wang, R.; Ren, C.; Xu, Y.; Lau, K.K.L.; Shi, Y. Mapping the local climate zones of urban areas by GIS-based and WUDAPT methods: A case study of Hong Kong. *Urban Clim.* **2018**, *24*, 567–576. [CrossRef]

29. Demuzere, M.; Kittner, J.; Bechtel, B. LCZ Generator: A web application to create Local Climate Zone maps. *Front. Environ. Sci.* **2021**, *9*, 637455. [CrossRef]
30. Mushore, T.D.; Dube, T.; Manjowe, M.; Gumindoga, W.; Chemura, A.; Rousta, I.; Odindi, J.; Mutanga, O. Remotely sensed retrieval of Local Climate Zones and their linkages to land surface temperature in Harare metropolitan city, Zimbabwe. *Urban Clim.* **2019**, *27*, 259–271. [CrossRef]
31. Lei, N.; Masanet, E. Climate-and technology-specific PUE and WUE estimations for US data centers using a hybrid statistical and thermodynamics-based approach. *Resour. Conserv. Recycl.* **2022**, *182*, 106323. [CrossRef]
32. Dimitrov, S.; Popov, A.; Iliev, M. Mapping and Assessment of Urban Heat Island Effects in the City of SOFIA, Bulgaria Through Integrated Application of Remote Sensing, Unmanned Aerial Systems (UAS) and GIS. In Proceedings of the Eighth International Conference on Remote Sensing and Geoinformation of the Environment (RSCy2020), Paphos, Cyprus, 16–18 March 2020; International Society for Optics and Photonics: Bellingham, WA, USA, 2020; Volume 11524, p. 115241A.
33. Zonato, A.; Martilli, A.; Di Sabatino, S.; Zardi, D.; Giovannini, L. Evaluating the performance of a novel WUDAPT averaging technique to define urban morphology with mesoscale models. *Urban Clim.* **2020**, *31*, 100584. [CrossRef]
34. Rodríguez-Carrión, N.M.; Hunt, S.D.; Goenaga-Jimenez, M.A.; Vélez-Reyez, M. June. Determining optimum pixel size for classification. In *Algorithms and Technologies for Multispectral, Hyperspectral, and Ultraspectral Imagery XX*; International Society for Optics and Photonics: Bellingham, WA, USA, 2014; Volume 9088, p. 90880X.
35. Pascale, S.; Kapnick, S.B.; Delworth, T.L.; Cooke, W.F. Increasing risk of another Cape Town "Day Zero" drought in the 21st century. *Proc. Natl. Acad. Sci. USA* **2020**, *117*, 29495–29503. [CrossRef]
36. Van der Walt, A.J.; Fitchett, J.M. Statistical classification of South African seasonal divisions on the basis of daily temperature data. *S. Afr. J. Sci.* **2020**, *116*, 1–15. [CrossRef]
37. Chikoore, H.; Bopape, M.J.M.; Ndarana, T.; Muofhe, T.P.; Gijben, M.; Munyai, R.B.; Manyanya, T.C.; Maisha, R. Synoptic structure of a sub-daily extreme precipitation and flood event in Thohoyandou, north-eastern South Africa. *Weather Clim. Extrem.* **2021**, *33*, 100327. [CrossRef]
38. Worden, N.; Van Heyningen, E.; Bickford-Smith, V. *Cape Town: The Making of a City: An Illustrated Social History*; Uitgeverij Verloren: Hilversum, The Netherlands, 1998.
39. Bickford-Smith, V. Creating a city of the tourist imagination: The case of Cape Town, The fairest Cape of them all'. *Urban. Stud.* **2009**, *46*, 1763–1785. [CrossRef]
40. Thornberry, E. Rape, Race, and Respectability in a South African Port City: East London, 1870-1927. *J. Urban. Hist.* **2016**, *42*, 863–880. [CrossRef]
41. Hangwelani, H.M.; Lovemore, C. The Impact of Island City in the Post-Apartheid South Africa: Focus on Bantustans. In Proceedings of the Urban Form and Social Context: From Traditions to Newest Demands, Krasnoyarsk, Russia, 5–9 July 2018; pp. 118–128.
42. Kruger, J. Singing psalms with owls: A Venda twentieth century musical history. *Afr. Music. J. Int. Libr. Afr. Music.* **1999**, *7*, 122–146. [CrossRef]
43. Sykas, D.; Papoutsis, I.; Zografakis, D. Sen4AgriNet: A Harmonized Multi-Country, Multi-Temporal Benchmark Dataset for Agricultural Earth Observation Machine Learning Applications. In Proceedings of the 2021 IEEE International Geoscience and Remote Sensing Symposium IGARSS, Brussels, Belgium, 11–16 July 2021; pp. 5830–5833.
44. Main-Knorn, M.; Pflug, B.; Louis, J.; Debaecker, V.; Müller-Wilm, U.; Gascon, F. Sen2Cor for sentinel-2. *Image Signal Process. Remote Sens. XXIII* **2017**, *10427*, 37–48.
45. Lillesand, T.; Kiefer, R.W.; Chipman, J. *Remote Sensing and Image Interpretation*; John Wiley & Sons: New York, NY, USA, 2015.
46. Sekertekin, A.; Abdikan, S.; Marangoz, A.M. The acquisition of impervious surface area from LANDSAT 8 satellite sensor data using urban indices: A comparative analysis. *Environ. Monit. Assess.* **2018**, *190*, 1–13. [CrossRef]
47. Bechtel, B.; Daneke, C. Classification of local climate zones based on multiple earth observation data. *IEEE J. Sel. Top. Appl. Earth Obs. Remote Sens.* **2012**, *5*, 1191–1202. [CrossRef]
48. Foody, G.M. Status of land cover classification accuracy assessment. *Remote Sens. Environ.* **2002**, *80*, 185–201. [CrossRef]
49. Weih, R.C.; Riggan, N.D. Object-based classification vs. pixel-based classification: Comparative importance of multi-resolution imagery. *Int. Arch. Photogramm. Remote Sens. Spat. Inf. Sci.* **2010**, *38*, C7.
50. Tabi, K.A. Coping with Weather in Cape Town: Use, Adaptation & Challenges in an Informal Settlement. Master's Thesis, University of the Western Cape, Cape Town, South Africa, 2013.
51. Plantier, T.; Loureiro, M.; Marques, P.; Caetano, M. Spectral analyses and classification of ikonos images for forest cover characterisation. *Cent. Remote Sens. Land Surf.* **2006**, *28*, 260–268.
52. Jin, L.; Pan, X.; Liu, L.; Liu, L.; Liu, J.; Gao, Y. Block-based local climate zone approach to urban climate maps using the UDC model. *Build. Environ.* **2020**, *186*, 107334. [CrossRef]
53. Abrams, F. *Mapping Local Climate Zones with SENTINEL Imagery in Ha Tinh and Hanoi, Vietnam*; Katholieke Universiteit Leuven: Leuven, Belgium, 2019.
54. Mora, C.; Vieira, G.; Pina, P.; Lousada, M.; Christiansen, H.H. Land cover classification using high-resolution aerial photography in adventdalen, Svalbard. *Geogr. Ann. Ser. A Phys. Geogr.* **2015**, *97*, 473–488. [CrossRef]
55. Geletič, J.; Lehnert, M.; Savić, S.; Milošević, D. Inter-/intra-zonal seasonal variability of the surface urban heat island based on local climate zones in three central European cities. *Build. Environ.* **2019**, *156*, 21–32. [CrossRef]

56. Bechtel, B.; Demuzere, M.; Mills, G.; Zhan, W.; Sismanidis, P.; Small, C.; Voogt, J. SUHI analysis using Local Climate Zones—A comparison of 50 cities. *Urban. Clim.* **2019**, *28*, 100451. [CrossRef]
57. Demuzere, M.; Bechtel, B.; Middel, A.; Mills, G. Mapping Europe into local climate zones. *PLoS ONE* **2019**, *14*, e0214474. [CrossRef] [PubMed]
58. Demuzere, M.; Hankey, S.; Mills, G.; Zhang, W.; Lu, T.; Bechtel, B. Combining expert and crowd-sourced training data to map urban form and functions for the continental US. *Sci. Data* **2020**, *7*, 264. [CrossRef]
59. Van damme, S.; Demuzere, M.; Verdonck, M.L.; Zhang, Z.; Van Coillie, F. Revealing kunming's (china) historical urban planning policies through local climate zones. *Remote Sens.* **2019**, *11*, 1731. [CrossRef]
60. Verdonck, M.L.; Okujeni, A.; van der Linden, S.; Demuzere, M.; De Wulf, R.; Van Coillie, F. Influence of neighbourhood information on 'Local Climate Zone' mapping in heterogeneous cities. *Int. J. Appl. Earth Obs. Geoinf.* **2017**, *62*, 102–113. [CrossRef]
61. Keany, E.; Bessardon, G.; Gleeson, E. Using machine learning to produce a cost-effective national building height map of Ireland to categorise local climate zones. *Adv. Sci. Res.* **2022**, *19*, 13–27. [CrossRef]
62. Qiu, C.; Mou, L.; Schmitt, M.; Zhu, X.X. Local climate zone-based urban land cover classification from multi-seasonal Sentinel-2 images with a recurrent residual network. *ISPRS J. Photogramm. Remote Sens.* **2019**, *154*, 151–162. [CrossRef] [PubMed]
63. Kabano, P.; Lindley, S.; Harris, A. Evidence of urban heat island impacts on the vegetation growing season length in a tropical city. *Landsc. Urban Plan.* **2021**, *206*, 103989. [CrossRef]
64. Zhao, W.; Qu, Y.; Zhang, L.; Li, K. Spatial-aware SAR-optical time-series deep integration for crop phenology tracking. *Remote Sens. Environ.* **2022**, *276*, 113046. [CrossRef]
65. Zhou, Y. Understanding urban plant phenology for sustainable cities and planet. *Nat. Clim. Chang.* **2022**, *12*, 302–304. [CrossRef]
66. Cai, M.; Ren, C.; Xu, Y.; Lau, K.K.L.; Wang, R. Investigating the relationship between local climate zone and land surface temperature using an improved WUDAPT methodology—A case study of Yangtze River Delta, China. *Urban. Clim.* **2018**, *24*, 485–502. [CrossRef]
67. Shi, L.; Ling, F.; Foody, G.M.; Yang, Z.; Liu, X.; Du, Y. Seasonal SUHI Analysis Using Local Climate Zone Classification: A Case Study of Wuhan, China. *Int. J. Environ. Res. Public Health* **2021**, *18*, 7242. [CrossRef]
68. Rosentreter, J.; Hagensieker, R.; Waske, B. Towards large-scale mapping of local climate zones using multitemporal Sentinel 2 data and convolutional neural networks. *Remote Sens. Environ.* **2020**, *237*, 111472. [CrossRef]
69. Yoo, C.; Han, D.; Im, J.; Bechtel, B. Comparison between convolutional neural networks and random forest for local climate zone classification in mega urban areas using Landsat images. *ISPRS J. Photogramm. Remote Sens.* **2019**, *157*, 155–170. [CrossRef]

 *remote sensing*

MDPI

Article

Determining the Influence of Long Term Urban Growth on Surface Urban Heat Islands Using Local Climate Zones and Intensity Analysis Techniques

Terence Darlington Mushore [1,2,*], Onisimo Mutanga [1] and John Odindi [1]

1 Discipline of Geography, School of Agricultural, Earth and Environmental Sciences, University of KwaZulu-Natal, Scottsville, Pietermaritzburg 3209, South Africa; mutangao@ukzn.ac.za (O.M.); odindi@ukzn.ac.za (J.O.)
2 Department of Space Science and Applied Physics, Faculty of Science, University of Zimbabwe, 630 Churchill Avenue, Mt Pleasant, Harare 00263, Zimbabwe
* Correspondence: tdmushore@science.uz.ac.zw

Abstract: Urban growth, typified by conversion from natural to built-up impervious surfaces, is known to cause warming and associated adverse impacts. Local climate zones present a standardized technique for evaluating the implications of urban land use and surface changes on temperatures of the overlying atmosphere. In this study, long term changes in local climate zones of the Bulawayo metropolitan city were used to assess the influence of the city's growth on its thermal characteristics. The zones were mapped using the World Urban Database and Access Portal Tool (WUDAPT) procedure while Landsat data were used to determine temporal changes. Data were divided into 1990 to 2005 and 2005 to 2020 temporal splits and intensity analysis used to characterize transformation patterns at each interval. Results indicated that growth of the built local climate zones (LCZ) in Bulawayo was faster in the 1990 to 2005 interval than the 2005 to 2020. Transition level intensity analysis showed that growth of built local climate zones was more prevalent in areas with water, low plants and dense forest LCZ in both intervals. There was a westward growth of light weight low rise built LCZ category than eastern direction, which could be attributed to high land value in the latter. Low plants land cover type experienced a large expansion of light weight low rise buildings than the compact low rise, water, and open low-rise areas. The reduction of dense forest was mainly linked to active expansion of low plants in the 2005 to 2020 interval, symbolizing increased deforestation and vegetation clearance. In Bulawayo's growth, areas where built-up LCZs invade vegetation and wetlands have increased anthropogenic warming (i.e., Surface Urban Heat Island intensities) in the city. This study demonstrates the value of LCZs in among others creating a global urban land use land cover database and assessing the influence of urban growth pattern on urban thermal characteristics.

Keywords: WUDAPT; thermal environment; urban climate; local climate zones; intensity analysis; urban growth

Citation: Mushore, T.D.; Mutanga, O.; Odindi, J. Determining the Influence of Long Term Urban Growth on Surface Urban Heat Islands Using Local Climate Zones and Intensity Analysis Techniques. *Remote Sens.* **2022**, *14*, 2060. https://doi.org/10.3390/rs14092060

Academic Editor: Xuecao Li

Received: 28 January 2022
Accepted: 18 April 2022
Published: 25 April 2022

Publisher's Note: MDPI stays neutral with regard to jurisdictional claims in published maps and institutional affiliations.

1. Introduction

Urban areas continue to expand in population and built-up extent, with faster rates in developing countries [1–6]. Whereas urban growth is often considered as a sign of economic vitality [3], its adverse impacts that include increased air pollution, Surface Urban Heat Islands, dust and haze, significantly influence urban micro- and macro-climate and affect urban environmental quality and human health [7,8]. Globally, urbanization has caused climate modifications, most evident in higher temperatures in urbanized than the non-urbanized surroundings [9–13]. Such growth often exacerbates heat stress in the already warming (due to global climate change) urban areas, leading to deterioration of outdoor thermal comfort [14]. Urban growth and associated surface changes induce near-surface warming, which increases energy and water demand due to search for indoor and outdoor thermal comfort [15–18]. Hence, urban growth assessment techniques that consider both surface and near surface characteristics and atmospheric gas/pollutant emissions are valuable in determining the influence of anthropogenic processes on the urban thermal emissions to inform sustainable urban growth.

Remote sensing offers a variety of data for analyzing spatial and temporal effects of urban land surface changes on the urban thermal environment. This has largely been facilitated by advances in sensor development that has improved the quality and availability of remotely sensed data. For instance, missions such as Landsat offer large archival data spanning as far back as 1972 at reasonable and improved radiometric, spectral and spatial resolution, valuable for assessing both large scale and localized temporal and multitemporal landscape and environmental patterns [19–23]. A large body of literature has investigated the impact of land use land cover (LULCs) changes on surface and near surface temperature, e.g., Kumar and Shekar [24] and Uddin et al. [25]. These studies have indicated that impervious, bare and built-up surfaces result in urban warming while vegetation and water bodies act as thermal sinks. However, whereas LULC-based techniques and surface characterization schemes account for the contributions of land surface changes to temperature dynamics, they often exclude other anthropogenic contributions such as emission, which are key drivers of temperature changes associated with urban growth. Hence, schemes that account for both land surface characteristics and gas emissions/pollutants as drivers of changes in the thermal environment are necessary to adequately explain climatic changes in such complex environments.

Due to the dependence on LULC schemes, studies on effects of urbanization on temperature have mostly defined Surface Urban Heat Island (SUHI) as the difference between "urban" and "rural" temperature [26–31]. In such studies, rural and urban are vaguely defined by differences in population and built-up extent in a manner that is not universal [32–34]. However, this separation is no longer always clear cut as traditional and non-traditional urban and rural land uses increasingly continue to coexist [33]. The traditional classification scheme is area specific, making it difficult to make global comparisons as LULC characteristics vary between cities of the same country and between countries. However, analysis based on Local Climate Zones (LCZ) provide understanding of urban structures and land uses in a globally standardized manner, useful for understanding the influence of urbanization on urban climate [33–35]. The LCZ scheme is local and climatic in nature considering surface cover, three dimensional surface structures (such as height and density of buildings and vegetation) as well as anthropogenic thermal emissions [8,35–39]. LCZs provide useful information for assessing adherence of cities to the 2030 agenda for sustainable development Goal 11 [40] both in the form of LULC transitions and anthropogenic emissions effects on climate. Hence, LCZ provide a standard basis upon which the impacts of urban growth on the thermal environment can be monitored and assessed.

The LCZ scheme emphasizes the difference in temperature among the categories within and between cities, thus directly linked to climate of an area, while contributing to the creation of the global urban database [11,32]. Close association between LCZ and LST shows that LCZs are helpful for examination of evolution of SUHI over time [41]. LCZs are more conducive to analysis and less prone to confusion because they highlight common exposure characteristics and invite physically based explanations of SUHI magnitude [42–49]. Studies which used LCZs in SUHI analysis mostly focused on short temporal scales such as diurnal, seasonal and annual, e.g., United Nations General Assembly [40] and Ardiyansyah et al. [49]. Focus on long term interactions between LCZs and SUHI have remained understudied, especially in Africa. Furthermore, although Stewart and Oke [33] showed the effectiveness of LCZs in defining SUHI in cities, application of inter-LCZ temperature difference to define SUHI has remained minimal, especially in the analysis of long-term changes. Most of the studies that analyzed the relationship between LCZs and SUHI either used LST to directly represent SUHI [33,46,49], reclassified LST into different SUHI categories [45], converted LST into other forms such as Distribution Index [41] or normalized temperatures [47,50], or used urban to rural temperature gradient [29,48,51,52].

Of the few studies that used temperature difference between LCZ types to quantify SUHI, most of them used the Low plants LCZ as a reference against which LSTs of other LCZs were compared [50,53–55]. A number of studies on LCZ have mainly focused on the current state and short-term variations as well as contributing to World Urban Database and Access Portal Tool (WUDAPT), e.g., Cai et al. [32], Cai et al. [36], Danylo et al. [37], Qiu et al. [39], and Demuzere et al. [56], with little effort towards understanding and explaining their long term changes that affect a city's SUHI intensity. In long term analysis, single date imageries are commonly used to develop static LCZ for each year, ignoring the value of combining multi-seasonal data for the same analysis. Long term changes provide a better understanding of the contribution of human activities to local climate change and assessment of adherence of city growth patterns to Agenda 2030 for Sustainable Development Goal 11 [39]. Hence, there is a need to use multi-season imageries to generate LCZ and LCZ-based SUHI to determine long term effects of urban growth on a city's thermal environment.

Cities in developing countries have been experiencing growth characterized by massive changes in land surface characteristics and intensification of activities, which have the potential to pollute the atmosphere and exacerbate changes in local thermal environments. Although the trends have been observed in different parts of the world, actual changes vary between and within countries, triggering the need for detailed and city-specific assessments. Hence, in-depth understanding of a city's specific influences on LCZ is important for ensuring that further development is climate smart and sustainable. However, available literature on long term LCZ changes [52,57–59] uses the traditional "from to" change detection approach which lacks in depth analysis to provide detailed understanding of long term LCZ transitions and their potential impacts on local climate. Although not yet applied to understand long term LCZ transitions, in depth analysis of changes based on transition matrices of different periods is better done using intensity analysis than the traditional "from to" change detection approach. Intensity analysis is useful for effecting classifications of different time intervals to understand sizes and intensities of temporal changes among categories [60–63], as it provides details on whether transition from one category to another deviates from a uniform process [61]. It also identifies time intervals when rate of change was fast or slow, identifies whether category changes were active or dormant in a time interval and whether a category was targeted or avoided by changes during an interval [62,64,65]. Furthermore, it analyzes land changes relative to size of category to identify systematic transitions over time [62]. As such, it reveals information such as underlying processes associated with changes which ordinary change detection conceals [66]. For instance, Alo and Pontius [67] revealed that protected areas in Ghana experienced systematic transitions from closed forest to bare and bush fires, while Ekumah et al. [66] revealed that between 1985 and 2017, human induced LULCs grew at the expense of natural categories in the Densu Delta, Sukumo II and Muni Pomodze Ramssar sites in Ghana. Intensity analysis is thus valuable for obtaining an in-depth and detailed analysis and understanding of long term LCZ changes, especially for cities such as Bulawayo where spatial and temporal temperature patterns are not yet documented. Combining intensity analysis and SUHI retrieval in the context of LCZ will therefore provide a detailed understanding of the effect of urban growth patterns on the thermal environment of cities.

Hence, the aim of this study was to integrate intensity analysis and LCZ-based SUHI retrieval to provide detailed analysis of the impact of urban growth on the thermal environment in Bulawayo metropolitan city in Zimbabwe. Specifically, this study sought to determine long term effect of urban growth on the thermal environment using LCZ between 1990 and 2020 in Bulawayo city, Zimbabwe. The study also utilized intensity analysis for an in-depth assessment of the changes in LCZ in Bulawayo between 1990 and 2020. Additionally, and contrary to the broad literature that uses the between rural–urban difference in temperature, this study enhanced the use of inter-LCZ temperature difference approach to quantify SUHI intensity and their long term changes.

2. Methodology

2.1. Description of the Study Area

Bulawayo is the second largest city in Zimbabwe (Figure 1). It is located to the southeast of the country (Figure 1a) at an elevation of approximately 1358 m above sea level. The period between October and March is hot and wet with a minimum of 16 °C and maximum of 30 °C, with an average temperature of 25 °C, while the rest of the year is cool and dry, with a minimum of 10 °C, maximum of 25 °C and average temperature of 15 °C [68]. Generally, the area receives erratic rainfall, with annual average precipitation of 600 mm that ranges from 199.3 mm to 1258.8 mm, typical of a semi-arid climate. Bulawayo (Figure 1b) lies in the subtropical steppe (Bsh) according to Koppen climate classification. The period from December to February is the wettest. Most rain falls from December to February and the area is vulnerable to droughts due to proximity to the Kalahari Desert [69,70].

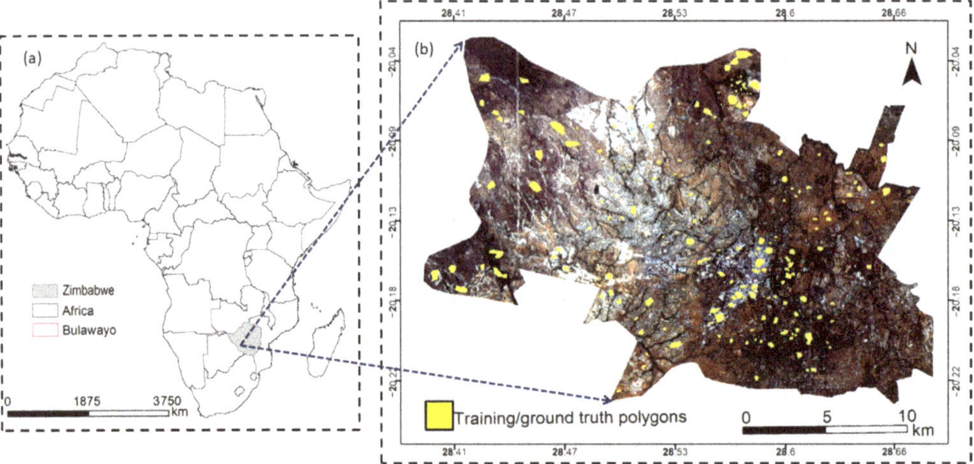

Figure 1. Map of Africa showing the location of Zimbabwe and Bulawayo (**a**) and map of Bulawayo showing distribution of training areas (**b**)—training sites not visible.

2.2. Field Observations of Local Climate Zones in Bulawayo

Since the WUDAPT places little emphasis on field data collection, a survey to identify and obtain ground truth samples of LCZ categories in the study area was important to guide digitizing of training polygons in Google Earth Field observations, which allowed for identification of inter- and intra-category variabilities that could not be adequately captured from Google Earth. The sample coordinates of each LCZ category (ground truth data) were obtained between 18 and 27 October in 2020. This experience also guided selection of training areas for the historical periods using Google Earth in the absence of field measurements for that period. Field observations increased the validity of the analysis instead of exclusive reliance on Google Earth retrievals. Generally, 8 LCZ categories were identified in the study area that fit into the description of LCZs provided by Stewart and Oke [33]. The categories were three land use-based LCZs, namely Compact low rise (LCZ3), Open low rise (LCZ6) and Light weight low rise (LCZ7), as well as three land cover-based LCZs, which were Dense forest (LCZA), Low plants (LCZD) and Water (LCZ G). The study used LCZs definitions and pictorials provided by Stewart and Oke [33] as reference to identify similar classes in the study area for global comparability.

2.3. Multi-Temporal Remotely Sensed Datasets

Multi-temporal Landsat 5, Landsat 7 and Landsat 8 Operational Land Imager datasets were downloaded from the United States Geological Survey's (USGS) earth explorer website for analysis. Cloud free imageries for the wet and dry vegetation periods were downloaded to minimize compromising effects of atmospheric noise on image radiometric values and LCZ mapping accuracy. Table 1 shows the imageries used for 1990, 2005 and 2020. In this study, Landsat thermal, panchromatic as well as bands for monitoring cirrus clouds and coastal aerosols were not used for analysis. For each year, dry and wet periods were selected in order to capture seasonal variations in LCZ, especially in areas with vegetation. The post rain period was chosen to represent the wet biomass period because during that period, trees and grasses are vibrant after a rainy season. The rainy season was avoided to attain both temporal and multi-temporal cloud free imagery. Amorim [71] indicated that Surface Urban Heat Island intensity is influenced by responses of vegetation to rainfall patterns before the imagery date. Amorim [71] observed that Heat Island Intensities increased during periods of high biomass, which reduce temperatures in vegetation areas. Therefore, precipitation patterns of up to 10 days prior to overpass (amounts and rainy days) are shown in Table 1. Generally, the number of rainy days prior to overpass was higher in the post-rain than other seasons, with the lowest number in the cool season. Cumulative rainfall amounts in 10 days before overpass were also low in all seasons (less than 20 mm).

Table 1. Multi-temporal and multi-spectral remote sensing imagery used in the study.

Imagery	Date	Season	Days to Recent Precipitation before Overpass (Days)	Rainy Days in 10 Days before Overpass (Days)	Precipitation in 10 before Overpass (mm)
Landsat 5	27 April 1990	Post rain	1.0	1.0	4.8
Landsat 7	12 April 2005	Post rain	4.0	6.06	13.1
Landsat 7	21 April 2020	Post rain	1.0	10.0	16.9
Landsat 5	14 June 1990	Cool	27.0	0.0	10.0
Landsat 5	7 June 2005	Cool	22.0	0.0	0.0
Landsat 7	24 June 2020	Cool	2.0	4.0	1.4
Landsat 5	20 October 1990	Hot	1.0	1.0	7.0
Landsat 7	21 October 2005	Hot	99.0	0.0	0.0
Landsat 8 OLI	15 October 2020	Hot	3.0	5.0	6.5

2.4. Mapping of LCZ Using Dry and Wet Season Imagery

The advantage of the LCZ scheme is that their mapping follows an easy and standardized approach for mapping LCZ, which involves downloading of imagery, digitizing of training sites (for classification and accuracy assessment) on Google Earth and supervised classification using the random forest (RF) classifier [37,72,73]. Local Climate Zones for Bulawayo were thus mapped following the WUDAPT L0 procedure [8,36,56,73,74]. The procedure involves downloading suitable imagery of the study area, on-screen selection and digitizing of training areas on Google Earth, supervised image classification using the Random Forest (RF) Classifier and post classification accuracy assessment in SAGA GIS. The procedure was adopted due to its use of readily available and freely downloadable imageries as well as easy access to the SAGA GIS software for implementation. Additionally, the steps have been followed in different parts of the world, making it very easy to follow. Use of the RF classifier makes the procedure attractive as it can perform bootstrapping analysis which is used to assess quality of the LCZ database [56,75]. The RF model is a collection of decision trees and each tree is made up of a subset of training dataset for a subset of predictors [73,76,77]. It is termed RF because its subsets are randomly formed. In RF classification, predicted value is the mode of the predictions from all trees. Main advantages of RF are the use of both categorical and numerical values, the evaluation of the precision of prediction, the robustness in the presence of outliers, noise, and overfitting. The RF model can quantify the contribution of each predictor to the total spatial variability of the target and assigns a variable importance score to each predictor. Random forest requires a small amount of training data, yet provides competitive results and can handle a large volume of input data without deletion, while still capable of identifying important variables for classification [78–83]. In addition, RF is not sensitive to over-training or noise and is desirable for multi-source remote sensing and geographical information systems data [75,81]. LCZ maps were produced for the years 1990, 2005 and 2020 using the same training areas. These were collected from locations whose LULC category did not change over time in order to eliminate the effect of differences in ground truth data on mapping accuracy. The use of the same training areas was made possible by availability of historical Google Earth imagery where the areas could be clearly identified at different periods. In order to adhere to the definition of LCZ, which requires that they cover at least a hundred meters to several kilometers to influence temperature [8,11,41], the maps were resampled using a 5 by 5 pixels window. A LCZ must be large enough to influence temperature of an area. In order to quantify the effect of seasonality on mapping accuracy, LCZ maps were also produced using data for the hot and dry season for comparison with analysis based on a combination of data from the post rain, cool dry and hot dry periods. Figure 2 provides a summary of the procedure followed. The broken arrow in the figure shows the approach taken by previous studies which largely skip the iterative step of qualitative accuracy assessment, further improving accuracy before change detection.

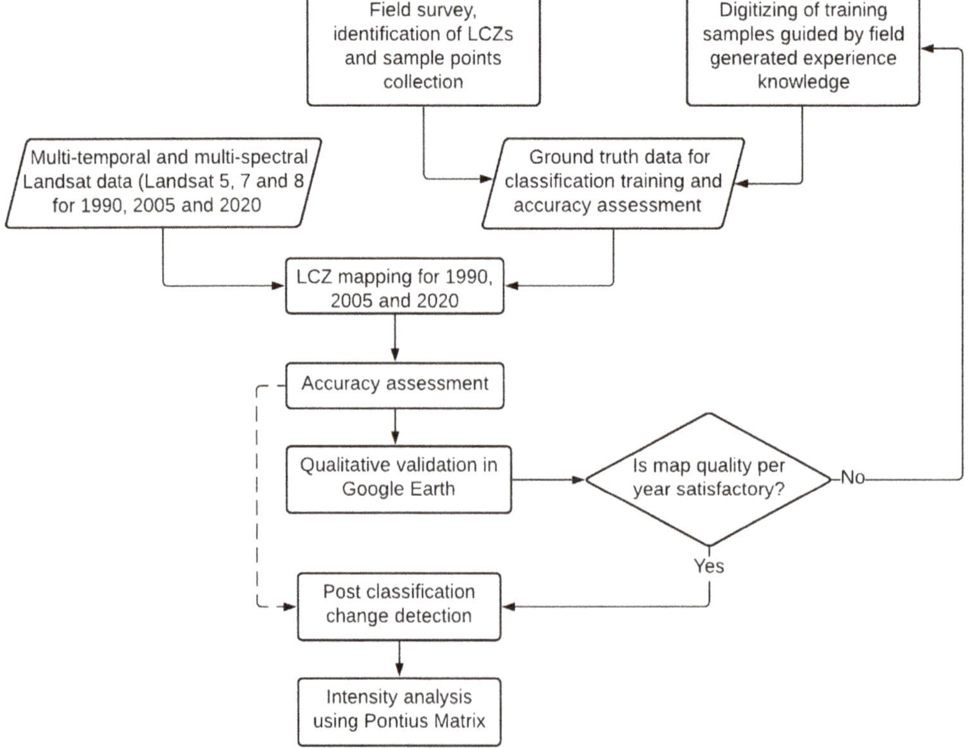

Figure 2. Flowchart showing summary of procedures used in this study.

2.5. Accuracy Assessment

The WUDAPT procedure automatically splits training areas into 50% for classification and 50% for accuracy assessment. A confusion matrix is formulated in a tabular form for the purposes of comparing reference class labels (ground truth) with labels for corresponding pixels on a derived classification of remote sensing imagery [83]. The diagonal values on the matrix indicate where categories assigned on the classified map correctly corresponded with ground truth. For example, the matrix shows the number of pixels which were assigned same LCZ value as observed on the ground as well as those that are misallocated following classification of remote sensing data. The confusion matrix was used to generate indicators of accuracy at class level (Producer Accuracy and User Accuracy) as well as at entire study area level (Overall Accuracy—OA—and Kappa—K). The use of OA and K with the inclusion of ground data is the most common and reliable way of assessing accuracy [84]. Accuracy assessment is important for the separation of real changes from changes due to errors for LCZ maps of different periods. Error analysis was done for all the study years. Accuracy was also assessed qualitatively by overlaying the produced LCZ maps with a corresponding Google Earth imagery in combination with expert judgement. The number of training areas was objectively and iteratively increased in areas where marked mismatches were observed to capture for intra- and inter-class variabilities using Google Earth.

2.6. Detection of Long-Term Changes in LCZ in Bulawayo

A 30-year period (1990 to 2020) was selected, as the study aimed at using LCZ dynamics as a proxy for temperature changes in Bulawayo. The World Meteorology Organization (WMO) recommends a minimum of 30 years for a representative climate change analysis. The period was further split into two 15 year periods (i.e., 1990 to 2005 and 2005 to 2020) for understanding of rapid changes that occur at local scale and for intensity analysis purposes. Additionally, Coppin and Bauer [85] recommended a period of at least 3 years for change detection involving forests and other vegetation types. In an urban setting, there is a mix of rapid and slow LCZ making a period of at least 15 years enough to detect effect of all change trajectories on the climate of an area. A post classification change detection approach was used. The approach produces a change matrix/table which shows the number of pixels which were converted to other LCZ types or remained in the same categories in the considered interval. For instance, the table indicated the number of pixels which were in LCZ1 at the beginning and remained in the same as well as those that were changed to other LCZ categories during the same time interval. Although the change matrix is useful in depicting the quantities and directions of change, it does not adequately explain the changes [60,85,86], hence the need for intensity analysis.

2.7. Intensity Analysis for In-Depth Characterization of LCZ Changes

Intensity analysis was used to obtain a better understanding of LCZ transitions between 1990 and 2020 in Bulawayo. It was used to assess locations and intensities of temporal changes among categories. The analysis was done at the interval, category and transition levels [87–90] using freely available software on Pontius Clarke University web page (https://www2.clarku.edu/faculty/rpontius/ (accessed on 15 April 2021)). The site provides an easy to use Excel sheet where the change matrix for a given time interval is entered. Varga et al. [91] provide descriptions and defining equations used in the analysis that were adopted in this study.

2.7.1. Interval Level Intensity Analysis

Interval level was used to determine overall changes per time period for the 1990 to 2005 and 2005 to 2020 time intervals. Overall changes in the interval 1990 to 2005 were compared with those for the interval 2005 and 2020. This was important to identify which interval had changes characterized as fast or slow. The change percentage for an interval t is defined as in Equation (1) [91]:

$$S_t = \frac{(size\ of\ changes\ during\ interval\ t) \times 100\%}{size\ of\ studyarea\ in\ which\ changes\ are\ occurring} \tag{1}$$

S_t is the uniform intensity in interval t. An interval change is fast if it exceeds uniform intensity and slow if otherwise.

2.7.2. Category Level Analysis

Changes in gross loss or gain in intensity among different categories was described by category level intensity analysis during time interval t (where t separately represents the two interval periods—1990 to 2005 and 2005 to 2020). Category level loss (L_{ti}) and gain (G_{tj}) during the interval t are obtained using Equations (2) and (3) as

$$L_{ti} = \frac{(size\ of\ loss\ of\ category\ i\ during\ interval\ t) \times 100}{size\ of\ i\ at\ the\ start\ of\ interval\ t} \tag{2}$$

$$G_{tj} = \frac{(size\ of\ gain\ of\ category\ j\ during\ interval\ t) \times 100}{size\ of\ category\ j\ at\ the\ end\ of\ interval\ t} \tag{3}$$

According to [64], calculation of category persistence (P_{ti}) is done using Equation (4):

$$P_{ti} = \frac{(size\ that\ has\ maintained\ category\ i\ during\ interval) \times 100\%}{size\ of\ spatial\ extent} \tag{4}$$

$S_t = L_{ti} = G_{tj}$ for all categories i and j if changes are uniformly distributed across spatial extent. Uniform transition assumes category i uniformly changes to other categories during the time interval. If $L_{it} > S_t$ then the loss of category i is active in the interval t while loss is dormant if $L_{it} < S_t$. Dormant implies that the loss of category i slowed down or stopped within the interval t. Similarly, gain of category j in the interval t is active if $G_{tj} > S_t$ and dormant if $G_{tj} > S_t$. When two intervals are considered and the status of a change as dormant or active is same in both intervals, then the category's loss or gain is said to be stationary.

2.7.3. Transition Level Analysis

The analysis describes the variation in intensity with which the gain of a particular category transitions from other categories within a time interval [60]. Transition level intensity analysis was used to determine whether change of a category avoids or targets other categories. If the intensity of the change from category i to category j exceeds uniform intensity, then category i targets j otherwise it avoids.

2.7.4. Retrieval of Changes in SUHI in Response to Long Term LCZ Dynamics

Thermal data of Landsat 5, 7 and 8 were used to compute land surface temperature (LST) for 1990, 2005 and 2020. Initially, the data were corrected of differences in solar zenith angles. While Landsat 8 has two thermal infrared bands, a single channel technique was applied for all the periods to minimize effects of differences in computation algorithms on LST variations between time periods. Digital numbers of thermal data were converted to radiances, which were then used to determine brightness temperature (T_b) and surface temperature (T_s) using Equations (4) and (6), respectively [91–94].

$$T_b = \frac{K_2}{\ln\left(\frac{K_1}{L_\lambda} + 1\right)} \tag{5}$$

where K_1 takes a values of 607.76, 666.09 and 774.89 W/(m^2 srμm), while K_2 has values of 1260.56, 1282.71 and 1321.08 W/(m^2 srμm), using Landsat 5, Landsat 7 and Landsat 8 data, respectively. A method based on spectral and blackbody radiance of the thermal infrared band was used to obtain pixel-based land surface emissivity map (ε) [95]. Emissivity correction was applied on brightness temperature to obtain actual land surface temperature using Equation (6) [96].

$$T_s = \frac{T_B}{1 + \left(\frac{\lambda T_B}{\rho}\right) \ln \varepsilon}$$ (6)

where λ is the central wavelength of emitted thermal radiance (11.5 μm for Landsat 5 and Landsat 7 and 10.9 μm for band 10 of Landsat 8) and ρ is equal to 1.438×10^{-2} mK. The procedure above was used to retrieve LST on two different dates so that independent sets were used for training and accuracy assessment of the developed estimation algorithm. The spatial structure of LST intensities were used for visual and quantitative analysis of changes which occurred between 1990 and 2020.

Stewart and Oke [33] defined SUHI as the difference in LST between LCZs. In this study, we adopted an approach by Dimitrov [97], which defined SUHI as the maximum temperature difference between LCZs. For each year, the Surface Urban Heat Island was computed as the difference between the average T_s of LCZ category and the mean surface temperature of the water LCZ (LCZ G), which was identified as giving the highest LST difference with other LCZs consistently in all years using Equation (7).

$$SUHI_{LCZ} = LST_{LCZ\ X} - LST_{LCZ\ Y}$$ (7)

$SUHI_{LCZ}$ is the SUHI derived from LST difference between other LCZs (X) and the Water LCZ (Y). Although studies such as Stewart and Oke [33] used LCZ D as reference, it was not applicable in this study due to the varying and opposing effects of the LCZ in different places and seasons for daytime analysis. For instance, agricultural areas had thermal values that vary in space and time and between seasons, rendering them as heat sinks in some instances and heat sources in others. As such, during the growing season, they acted as heat sinks while in the dry season, their heat mitigation value was reduced. In other areas, they were completely removed, as they turned to dry biomass or bare soil areas. Similarly, grasslands (also in LCZ D) vary in heat mitigation value depending on season and maintenance, making them another example of inconsistency of LCZ D. The LCZ G was chosen as a reference since it was the coolest and more stable than the vegetation based categories, which have broad seasonal and long term variations in characteristics. Mean LST per LCZ category was obtained using the Zonal Statistics overlay function in ArcGIS version 10.2 for each year. Changes in SUHI per LCZ strata were monitored and linked with observed changes in LCZ from 1990 to 2020 in 15-year intervals.

3. Results

3.1. LCZ Maps Based on Multi-Seasonal Image Analysis

The use of imagery for the wet and dry vegetation periods reduced the confusion between light weight low rise and low plants in the western areas (Figure 3). The overall classification accuracies were 98%, 98.2% and 95%, for 1990, 2005 and 2020, respectively. Visual inspection shows that between 1990 and 2020, light weight low rise LCZ was spread westwards into areas formerly occupied by low plants. All built local climate zones increased in spatial coverage while low plants and dense forest LCZ decreased in coverage. The water LCZ also decreased in coverage during the study period.

Compact low rise increased by 4.33 km^2 between 1990 and 2005 and by 2.50 km^2 between 2005 and 2020 (Table 2). Similarly, light weight low rise expanded by 21.27 km^2 between 1990 and 2005 and by 14.25 km^2 between 2005 and 2020. The open low rise LCZ also showed the same pattern of larger increase in the 1990 to 2005 interval than in the 2005 to 2020 interval. On the other hand, the expansion rate of dense forest between 1990 and 2005 (2.91 km^2 in 15 years) was smaller than the depletion rate of the LCZ between 2005 and 2020 (15.25 km^2 in 15 years). Low coverage diminished faster in the 1990 to 2005 interval than in the 2005 to 2020 interval.

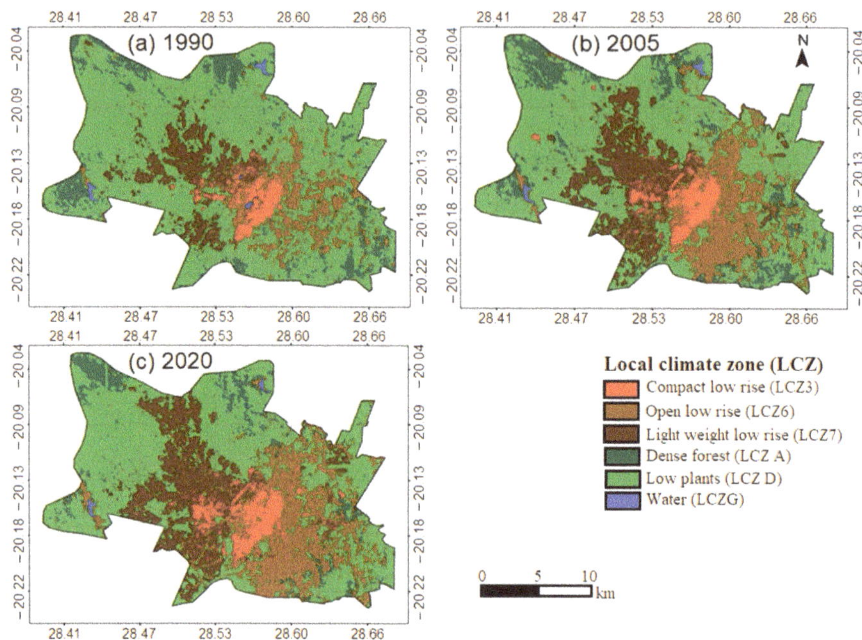

Figure 3. LCZ maps produced using multi-seasonal remotely sensed data.

Table 2. Coverage of LCZ categories in 1990, 2005 and 2020.

LCZ Category	Coverage of LCZ Categories in km^2 (% in Bracket)		
	1990	**2005**	**2020**
Compact low rise	12.84 (3.0)	17.17 (4.0)	19.67 (4.5)
Dense Forest	38.41 (8.9)	41.32 (9.5)	26.07 (6.0)
Light weight low rise	39.60 (9.1)	60.87 (14.0)	75.12 (17.3)
Open low rise	45.13 (10.4)	74.00 (17.1)	88.56 (20.4)
Water	2.06 (0.5)	1.45 (0.3)	1.26 (0.3)
Low plants	295.42 (68.2)	238.66 (55.1)	222.77 (51.4)

3.2. Changes in LCZs for Bulawayo Using Multi-Temporal (Dry and Wet) Datasets

Table 3 shows that Compact low rise increased by 1.5% between 1990 and 2005 and a further 2.2% between 2005 and 2020, giving a 30-year expansion of 3.7%. A significant decrease in coverage was observed in the dense forest LCZ, which experienced a net reduction of 42% between 1990 and 2020, with greater change in the 1990 to 2005 than 2005 to 2020 periods. Generally, all built-up LCZ increased in coverage between 1990 and 2020, with larger expansion in the light weight low rise than other built-up LCZs. Sustained contraction was recorded in dense forest and water LCZs. The low plants LCZ, which in this study included croplands, grasslands and parks increased by 4.3% over the entire period, except a 0.9% decrease recorded between 2005 and 2020.

3.3. Intensity Analysis

3.3.1. Category Level Intensity Analysis for 1990 to 2005 and 2005 to 2020 Intervals

All other LCZs except compact low rise were gainers or losers in the 1990 to 2005 and 2005 to 2020 intervals (Figure 4a,b). The gain in Compact low rise LCZ was dormant in the 1990 to 2005 interval, implying the gain stopped or slowed along the interval. In the interval 2005 to 2020, the gain of compact low rise was active in the 2005 to 2020 interval.

Low plants were active losers in the 1990 to 2005 interval and became dormant losers (the loss of the class was slow or stopped along the interval) in the 2005 to 2020. The water LCZ was an active loser in both intervals. The light weight low rise LCZ was a dormant gainer in the initial interval, but turned into an active gainer in the interval 2005 to 2020.

Table 3. LCZ changes from 1990 to 2020 in Bulawayo.

LCZ Category	LCZ Category Changes in km² (% in Bracket)		
	1990 to 2005	2005 to 2020	1990 to 2020
Compact low rise	4.33 (33.7)	2.50 (14.6)	6.83 (53.2)
Dense Forest	2.91 (7.6)	−15.26 (−37.0)	−12.35 (−32.1)
Light weight low rise	21.27 (53.7)	14.27 (23.5)	35.54 (89.7)
Open low rise	28.88 (64.0)	14.50 (19.7)	43.43 (96.2)
Water	−0.61 (−29.8)	−0.19 (−13.2)	−0.81 (−39.1)
Low plants	−56.77 (−19.2)	−15.89 (−6.7)	−72.66 (−24.6)

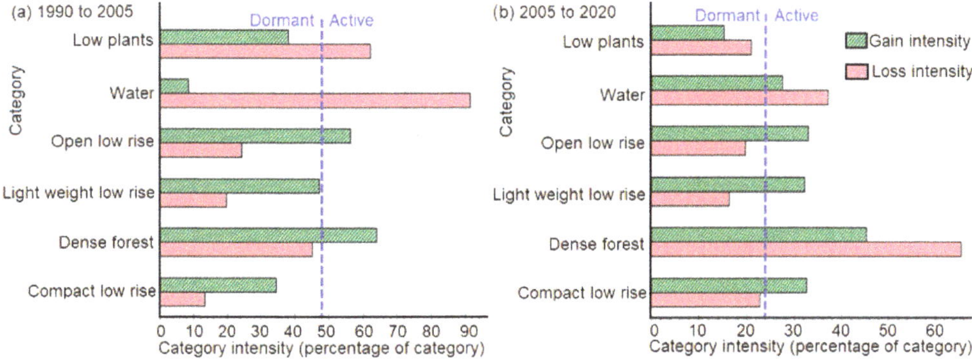

Figure 4. Category level intensity analysis for 1990 to 2005 (**a**) and 2005 to 2020 (**b**).

3.3.2. Transition Intensity of Gaining Categories Encroaching into Losing Categories

The expansion of compact low rise LCZ targeted water, open low rise and light weight low rise between 1990 and 2005 with open low rise LCZ as the most intensely targeted (Figure 5a). The growth of compact low rise avoided low plants and vegetation LCZs during the same period. In the 2005 to 2020 interval, the Compact low rise LCZ continued to target open low rise and lightweight low rise areas (Figure 5b). The intensity of transition of water to compact low rise was greater in the interval 1990 to 2005 than 2005 to 2020. In both intervals, the gain of compact low rise avoided low plants and dense forest LCZ areas.

The gain of light weight low rise between 1990 and 2005 targeted low plants and avoided dense forests and open low rise water and compact low rise LCZ areas (Figure 6a). Between 2005 and 2020, the gain of Light weight low rise LCZ continued to target low plants while avoiding other LCZs (Figure 6b).

The gain of the dense forest LCZ in the 1990 to 2005 interval targeted low plants (Figure 7a). This could imply growth of trees in grasslands such as parks in addition to other tree planting efforts. The expansion of low plants LCZ targeted dense forest, open low rise and light weight low rise, while it avoided water and compact low rise in the 2005 to 2020 interval (Figure 7b). The study also noticed slight spectral confusion between light weight low rise and low plants, especially in the western parts of the study area.

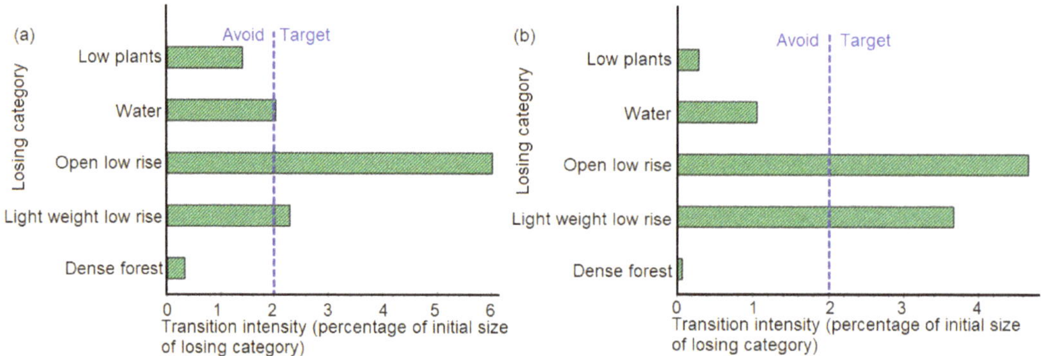

Figure 5. Transition intensity given gain of Compact low rise from (**a**) 1990 to 2005 and (**b**) 2005 to 2020.

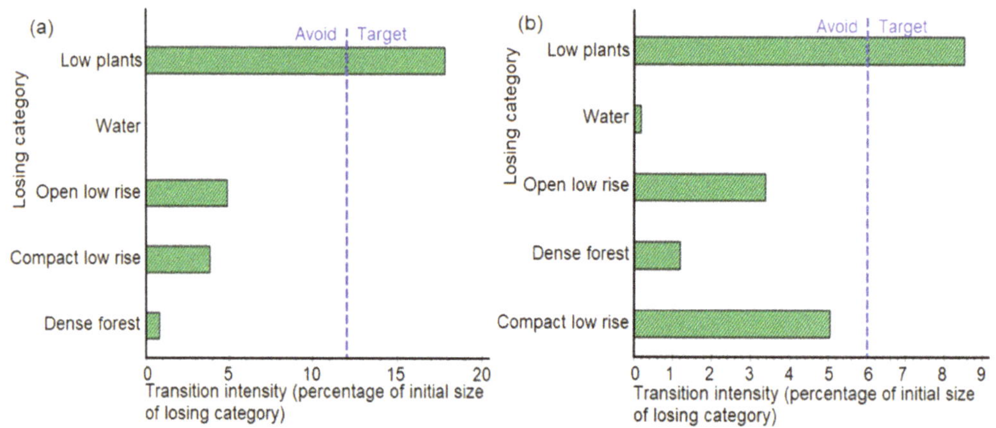

Figure 6. Transition intensity given the gain of Light weight low rise form (**a**) 1990 to 2005 and (**b**) 2005 to 2020.

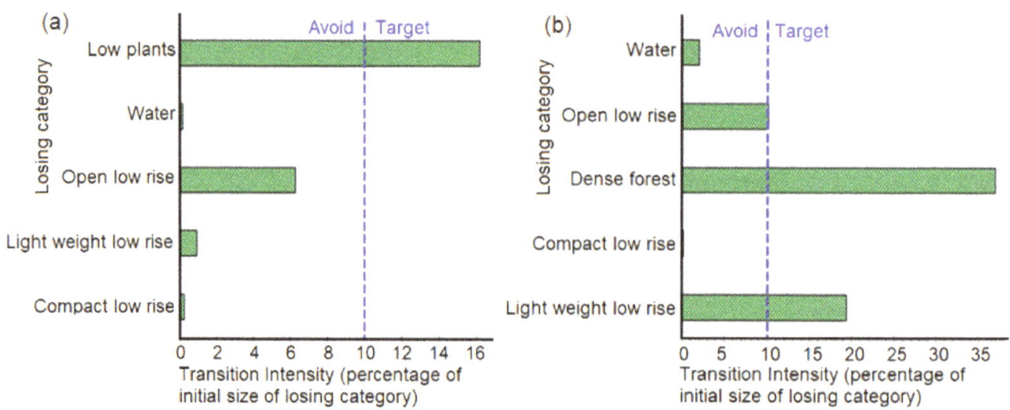

Figure 7. Transition intensity given (**a**) gain of dense forest from 1990 to 2005 and (**b**) low plants from 2005 to 2020.

The gain of open low rise LCZ in the 1990 to 2005 interval targeted low plants (Figure 8a), while in the interval 2005 to 2020 it targeted low plants, water, dense forest and compact low rise (Figure 8b). The gain targeted dense forest more than low plants LCZ, which could be due to the nature of the LCZ that consists of a few and well-spaced buildings surrounded by trees and grass. Expansion into water reveals an adverse environmental impact, with growth intrusion into wetland areas.

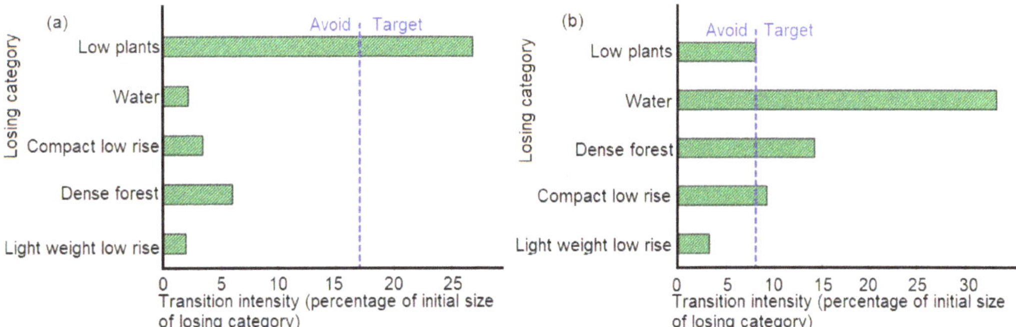

Figure 8. Transition intensity given gain of Open low rise from (**a**) 1990 to 2005 and (**b**) 2005 to 2020.

3.4. Long Term Changes in the Two Dimensional LST in Response to LCZ Changes

Figure 9 shows expansion of high temperature surfaces between 1990 and 2015. In 1990, LSTs in the 37.8 to 43.8 °C range (Figure 9a) dominated the city, while LSTs below 43.8 °C became uncommon in 2005 (Figure 9b). Visual inspection shows that in 2020, the LSTs became even higher, with most areas recording values above 46.8 °C (Figure 9c). The water areas were the most stable, with LSTs in the 16.8 to 37.8 °C range in 1990, 2005 and 2020.

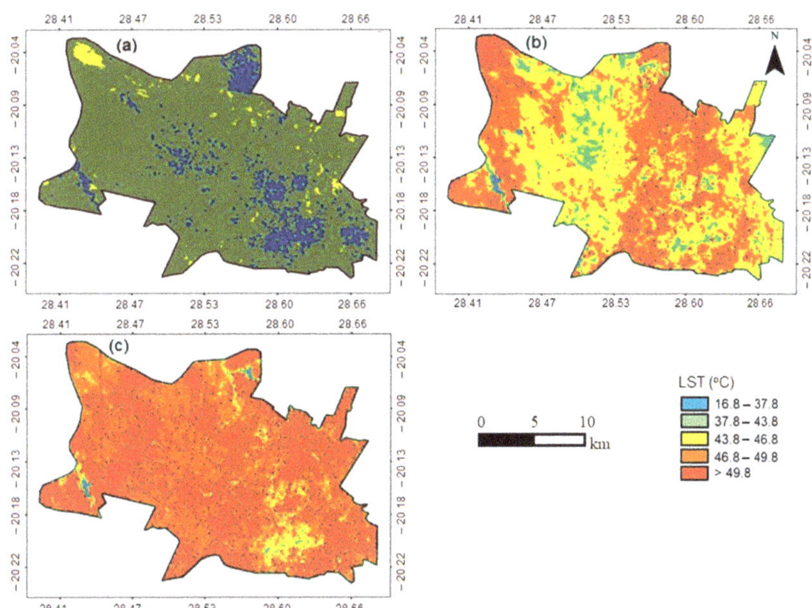

Figure 9. Spatial structure of LSTs in Bulawayo in (**a**) 1990, (**b**) 2005 and (**c**) 2020.

In each year, SUHI intensities between the Compact low rise and Low plants LCZs were comparable, although in 2020 the Low plants were slightly warmer (Figure 10). SUHI intensities increased with built-up density as well as density of tall buildings evidenced by largest intensity in Compact low rise (e.g., 11.7 °C in 2020) and lowest in Open low rise (10 °C). Dense vegetation LCZ areas were cooler than built LCZ in all the periods.

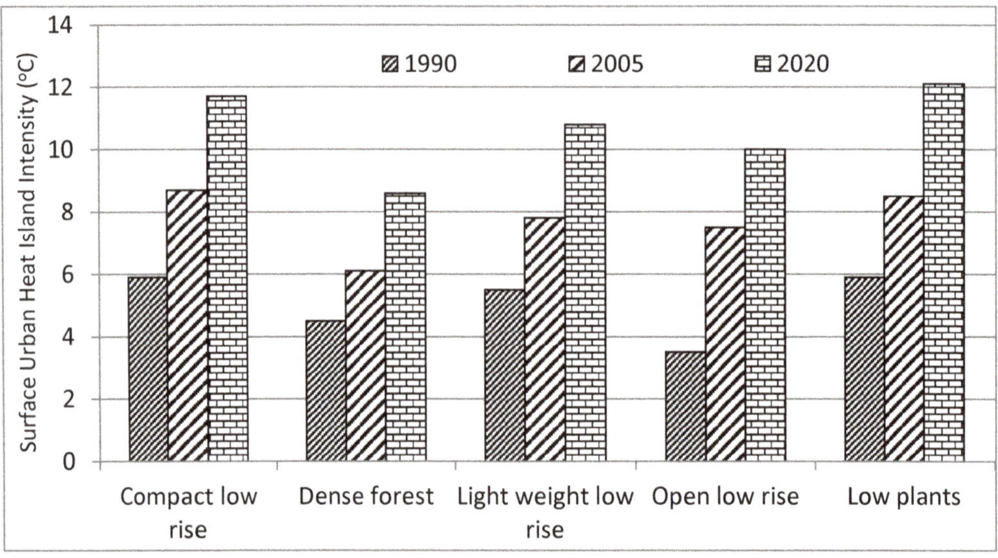

Figure 10. Changes in SUHI intensities across LCZs from 1990 to 2020 in Bulawayo.

4. Discussion

Mapping accuracy was very high, exceeding 95% for 1990, 2005 and 2020 LCZ maps. Accuracy of at least 95% in LCZ mapping were also recorded by Danylo et al. [37] in Kyiv and Lviv in Ukraine for city specific analysis, which was higher than the below 75% accuracy they achieved using training data for multiple city mapping. The differences in accuracy between city level and large scale LCZ mapping approaches demonstrate that better LCZ maps are generated with city specific efforts than when training areas of another city are used for LCZ mapping [55]. The high accuracy achieved stresses the high performance of random forest classifier in comparison to other classifiers such as support vector machine [75,80,81,83]. RF uses a set of classifiers that make it superior to individual classifier based approaches [81]. The use of multi-seasonal data also enhanced discriminability of LCZs in this study, which resulted in high yearly accuracies. LCZ maps did not show most of the linear features such as roads and rivers due to the large filter used. As noted by Kotharkar and Bagade [8] and Gal et al. [72], an LCZ must be large enough to affect an area's temperature, while use of a large filter removes linear features and expands urban area. According to Kotharkar and Bagade [8] and Gal et al. [72], shifting to a coarser scale results in loss of a number of LCZs.

Consistent with global trends, there was a general increase in built LCZ in Bulawayo between 1990 and 2020. This is a characteristic of city growth globally typified by conversion from natural land covers to urban fabric. For instance, in Kigali, Rwanda, Akinyemi et al. [60] observed an increase in built area from 1% in 1981 to 19% in 2002, followed by a slight decrease to 18% in 2014. Similar to Bulawayo, and indeed other global cities trends, Akinyemi et al. [60] showed a general increase in built-up area over a 33 year period. Whereas the expansion of Built LCZ continued throughout the study period, it was faster between 1990 and 2005 than between 2005 and 2020. For instance, expansion of compact low rise, which coincides with the Central Business District, was slower than other built LCZs in both intervals. Between 1990 and 2020, lightweight low rise LCZ expanded faster than other built LCZ. Generally, in Bulawayo, other built LCZ grow around the Central Business District with light weight low rise spreading to the west and spacious built-up LCZ spreading to the east. This limits space for further expansion of the Central Business District (CBD), which explains slowed growth of the corresponding LCZ between 2005 and 2020.

All other LCZs were gainers or losers in the 1990 to 2005 and 2005 to 2020 intervals, except compact low rise LCZ that was dormant in the 1990 to 2005 interval, implying the gain stopped or slowed along the interval. In the 2005 to 2020 interval, the gain of compact low rise was active. Low plants were active losers in the 1990 to 2005 interval and became dormant losers (the loss of the class was slow or stopped along the interval) in the 2005 to 2020. The water LCZ was an active loser in both intervals. The light weight low rise LCZ was a dormant gainer in the initial interval, but was an active gainer in the 2005 to 2020 interval. Sustained contraction was recorded in dense forest and water LCZs. The low plants LCZ, which in this study included croplands, grasslands and parks increased over the entire period (by 4.3%), although they experienced a decrease (by 0.9%) between 2005 and 2020. This shows that although the LCZ had expanded due to activities such as deforestation, it was affected by the expansion of built-up between 2005 and 2020.

The expansion of built LCZs between 1990 and 2005 was most prevalent in the open low rise LCZ. The growth of compact low rise avoided low plants and vegetation LCZs during the 1990 to 2005 period. In the interval 2005 to 2020, the Compact low rise LCZ continued to target open low rise and lightweight low rise areas. This may signify growth of the industrial and CBD to serve the surrounding residential areas, which were also expanding. The intensity of transition of water to compact low rise was greater in the interval 1990 to 2005 than 2005 to 2020. In both intervals, the gain of compact low rise also avoided low plants and dense forest LCZ areas. This may signify adherence to the Environmental Management Act, which was signed into law in 2013 [98]. The Act has increased protection of wetlands and the general environment with non-compliance, especially of business enterprises attracting heavy fines. Avoidance of water areas by expansion of compact low rise may also be a result of costs associated with construction in wetlands.

The gain of light weight low rise between 1990 and 2005 targeted low plants and avoided dense forests, open low rise water and compact low rise LCZ areas. Between 2005 and 2020, the gain of Light weight low rise LCZ continued to target low plants while avoiding other LCZ. The shift from low plant to light weight low rise indicates the change from primary production to industry based economy as the city grows. This may have resulted in expansion of low-income residential areas, which comprise the Light weight low rise occupied by most of the people who work in the industries and Central Business District (Compact low rise). The expansion of light weight low rise affecting low plant LCZ in the western direction may also indicate spreading of low income residential areas from the Central Business District (compact low rise area), where cost of land is high. Beside the agricultural land, grasslands are also part of the low plants that were targeted by the expansion of the densely built-up light weight low rise. Although transition level shows avoidance, the expansion of light weight low rise into open low rise areas during both intervals (1990 to 2005 and 2005 to 2020) could indicate increase in built-up density in an area which previously had few buildings surrounded by enough vegetation to be classified as open low rise. Such areas have few buildings during early stages of land allocation with densities increasing with time, thus changing from an open to a densely packed built-up setting.

The gain of the dense forest LCZ in the 1990 to 2005 interval which targeted low plants could imply growth of trees in grasslands such as parks in addition to other tree planting efforts. Over a period of 15 years, the planted and naturally growing trees can increase in canopy size, density and leaf area, enough to be separable from low plants. The expansion of low plants LCZ targeted dense forest, open low rise and light weight low rise, while it avoided water and compact low rise in the 2005 to 2020 interval. The spread into dense forests indicates deforestation, which turns formerly dense forest LCZ into low plants areas such as grasslands and croplands. As the light weight low rise occupied formerly low plant areas, demand for areas such as peri-urban agriculture increased. This may cause communities to spread activities into unused areas by clearing some of the dense forests. Low plants targeting light weight low rise areas could be associated with growth of vegetation within the built-up LCZ. The vegetation includes edible vegetation in small gardens as well as lawns around households. The study also noticed slight spectral confusion between light weight low rise and low plants, especially in the western parts of the study area. However, most of the confusion was eliminated through the use of multi-season data, which increased inter-class discriminability in supervised classification.

The gain of open low rise LCZ in the 1990 to 2005 interval targeted low plants while in the 2005 to 2020 interval it targeted low plants, water, dense forest and compact low rise. The gain targeted dense forest more than low plants LCZ, which could be due to the nature of the LCZ that consists of a few well-spaced buildings surrounded by trees and grass. Expansion into water demonstrates adverse environmental impact with growth intrusion into wetlands. The expansion of open low rise into low plant areas could indicate development of spacious settlements into formerly grassland, bare and agricultural areas. Furthermore, the expansion into other LCZs could indicate the advantage of wealth, as this LCZ is mostly occupied by medium to high income strata which can afford land and develop in any area. The expansion of this spacious LCZ could also be a sign of economic emancipation which enables residence of the city to purchase tracts of large land. In Zimbabwe, this includes existing land owners in densely built-up low income areas but who prefer accommodation in low density built-up areas. Due to increased demand for such spacious settings, developments encroach into formerly protected LCZs such as wetlands and dense forests.

Intensity analysis provided details of LCZ transitions in Bulawayo between 1990 and 2020 beyond the usual "from to" change detection analysis. It revealed the tendency of built-up growth, which was at the expense of vegetation areas, especially low plants. The analysis also showed that the growth of light weight low rise targeted low plant areas and completely avoided compact low rise and open low rise areas. This could be associated with the cost of land in the compact low rise and open low rise, mainly found in the central business district and low-medium density residential (largely occupied by medium to high income strata), respectively. The findings of this study emphasize the argument by Niya et al. [63] that intensity analysis clarifies substantial causes and processes of land use changes. Additionally, in agreement with Huang et al. [66], intensity analysis can assess evidence of a particular change and help develop hypothesis concerning processes of change.

High temperature surfaces expanded while LST temperatures increased between 1990 and 2015. According to Nayak and Mandal [99], urbanization causes temperature change due to both alteration of land use land cover and greenhouse gas concentrations. Similarly, in this study, we attributed surface warming to both LCZ transitions and background cause by anthropogenic activities such as industrial emissions. Blake et al. [100] also reported that Harare was warming, despite cooling in the decade from 1900 to 2002. The expansion of high LST areas and increase in LST intensities was largely due to replacement of natural land covers with built-up LCZs. Buildings and impervious surfaces have high heat absorption capacities which cause elevation of LSTs, especially where vegetation fraction and surface wetness are low. High LSTs were recorded in compact low rise areas especially as their coverage increased over time. This is in agreement with the sentiment that large areas of densely packed buildings create homogeneous areas with high LST [49]. Therefore, the growth patterns of Bulawayo have caused warming due to massive replacement of natural surfaces with buildings and impervious surfaces.

Each year, SUHI intensities between the Compact low rise and Low plants LCZs were comparable, although in 2020, the Low plants were slightly warmer. This is because field observations showed that during the hot and dry seasons, low plant LCZ areas are characterized by dry vegetation and even bare or close to bare ground in areas that are cleared at the end of farming seasons. On the other hand, the link between heat stress and built-up extent is also complex, as buildings also provide shading while on the other hand reducing ventilation leading to opposite effects on thermal comfort [100,101]. Additionally, vegetation within built-up LCZs has a heat mitigation effect, which can reduce differences in SUHI intensities between built-up and natural (land cover based) LCZs.

Surface Urban Heat Island intensities increased with built-up density as well as density of tall buildings evidenced by largest intensity in Compact low rise. According to Stewart and Oke [33], built-up LCZs vary in air temperature depending on factors that include density and height of buildings as well as type and density of vegetation within built-up areas. Consistent with this study, Lelovics et al. [11] stressed that LCZ2 is warmer than LCZ3, which is warmer than LCZ6. They also concluded that LCZ maps can distinguish areas based on degree of LULC modification. Contrasts in temperature between classes with differences in geometry/cover can exceed 10 °C, while classes with few physical differences can be less than 2 °C [33]. Similarly, Lau et al. [14] recorded the highest temperature (38.9 °C) and lowest (29.9 °C) in land cover LCZs in Hong Kong. According to Qiu et al. [39], LCZ scheme considers three-dimensional surface structure and anthropogenic parameters such as heat from human activities that influence temperature. Based on this understanding, comparatively very high land surface temperatures and SUHI were observed in the compact low rise area of the city, followed by the densely packed light weight low rise while the open low rise LCZ was the coolest of the built LCZ.

Dense vegetation LCZ areas were cooler and had lower SUHI intensities than built LCZ in all the periods. Even as vegetation cover declined, their heat mitigation value remained remarkably high within artificial LCZs that cause SUHI intensification as the city grows. Low plants LCZ areas were warmer than Open low rise areas in all the periods. This could be due to the fact that the Open low rise areas constitute residents of the middle to high income strata who have resources to ensure that vegetation around their homes is well-maintained and healthy throughout the year. This is because vegetation in urban areas serves as temperature refugees in streets and parks providing cooling effect through evapotranspiration and shading [14]. According to Lu et al. [59], vegetation within buildings reduces patch sizes of built-up LCZs thus lowering their thermal effect on the surrounding LCZs. High SUHI in low plants LCZ contradicted with other studies such as Shi et al. [102] and Lu et al. [59] which had low plants as heat sinks. The disparity was because open low rise includes natural grassland areas, which in Bulawayo experienced drying of vegetation during the hot dry season. Low plants also include bare areas and croplands whose cover during the dry season could be bare or dry vegetation, reducing the surface cooling effect of latent heat transfer. Therefore, spatial and temporal variations in the thermal characteristics of low plants significantly reduced their heat mitigation value in Bulawayo. Higher UHI in 2005 and 2020 may partly be explained by higher average precipitation around satellite overpass dates in those years than in 1990. This is also in tandem with Amorim [71], that the heat mitigation value of urban greenery and SUHI varies with seasons and are increased during wet periods around overpass when the vegetation biomass is increased.

5. Conclusions

In order to understand the effect of urban growth on the thermal environment, the study used multi-temporal Landsat data to map LCZ with very high accuracy and retrieve SUHI for Bulawayo metropolitan city in Zimbabwe for 1990, 2005 and 2020. LST of the water LCZ were used as reference for quantifying SUHI intensities instead of the subjective traditional "rural-urban" LST difference. The high LCZ mapping accuracies were attributed to precise generation of training data and the robustness of the RF classifier, an ensemble based technique, compared to single classifier-based approaches. Furthermore, the use of dry and wet biomass periods significantly improved LCZ mapping accuracy in all years. In both intervals, i.e., 1990 to 2005 and 2005 to 2020, built LCZs monotonically expanded at the cost of vegetation- and water-based LCZs. Intensity analysis showed that the growth of lightweight low rise mainly targeted low plant areas. Deforestation in the city was expressed by the gain of low plants, which targeted dense forests. Intensity analysis also showed that the growth of compact low rise occupied mostly by low income strata avoided eastern direction where there are compact low rise and open low rise generally characterized by high cost per land unit. Due to expansion of built and polluting LCZs, the SUHI intensities rose monotonically during the study period. SUHI intensities varied between LCZs as they intensified with built-up proportion and density of tall building while decreasing with abundance of healthy vegetation. Based on the findings, the study concluded that human activities and growth induced LCZ changes have continued to trigger warming in Bulawayo. SUHI retrieval based on LCZ scheme proved effective in determining effects of urban growth on the thermal environment.

Author Contributions: Conceptualization, T.D.M., O.M. and J.O.; methodology, T.D.M.; software, T.D.M.; validation, T.D.M., O.M. and J.O.; formal analysis, T.D.M.; investigation, T.D.M., O.M. and J.O.; resources, T.D.M., O.M. and J.O.; data curation, T.D.M., O.M. and J.O.; writing—original draft preparation, T.D.M.; writing—review and editing, T.D.M., O.M. and J.O.; visualization, T.D.M.; supervision, O.M. and J.O.; project administration, O.M. and J.O.; funding acquisition, T.D.M., O.M. and J.O. All authors have read and agreed to the published version of the manuscript.

Funding: We acknowledge the German DAAD climapAfrica and the National Research Foundation of South Africa for funding the research. The research of this article was supported by DAAD within the framework of the climapAfrica program of the Federal Ministry of Education and Research. The publisher is fully responsible for the content. The work and article processing charge was also funded by the National Research Foundation of South Africa (NRF) Research Chair in Land Use Planning and Management (Grant Number: 84157).

Data Availability Statement: Remotely sensed data used in this study can be freely downloaded from United States Geological Survey (USGS) Earth Explorer website (www.earthexplorer.usgs.gov (accessed on 27 January 2022)). The training data used to map LCZ have been uploaded on the WUDAPT website (https://lcz-generator.rub.de/factsheets/3eb90c0ab4210886bdd0b23f1fc7dccae8 93c005/3eb90c0ab4210886bdd0b23f1fc7dccae893c005_factsheet.html) (accessed on 27 January 2022).

Acknowledgments: We acknowledge the Climate Modeling Group of the climapAfrica fellowship for inputs during virtual presentations which contributed to the quality of this manuscript. We thank the Discipline of Geography, School of Agricultural, Earth and Environmental Sciences in Pietermaritzburg, South Africa for availing a fruitful research environment. The Department of Space Science and Applied Physics at University of Zimbabwe also provided a working environment for this research.

Conflicts of Interest: The authors declare no conflict of interest.

References

1. Ahmed, B.; Ahmed, R. Modeling urban land cover growth dynamics using multioral satellite images: A case study of Dhaka, Bangladesh. *ISPRS Int. J. Geo-Inf.* **2012**, *1*, 3–31. [CrossRef]
2. Grimmond, S. Urbanization and global environmental change: Local effects of urban warming. *Geogr. J.* **2007**, *173*, 83–88. [CrossRef]
3. Hegazy, I.R.; Kaloop, M.R. Monitoring urban growth and land use change detection with GIS and remote sensing techniques in Daqahlia governorate Egypt. *Int. J. Sustain. Built Environ.* **2015**, *4*, 117–124. [CrossRef]
4. McDonald, R.I.; Green, P.; Balk, D.; Fekete, B.M.; Revenga, C.; Todd, M.; Montgomery, M. Urban growth, climate change, and freshwater availability. *Proc. Natl. Acad. Sci. USA* **2011**, *108*, 6312–6317. [CrossRef]
5. Nyamekye, C.; Kwofie, S.; Ghansah, B.; Agyapong, E.; Appiah, L. Land Use Policy Assessing urban growth in Ghana using machine learning and intensity analysis: A case study of the New Juaben Municipality. *Land Use Policy* **2020**, *99*, 105057. [CrossRef]
6. Seto, K.C.; Güneralp, B.; Hutyra, L.R. Global forecasts of urban expansion to 2030 and direct impacts on biodiversity and carbon pools. *Proc. Natl. Acad. Sci. USA* **2012**, *109*, 16083–16088. [CrossRef]

7. Heaviside, C.; Vardoulakis, S.; Cai, X.M. Attribution of mortality to the urban heat island during heatwaves in the West Midlands, UK. *Environ. Health A Glob. Access Sci. Source* **2016**, *15*, 49–59. [CrossRef]
8. Kotharkar, R.; Bagade, A. Urban Climate Local Climate Zone classification for Indian cities: A case study of Nagpur. *Urban Clim.* **2018**, *24*, 369–392. [CrossRef]
9. Bektaş Balçik, F. Determining the impact of urban components on land surface temperature of Istanbul by using remote sensing indices. *Environ. Monit. Assess.* **2014**, *186*, 859–872. [CrossRef]
10. Chow WT, L.; Brennan, D.; Brazel, A.J. Urban heat island research in Phoenix, Arizona. *Bull. Am. Meteorol. Soc.* **2012**, *93*, 517–530. [CrossRef]
11. Lelovics, E.; Unger, J.; Gál, T.; Gál, C.V. Design of an urban monitoring network based on Local Climate Zone mapping and temperature pattern modelling. *Clim. Res.* **2014**, *60*, 51–62. [CrossRef]
12. Odindi, J.O.; Bangamwabo, V.; Mutanga, O. Assessing the value of urban green spaces in mitigating multi-seasonal urban heat using MODIS land surface temperature (LST) and Landsat 8 data. *Int. J. Environ. Res.* **2015**, *9*, 9–18.
13. Saaroni, H.; Ben-Dor, E.; Bitan, A.; Potchter, O. Spatial distribution and microscale characteristics of the urban heat island in Tel-Aviv, Israel. *Landsc. Urban Plan.* **2000**, *48*, 1–18. [CrossRef]
14. Lau, K.K.; Ching, S.; Ren, C. Outdoor thermal comfort in different urban settings of sub-tropical high- density cities: An approach of adopting local climate zone (LCZ) classification. *Build. Environ.* **2019**, *154*, 227–238. [CrossRef]
15. Ormandy, D.; Ezratty, V. Thermal discomfort and health: Protecting the susceptible from excess cold and excess heat in housing. *Adv. Build. Energy Res.* **2016**, *10*, 84–98. [CrossRef]
16. Radhi, H.; Sharples, S. Quantifying the domestic electricity consumption for air-conditioning due to urban heat islands in hot arid regions. *Appl. Energy* **2013**, *112*, 371–380. [CrossRef]
17. Radhi, H.; Fikry, F.; Sharples, S. Impacts of urbanisation on the thermal behaviour of new built up environments: A scoping study of the urban heat island in Bahrain. *Landsc. Urban Plan.* **2013**, *113*, 47–61. [CrossRef]
18. Zhou, Y.; Weng, Q.; Gurney, K.R.; Shuai, Y.; Hu, X. Estimation of the relationship between remotely sensed anthropogenic heat discharge and building energy use. *ISPRS J. Photogramm. Remote Sens.* **2012**, *67*, 65–72. [CrossRef]
19. Jianya, G.; Haigang, S.; Guorui, M.; Qiming, Z. A review of multi-temporal remote sensing data change detection algorithms. *Int. Arch. Photogramm. Remote Sens. Spat. Inf. Sci.* **2008**, *37*, 757–762.
20. Lee, L.; Chen, L.; Wang, X.; Zhao, J. Use of Landsat TM/ETM+ data to analyze urban heat island and its relationship with land use/cover change. In Proceedings of the 2011 International Conference on Remote Sensing, Environment and Transportation Engineering, Nanjing, China, 24–26 June 2011; pp. 922–927. [CrossRef]
21. Gobakis, K.; Kolokotsa, D.; Synnefa, A.; Saliari, M.; Giannopoulou, K.; Santamouris, M. Development of a model for urban heat island prediction using neural network techniques. *Sustain. Cities Soc.* **2011**, *1*, 104–115. [CrossRef]
22. Keramitsoglou, I.; Kiranoudis, C.T.; Ceriola, G.; Weng, Q.; Rajasekar, U. Identification and analysis of urban surface temperature patterns in Greater Athens, Greece, using MODIS imagery. *Remote Sens. Environ.* **2011**, *115*, 3080–3090. [CrossRef]
23. Kolokotsa, D.; Psomas, A.; Karapidakis, E. Urban heat island in southern Europe: The case study of Hania, Crete. *Sol. Energy* **2009**, *83*, 1871–1883. [CrossRef]
24. Kumar, D.; Shekhar, S. Statistical analysis of land surface temperature-vegetation indexes relationship through thermal remote sensing. *Ecotoxicol. Environ. Saf.* **2015**, *121*, 39–44. [CrossRef]
25. Uddin, S.; Al Ghadban, A.N.; Al Dousari, A.; Al Murad, M.; Al Shamroukh, D. A remote sensing classification for land-cover changes and micro-climate in Kuwait. *Int. J. Sustain. Dev. Plan.* **2010**, *5*, 367–377. [CrossRef]
26. Li, H.; Liu, Q. Comparison of NDBI and NDVI as indicators of surface urban heat island effect in MODIS imagery. *Int. Conf. Earth Obs. Data Process. Anal.* **2008**, *7285*, 728503.
27. Jamei, Y.; Rajagopalan, P.; Sun, Q. Spatial structure of surface urban heat island and its relationship with vegetation and built-up areas in Melbourne, Australia. *Sci. Total Environ.* **2019**, *659*, 1335–1351. [CrossRef] [PubMed]
28. Sung, C.Y. Mitigating surface urban heat island by a tree protection policy: A case study of The Woodland, Texas, USA. *Urban For. Urban Green.* **2013**, *12*, 474–480. [CrossRef]
29. Rasul, A.; Balzter, H.; Smith, C. Spatial variation of the daytime Surface Urban Cool Island during the dry season in Erbil, Iraqi Kurdistan, from Landsat 8. *Urban Clim.* **2015**, *14*, 176–186. [CrossRef]
30. Sheng, L.; Lu, D.; Huang, J. Impacts of land-cover types on an urban heat island in Hangzhou, China. *Int. J. Remote Sens.* **2015**, *36*, 1584–1603. [CrossRef]
31. Wu, C.D.; Lung, S.C.C.; Jan, J.F. Development of a 3-D urbanization index using digital terrain models for surface urban heat island effects. *ISPRS J. Photogramm. Remote Sens.* **2013**, *81*, 1–11. [CrossRef]
32. Cai, M.; Ren, C.; Xu, Y.; Dai, W.; Wang, X.M. Local Climate Zone Study for Sustainable Megacities Development by Using Improved WUDAPT Methodology–A Case Study in Guangzhou. *Procedia Environ. Sci.* **2016**, *36*, 82–89. [CrossRef]
33. Stewart, I.D.; Oke, T.R. Local climate zones for urban temperature studies. *Bull. Am. Meteorol. Soc.* **2012**, *93*, 1879–1900. [CrossRef]
34. Fenner, D.; Meier, F.; Bechtel, B.; Otto, M.; Scherer, D. Intra and inter 'local climate zone' variability of air temperature as observed by crowdsourced citizen weather stations in Berlin, Germany. *Meteorol. Zeitscrift* **2017**, *26*, 525–547. [CrossRef]
35. Bechtel, B.; Alexander, P.J.; Böhner, J.; Ching, J.; Conrad, O.; Feddema, J.; Mills, G.; See, L.; Stewart, I. Mapping local climate zones for a worldwide database of the form and function of cities. *ISPRS Int. J. Geo-Inf.* **2015**, *4*, 199–219. [CrossRef]

36. Cai, M.; Ren, C.; Xu, Y.; Lau, K.K.; Wang, R. Urban Climate Investigating the relationship between local climate zone and land surface temperature using an improved WUDAPT methodology–A case study of Yangtze River Delta, China. *Urban Clim.* **2018**, *24*, 485–502. [CrossRef]

37. Danylo, O.; See, L.; Bechtel, B.; Schepaschenko, D.; Fritz, S. Contributing to WUDAPT: A Local Climate Zone Classification of Two Cities in Ukraine. *IEEE J. Sel. Top. Appl. Earth Obs. Remote Sens.* **2016**, *9*, 1841–1853. [CrossRef]

38. Perera NG, R.; Emmanuel, R. Urban Climate A "Local Climate Zone" based approach to urban planning in Colombo, Sri Lanka. *Urban Clim.* **2018**, *23*, 188–203. [CrossRef]

39. Qiu, C.; Mou, L.; Schmitt, M.; Xiang, X. Local climate zone-based urban land cover classification from multi-seasonal Sentinel-2 images with a recurrent residual network. *ISPRS J. Photogramm. Remote Sens.* **2019**, *154*, 151–162. [CrossRef]

40. United Nations General Assembly. *Sustainable Development Goals. SDGs Transform Our World 2030;* United Nations General Assembly: New York, NY, USA, 2015.

41. Zhao, C.; Jensen JL, R.; Weng, Q.; Currit, N. Use of Local Climate Zones to investigate surface urban heat islands in Texas Use of Local Climate Zones to investigate surface urban heat islands in Texas. *GISci. Remote Sens.* **2020**, *57*, 1083–1101. [CrossRef]

42. Nassar, A.K.; Blackburn, G.A.; Whyatt, J.D. International Journal of Applied Earth Observation and Geoinformation Dynamics and controls of urban heat sink and island phenomena in a desert city: Development of a local climate zone scheme using remotely-sensed inputs. *Int. J. Appl. Earth Obs. Geoinf.* **2016**, *51*, 76–90. [CrossRef]

43. Zhou, X.; Xu, L.; Zhang, J.; Niu, B.; Luo, M.; Zhou, G.; Zhang, X. Data-driven thermal comfort model via support vector machine algorithms: Insights from ASHRAE RP-884 database. *Energy Build.* **2020**, *211*, 109795. [CrossRef]

44. Eldesoky AH, M.; Gil, J.; Berghauser, M. Urban Climate The suitability of the urban local climate zone classification scheme for surface temperature studies in distinct macroclimate regions. *Urban Clim.* **2021**, *37*, 100823. [CrossRef]

45. Dutta, K.; Basu, D.; Agrawal, S. Evaluation of seasonal variability in magnitude of urban heat islands using local climate zone classification and surface albedo. *Int. J. Environ. Sci. Technol.* **2021**, 1–22. [CrossRef]

46. Badaro-saliba, N.; Adjizian-gerard, J.; Zaarour, R.; Najjar, G. Urban Climate LCZ scheme for assessing Urban Heat Island intensity in a complex urban area (Beirut, Lebanon). *Urban Clim.* **2021**, *37*, 100846. [CrossRef]

47. Mandelmilch, M.; Ferenz, M.; Mandelmilch, N.; Potchter, O. Urban Spatial Patterns and Heat Exposure in the Mediterranean City of Tel Aviv. *Atmosphere* **2020**, *11*, 963. [CrossRef]

48. Lee, Y.; Lee, S.; Im, J.; Yoo, C. Analysis of Surface Urban Heat Island and Land Surface Temperature Using Deep Learning Based Local Climate Zone Classification: A Case Study of Suwon and Daegu, Korea. *Korean J. Remote Sens.* **2021**, *37*, 1447–1460.

49. Ardiyansyah, A.; Munir, A.; Gabric, A. The Utilization of Land Surface Temperature Information as an Input for Coastal City The Utilization of Land Surface Temperature Information as an Input for Coastal City. *IOP Conf. Ser. Earth Environ. Sci.* **2021**, *921*, 012004. [CrossRef]

50. Zhang, Y.; Li, D.; Liu, L.; Liang, Z.; Shen, J.; Wei, F.; Li, S. Spatiotemporal Characteristics of the Surface Urban Heat Island and Its Driving Factors Based on Local Climate Zones and Population in Beijing, China. *Atmosphere* **2021**, *12*, 1271. [CrossRef]

51. Fricke, C.; Pongrácz, R.; Gál, T.; Savić, S.; Unger, J. Using local climate zones to compare remotely sensed surface temperatures in temperate cities and hot desert cities. *Morav. Geogr. Rep.* **2020**, *28*, 48–60. [CrossRef]

52. Dian, C.; Pongrácz, R.; Dezső, Z.; Bartholy, J. Urban Climate Annual and monthly analysis of surface urban heat island intensity with respect to the local climate zones in Budapest. *Urban Clim.* **2020**, *31*, 100573. [CrossRef]

53. Budhiraja, B.; Gawuc, L.; Agrawal, G. Seasonality of Surface Urban Heat Island in Delhi City Region Measured by Local Climate Zones and Conventional Indicators. *IEEE J. Sel. Top. Appl. Earth Obs. Remote Sens.* **2019**, *12*, 5223–5232. [CrossRef]

54. Shi, L.; Ling, F.; Foody, G.M.; Yang, Z.; Liu, X.; Du, Y. Seasonal SUHI Analysis Using Local Climate Zone Classification: A Case Study of Wuhan, China. *Int. J. Environ. Res. Public Health* **2021**, *18*, 7242. [CrossRef]

55. Bechtel, B.; Demuzere, M.; Mills, G.; Zhan, W.; Sismanidis, P.; Small, C.; Voogt, J. Urban Climate SUHI analysis using Local Climate Zones—A comparison of 50 cities. *Urban Clim.* **2021**, *28*, 100451. [CrossRef]

56. Demuzere, M.; Bechtel, B.; Mills, G. Urban Climate Global transferability of local climate zone models. *Urban Clim.* **2019**, *27*, 46–63. [CrossRef]

57. Gusso, A.; Cafruni, C.; Bordin, F.; Veronez, M.R.; Lenz, L.; Crija, S. Multitemporal Analysis of Thermal Distribution Characteristics for Urban Heat Islands Management. In Proceedings of the 4th World Sustainability Forum, Basel, Switzerland, 1–30 November 2014. [CrossRef]

58. Wang, R.; Cai, M.; Ren, C.; Bechtel, B.; Xu, Y.; Ng, E. Urban Climate Detecting multi-temporal land cover change and land surface temperature in Pearl River Delta by adopting local climate zone. *Urban Clim.* **2019**, *28*, 100455. [CrossRef]

59. Lu, Y.; Yang, J.; Ma, S. Dynamic Changes of Local Climate Zones in the Guangdong–Hong Kong–Macao Greater Bay Area and Their Spatio-Temporal Impacts on the Surface Urban Heat Island Effect between 2005 and 2015. *Sustainability* **2021**, *13*, 6374. [CrossRef]

60. Akinyemi, F.O.; Pontius, R.G., Jr.; Braimoh, A.K. Land change dynamics: Insights from Intensity Analysis applied to an African emerging city. *J. Spat. Sci.* **2017**, *62*, 69–83.

61. Aldwaik, S.Z.; Pontius, R.G., Jr. Map errors that could account for deviations from a uniform intensity of land change. *Int. J. Geogr. Inf. Sci.* **2013**, *27*, 1717–1739. [CrossRef]

62. Aldwaik, S.Z.; Gilmore, R.P., Jr. Landscape and Urban Planning Intensity analysis to unify measurements of size and stationarity of land changes by interval, category, and transition. *Landsc. Urban Plan.* **2012**, *106*, 103–114. [CrossRef]

63. Niya, A.K.; Huang, J.; Karimi, H.; Keshtkar, H.; Naimi, B. Use of Intensity Analysis to Characterize Land Use/Cover Change in the Biggest Island of Persian. *Sustainability* **2019**, *11*, 4396. [CrossRef]
64. Feng, Y.; Lei, Z.; Tong, X.; Gao, C.; Chen, S.; Wang, J.; Wang, S. Spatially-explicit modeling and intensity analysis of China's land use change 2000–2050. *J. Environ. Manag.* **2020**, *263*, 110407. [CrossRef]
65. Huang, B.; Huang, J.; Gilmore, R.P., Jr.; Tu, Z. Comparison of Intensity Analysis and the land use dynamic degrees to measure land changes outside versus inside the coastal zone of Longhai, China. *Ecol. Indic.* **2018**, *89*, 336–347. [CrossRef]
66. Ekumah, B.; Armah, F.A.; Afrifa, E.K.; Aheto, D.W.; Odoi, J.O.; Afitiri, A.R. Assessing land use and land cover change in coastal urban wetlands of international importance in Ghana using Intensity Analysis. *Wetl. Ecol. Manag.* **2020**, *28*, 271–284. [CrossRef]
67. Alo, C.A.; Pontius, R.G. Identifying systematic land-cover transitions using remote sensing and GIS: The fate of forests inside and outside protected areas of Southwestern Ghana. *Environ. Plan. B Plan. Des.* **2008**, *35*, 280–295. [CrossRef]
68. Mutengu, S.; Hoko, Z.; Makoni, F.S. An assessment of the public health hazard potential of wastewater reuse for crop production. A case of Bulawayo city, Zimbabwe. *Phys. Chem. Earth* **2007**, *32*, 1195–1203. [CrossRef]
69. Gumbo, B.; Mlilo, S.; Broome, J.; Lumbroso, D. Industrial water demand management and cleaner production potential: A case of three industries in Bulawayo, Zimbabwe. *Phys. Chem. Earth* **2003**, *28*, 797–804. [CrossRef]
70. Muchingami, I.; Hlatywayo, D.J.; Nel, J.M.; Chuma, C. Electrical resistivity survey for groundwater investigations and shallow subsurface evaluation of the basaltic-greenstone formation of the urban Bulawayo aquifer. *Phys. Chem. Earth* **2012**, *50–52*, 44–51. [CrossRef]
71. Amorim, M.C.D.C.T. Spatial variability and intensity frequency of surface heat island in a Brazilian city with continental tropical climate through remote sensing. *Remote Sens. Appl. Soc. Environ.* **2018**, *9*, 10–16. [CrossRef]
72. Bechtel, B.; Langkamp, T.; Böhner, J.; Daneke, C.; Oßenbrügge, J.; Schempp, S. Classification and Modelling of Urban Micro-Climates Using Multisensoral and Multitemporal Remote Sensing Data. *ISPRS-Int. Arch. Photogramm. Remote Sens. Spat. Inf. Sci.* **2012**, *39*, 463–468. [CrossRef]
73. Gal, T.; Bechtel, B.; Unger, J. Comparison of two different Local Climate Zone mapping methods. In Proceedings of the ICUC9—9th International Conference on Urban Climate Jointly with 12th Symposium on the Urban Environment, Toulouse, France, 20–24 July 2015; pp. 1–6.
74. Shi, Y.; Lau, K.K.; Ren, C.; Ng, E. Urban Climate Evaluating the local climate zone classi fi cation in high-density heterogeneous urban environment using mobile measurement. *Urban Clim.* **2018**, *25*, 167–186. [CrossRef]
75. Verdonck, M.L.; Okujeni, A.; van der Linden, S.; Demuzere, M.; De Wulf, R.; Van Coillie, F. Influence of neighbourhood information on 'Local Climate Zone' mapping in heterogeneous cities. *Int. J. Appl. Earth Obs. Geoinf.* **2017**, *62*, 102–113. [CrossRef]
76. Gislason, P.O.; Benediktsson, J.A.; Sveinsson, J.R. Random Forest Classification of Multisource Remote Sensing and Geographic Data. In Proceedings of the IGARSS 2004. 2004 IEEE International Geoscience and Remote Sensing Symposium, Anchorage, AK, USA, 20–24 September 2004; pp. 1049–1052.
77. Novack, T.; Stilla, U. Classification of Urban Settlements Types based on space-borne SAR datasets. *ISPRS Ann. Photogramm. Remote Sens. Spat. Inf. Sci.* **2014**, *2*, 55–60. [CrossRef]
78. Choe, Y.J.; Yom, J.H. Improving accuracy of land surface temperature prediction model based on deep-learning. *Spat. Inf. Res.* **2020**, *28*, 377–382. [CrossRef]
79. Meinel, G.; Winkler, M. Long-term investigation of urban sprawl on the basis of remote sensing data-Results of an international city comparison. In Proceedings of the 24th EARSeL-Symposium: New Strategies for European Remote Sensing, Dubrovnik, Croatia, 25–27 May 2004.
80. Pal, M. Random forest classifer for remote sensing classification. *Int. J. Remote Sens.* **2005**, *26*, 217–222. [CrossRef]
81. Rodriguez-galiano, V.F.; Ghimire, B.; Rogan, J.; Chica-olmo, M.; Rigol-sanchez, J.P. An assessment of the effectiveness of a random forest classifier for land-cover classification. *ISPRS J. Photogramm. Remote Sens.* **2012**, *67*, 93–104. [CrossRef]
82. Gislason, P.O.; Benediktsson, J.A.; Sveinsson, J.R. Random Forests for land cover classification. *Pattern Recognit. Lett.* **2006**, *27*, 294–300. [CrossRef]
83. Sumaryono, S. Assessing Building Vulnerability to Tsunami Hazard Using Integrative Remote Sensing and GIS Approaches. Ph.D. Thesis, LMU München, Faculty of Geosciences, München, Germany, 2010.
84. Sheykhmousa, M.; Mahdianpari, M.; Ghanbari, H. Support Vector Machine vs. Random Forest for Remote Sensing Image Classification: A Meta-analysis and systematic review. *IEEE J. Sel. Top. Appl. Earth Obs. Remote Sens.* **2020**, *13*, 6308–6325. [CrossRef]
85. Coppin, P.R.; Bauer, M.E. Digital change detection in forest ecosystems with remote sensing imagery Digital Change Detection in Forest Ecosystems with Remote Sensing Imagery. *Remote Sens. Rev.* **1996**, *13*, 207–234. [CrossRef]
86. Feng, H.; Zhao, X.; Chen, F.; Wu, L. Using land use change trajectories to quantify the effects of urbanization on urban heat island. *Adv. Sp. Res.* **2014**, *53*, 463–473. [CrossRef]
87. Pontius, R.G.; Gao, Y.; Giner, N.M.; Kohyama, T.; Osaki, M.; Hirose, K. Design and Interpretation of Intensity Analysis Illustrated by Land Change in Central Kalimantan, Indonesia. *Land* **2013**, *2*, 351–369. [CrossRef]
88. Adu, I.K.; Tetteh, J.D.; Puthenkalam, J.J.; Antwi, K.E. Intensity Analysis to Link Changes in Land Use Pattern in the Abuakwa North and South Municipalities Ghana from 1986 to 2017. *World Acad. Sci. Eng. Technol. Int. J. Geol. Environ. Eng.* **2020**, *14*, 225–242.

89. Enaruvbe, G.O.; Gilmore, R.P., Jr. Influence of classification errors on Intensity Analysis of land changes in southern Nigeria. *Int. J. Remote Sens.* **2015**, *36*, 244–261. [CrossRef]
90. Yang, Y.; Liu, Y.; Xu, D.; Zhang, S. Use of Intensity Analysis to Measure Land Use Changes from 1932 to 2005 in Zhenlai County, Northeast China. *Chin. Geogra Sci.* **2017**, *27*, 441–455. [CrossRef]
91. Varga, O.G.; Pontius, R.G., Jr.; Kumar, S.; Szabó, S. Intensity Analysis and the Figure of Merit's components for assessment of a Cellular Automata–Markov simulation model. *Ecol. Indic.* **2019**, *101*, 933–942. [CrossRef]
92. Chen, X.L.; Zhao, H.M.; Li, P.X.; Yin, Z.Y. Remote sensing image-based analysis of the relationship between urban heat island and land use/cover changes. *Remote Sens. Environ.* **2006**, *104*, 133–146. [CrossRef]
93. Srivanit, M.; Hokao, K.; Phonekeo, V. Assessing the Impact of Urbanization on Urban Thermal Environment: A Case Study of Bangkok Metropolitan. *Int. J. Appl. Sci. Technol.* **2012**, *2*, 243–256.
94. Stathopoulos, T.; Wu, H.; Zacharias, J. Outdoor human comfort in an urban climate. *Build. Environ.* **2004**, *39*, 297–305. [CrossRef]
95. Yang, J.S.; Wang, Y.Q.; August, P.V. Estimation of Land Surface Temperature Using Spatial Interpolation and Satellite-Derived Surface Emissivity. *J. Environ. Inform.* **2015**, *4*, 40–47. [CrossRef]
96. Weng, Q.; Liu, H.; Lu, D. Assessing the effects of land use and land cover patterns on thermal conditions using landscape metrics in city of Indianapolis, United States. *Urban Ecosyst.* **2007**, *10*, 203–219. [CrossRef]
97. Dimitrov, S.; Popov, A.; Iliev, M. An Application of the LCZ Approach in Surface Urban Heat Island Mapping in Sofia, Bulgaria. *Atmosphere* **2021**, *12*, 1370. [CrossRef]
98. Chatira-Muchopa, B.; Tarisayi, K.S.; Chidarikire, M. Solid waste management practices in Zimbabwe: A case study of one secondary school. *TD J. Transdiscipl. Res. South. Afr.* **2019**, *15*, 1–5. [CrossRef]
99. Nayak, S.; Mandal, M. Impact of land use and land cover changes on temperature trends over India. *Land Use Policy* **2019**, *89*, 104238. [CrossRef]
100. Blake, R.; Grimm, A.; Ichinose, T.; Horton, R.; Gaffin, S.; Jiong, S.; Bader, D.; Cecil, L.D. Urban climate: Processes, trends and projections. In *Climate Change and Cities: First Assessment Report of the Urban Climate Change Research Network*; Cambridge University Press: Cambridge, UK, 2011; pp. 43–81.
101. Lin, T.P.; De Dear, R.; Hwang, R.L. Effect of thermal adaptation on seasonal outdoor thermal comfort. *Int. J. Climatol.* **2011**, *31*, 302–312. [CrossRef]
102. Shi, Y.; Xiang, Y.; Zhang, Y. Urban Design Factors Influencing Surface Urban Heat Island in the High-Density City of Guangzhou Based on the Local Climate Zone. *Sensors* **2019**, *19*, 3459. [CrossRef]

 *remote sensing*

Article

Local Climate Zones and Thermal Characteristics in Riyadh City, Saudi Arabia

Ali S. Alghamdi [1], Ahmed Ibrahim Alzhrani [1,*] and Humud Hadi Alanazi [1,2]

[1] Department of Geography, King Saud University, Riyadh 12372, Saudi Arabia; aorifi@ksu.edu.sa (A.S.A.); Hu.alanazi@uoh.edu.sa (H.H.A.)

[2] Department of Tourism and Archeology, University of Hail, Hail 55473, Saudi Arabia

* Correspondence: ahalzahrani@ksu.edu.sa

Abstract: Using the local climate zone (LCZ) framework and multiple Earth observation input features, an LCZ classification was developed and established for Riyadh City in 2017. Four land-cover-type and four urban-type LCZs were identified in the city with an overall accuracy of 87%. The bare soil/sand (LCZ-F) class was found to be the largest LCZ class, which was within the nature of arid climate cities. Other land-cover LCZs had a lower coverage percentage (each class with <7%). The compact low-rise (LCZ-3) class was the largest urban type, as urban development in arid climate cities tends to extend horizontally. The daytime surface thermal characteristics of the developed LCZs were analyzed at seasonal timescales using land surface temperature (LST) estimated from multiple Landsat 8 satellite images (June 2017–May 2018). The highest daytime mean LST was found over large low-rise (LCZ-8) class areas throughout the year. This class was the only urban-type LCZ class that demonstrated a positive LST departure from the overall mean LST across seasons. Other urban-type LCZ classes showed lower LSTs and negative deviations from the overall mean LSTs. The overall thermal results suggested the presence of the surface urban heat island sink phenomenon as urban areas experienced lower LSTs than their surroundings. Thermal results demonstrated that the magnitudes of LST differences among LCZs were considerably dependent on the way the region of interest/analysis was defined. This was related to the types of LCZ classes presented in the study area and the spatial distribution and abundance of these LCZ classes. The developed LCZ classification and thermal results have several potential applications in different areas including planning and urban design strategies and urban health-related studies.

Keywords: land surface temperature; urban heat island; surface urban heat island; local climate zones; local climate zone generator

Citation: Alghamdi, A.S.; Alzhrani, A.I.; Alanazi, H.H. Local Climate Zones and Thermal Characteristics in Riyadh City, Saudi Arabia. *Remote Sens.* **2021**, *13*, 4526. https://doi.org/10.3390/rs13224526

Academic Editors: Prasad S. Thenkabail, Elhadi Adam, John Odindi, Elfatih Abdel-Rahman and Yuji Murayama

Received: 27 July 2021
Accepted: 8 November 2021
Published: 11 November 2021

Publisher's Note: MDPI stays neutral with regard to jurisdictional claims in published maps and institutional affiliations.

1. Introduction

One of the major consequences of rapid urbanization and urban expansion in the 21st century is the modification of urban climate [1] by replacing natural covers and structures (e.g., vegetation and open areas) with urbanized formations [2], e.g., compact and high buildings. Urbanization processes alter local landscapes in which urban/local climate characteristics (e.g., airflow, and energy and water exchanges) are modified [3]. This creates environments within cities that are warmer than their surroundings, which are referred to as urban heat islands (UHIs). UHIs constitute an essential research topic in urban climate studies and have received increasing attention as the UHI effect involves serious environmental concerns, two of which are population growth and climate change [4].

Commonly, a UHI is quantified by the temperature difference between urban and rural areas. One of the challenges in UHI studies is urban–rural representation because of the lack of clear boundaries between urban and non-urban (rural) areas [5]. To overcome this challenge, several urban climate classification schemes have been developed, including urban terrain zones [6]; the Davenport Roughness Classification [7]; urban climate zones,

UCZs [8]; and local climate zones, LCZs [4]. One of the main goals of these climate-based classification frameworks is to classify local landscapes (urban and rural) based on their surface properties (e.g., land-cover and urban geometry parameters) in such a way that different local thermal climate characteristics within and around cities can be adequately captured. In addition, these schemes were introduced to allow for standardized definitions and to simplify worldwide communication on different urban studies [2,9]. The LCZ scheme is the most recent, which extends the UCZ scheme, and is considered to be an international standard framework for urban climate classification because of its wide range of applications [2,9,10] and its effectiveness in urban thermal analysis [9].

Although the LCZ scheme was originally designed for UHI studies [4], it has several potential uses in many research and application fields, including weather and climate modeling, climate change, ecology, and urban planning [4,9]. As Zheng et al. [2] stated, the LCZ classification is an effective means for facilitating communication among urban planners, who are not familiar with local climate processes, and climatic researchers, who are not familiar with urban planning practices, for a better understanding of local climate and urban management strategies. The LCZ scheme has been applied in a wide range of urban studies including thermal comfort, energy consumption, human health, urban planning, and carbon emission [11,12]. For example, Verdonck et al. [13] evaluated the LCZ classification as an assessment tool for heat stress in Belgium and showed that the LCZ scheme is valuable when used a heat stress indicator and will help in urban planning for extreme weather events. Kotharkar et al. [14] demonstrated similar findings by using the LCZ scheme to explore population vulnerability to heat stress. Yang et al. [15] demonstrated that cooling and heating loads differed among different LCZs. As Xue et al. [11] explained, the LCZs is an important tool that can be employed in many research areas related to urban climate. This is because the LCZ scheme provides a means for classifying urban areas into different local climate zones, and allowing studies across different applications to document and compare results in a standardized way [11]. LCZ classification has extended the concept of urban–rural temperature differences to the neighborhood level (i.e., zone differences) to provide more details within a city and among cities [12].

LCZ mapping/classification approaches can be grouped into four techniques [1,2,9]: in situ measurements/sampling, satellite-image-based, GIS-based, and integrated approaches. In balancing the advantages and disadvantages of different mapping methods, the satellite-image-based technique is considered a fast and low-cost method for LCZ classification since it is based on readily available data with continuous spatiotemporal coverage at reasonable spatiotemporal resolutions [2,9]. A commonly used satellite-image-based LCZ classification protocol is the World Urban Database and Access Portal Tools (WUDAPT) scheme [16]. WUDAPT is a community-based project to develop, store, and share consistent urban climate datasets around the world. WUDAPT has established a protocol to classify the urban landscape into components using the LCZ classification scheme. This protocol is recognized as a framework for urban climate/weather research, urban planning, and public health worldwide as it provides a consistent, fast, reliable, and low-cost framework (e.g., [1,2,17]). For instance, because the LCZ scheme allows many studies to examine the spatiotemporal thermal patterns of cities using the WUDAPT framework and land surface temperature (LST) (e.g., [1,17–19]), the WUDAPT framework enables direct exploration of how these thermal patterns differ among cities as the urban morphology/texture is delineated in a standardized manner. WUDAPT contributes substantially to progress in the urban climate and planning field, as it provides standardized and detailed information on the fabric and structure of cities globally. However, global demands for more inputs to WUDAPT and verification of the framework's applicability are still growing [1,17]. Nevertheless, WUDAPT involves three time-consuming steps: obtaining data, pre-processing, and accuracy assessment [20]. Recently, Demuzere et al. [20] developed an LCZ Generator tool that addresses these issues as it requires only a training sample and its reference date. The LCZ Generator is a new online platform that implements the WUDAPT framework in a reasonable time frame by using Google Earth's engine computing environment to perform random forest classification and automated accuracy assessment.

Despite the importance of LCZs and the availability of LCZ maps for cities becoming an important step in facilitating urban studies [11], a climate-relevant database for the local climate in Riyadh City has not been established, as UHI studies in Riyadh City are limited, with only a few studies available (e.g., [21–23]). Furthermore, no studies have been conducted for the city's LSTs at the seasonal scale. Wang et al. [9] noted that studies related to LCZs in arid climate cities are limited, and the LCZ scheme offers greater potential for further understanding UHIs in these cities. Eldesoky et al. [24] showed that the suitability of the LCZ scheme varies from one climate region to another. For LCZ classification to be applied in arid climate cities, they recommended adding subclasses (i.e., combining two zones into one zone) and additional observation input features to the classification process. To contribute to the WUDAPT initiative and the promotion of LCZs for arid climate cities [9], this work aimed to advance our knowledge about Riyadh City's urban climate by mapping the local climate using the LCZ classification scheme and LCZ Generator for the first time. More specifically, the objectives were to (1) map the city LCZs and (2) evaluate how the LST (using Landsat-8) behaves seasonally (June 2017–May 2018) in different LCZs in order to investigate Riyadh City's surface UHI (SUHI) phenomenon. We further highlighted the importance of defining the region of interest, which can influence LST differences among LCZ classes and ultimately the intensity of SUHIs.

2. Data and Methods

2.1. Study Area

Riyadh City, the capital of Saudi Arabia (Figure 1), is positioned in an arid climate type according to the Koppen climate classification. The average air temperature of the city ranges from 15 °C in winter to 35.5 °C in summer (Table 1). Spring is the wettest season with an average precipitation of 22.4 mm, and summer is the driest season with an average precipitation of <1 mm (Table 1). The city is located at an elevation of approximately 700 m (Figure S1). During the last few decades, the city has experienced significant population growth from 80,000 in 1952 to 6,700,000 in 2015 [21,25]. Several studies have examined changes in land cover/use (LCLU) in Riyadh City (e.g., [26–29]) and a few have looked at future expansion (e.g., [25,30]). According to these studies, the urban area has substantially increased over the last few decades, with a higher likelihood of significant future expansions. For instance, population growth has led to spatial expansion of the built environment from less than 3.5 km^2 before the 1950s [21] to 1500 km^2 in 2014, and it is projected to reach 2161 km^2 by 2034 [25].

UHIs in Riyadh City are a weak phenomenon [31] and have the characteristics of UHIs in arid environments [21]. Recently, Sobrino and Irakulis (2020) [32] explored nighttime SUHIs in 71 urban agglomerations (including Riyadh) across different climate types and showed that arid climate cities have the lowest nighttime SUHIs. Studies have attributed this to the high thermal similarity between semi-bare lands and built-up areas in arid environments [33] and higher vegetation cover within urban areas [21]. Chen et al. [34] showed that temperature differences between urban and rural areas are reduced as semi-bare land around urban areas increases.

Table 1. Seasonal averages (1985–2020) of precipitation (Precip), mean temperature (Tmean), maximum temperature (Tmax), and minimum temperature (Tmin). Averages were derived from two weather stations (Figure 1) provided by the Saudi National Center of Meteorology as monthly means.

Season	Precip (mm)	Tmean (°C)	Tmax (°C)	Tmin (°C)
Winter	11.2	15.9	22.4	9.4
Spring	22.4	27.1	33.9	19.9
Summer	0.1	35.5	43.4	27.3
Fall	2.7	27.8	34.8	20.0

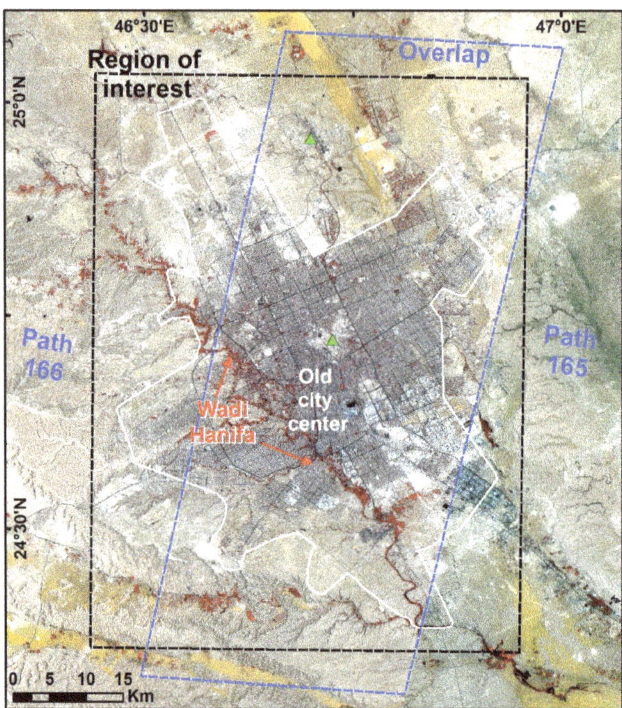

Figure 1. A map showing the study area along with weather stations (green triangles), urban growth limit phase II (white boundary), region of interest/research (black dashed box), and overlapped area (blue dashed box) covered by Landsat-8 path 166 and 165.

However, rural areas in Riyadh City have been shown to experience a negative daytime SUHI [21,23]. In arid desert cities, rural areas can display warmer daytime LSTs than urban areas, resulting in a SUHI sink phenomenon [9,21]. A SUHI sink in arid environments is due to the "oasis effect" [9], as urbanization in desert cities increases vegetation cover and water availability, which in turn lead to cooling effects by decreasing sensible heat flux and increasing latent heat flux and shading [21]. Alghamdi and Moore [21] and Aina et al. [23] used the urban growth boundary limit phase II (UG-II) to define the region of interest (Figure 1), yet Aina et al. [27] found that urbanization developments in the city do not precisely follow the UG-II. Accordingly, we defined the region of interest (i.e., research/analysis area) as the outer extent of the UG-II buffered by 3 km (black dashed box in Figure 1).

2.2. LCZ Definition

The LCZ classification includes 17 classes that can be divided into two major types: built-up/urban cover and land cover [4]. Not all LCZs can necessarily be found in a city due to differences in urban planning and strategies (for built cover types) and climate regimes (for land-cover types) (e.g., [1]). Given the climate and landscape of Riyadh City (i.e., arid environment), dense trees (LCZ-A), bush and scrub (LCZ-C), and water (LCZ-G) zones do not exist and were excluded. Some other original LCZ classes, namely compact high-rise (LCZ-1), compact midrise (LCZ-2), lightweight low-rise (LCZ-7), sparsely built (LCZ-9), and heavy industry (LCZ-10), had no or very few matches and were not generally identifiable at the local climate scale used for classification (i.e., 100 m) and thus were not included. The open midrise zone (LCZ-5) coexisted with the open high-rise zone (LCZ-4), and both were merged into one LCZ class (LCZ-5$_4$, open midrise with open high-rise).

As Stewart and Oke [4] explained, the LCZ system is flexible and can be adjusted to create new subclasses (e.g., LCZ-5_4) using proposed LCZ properties (e.g., sky view factor, surface cover fractions). However, Stewart and Oke (2012) [4] acknowledged that straight alignment between LCZ properties is not necessary. Recently, Wang et al. (2018b) [1] showed that LCZ property value ranges in desert environments do not align with the LCZ properties proposed in the literature. Alternatively, traffic flow and variability in building heights can be used to distinguish between adjacent/mixed LCZ classes and create a new subclassification [4]. Accordingly, Table 2 summarizes the LCZ classification developed and employed in this work based on the nature of Riyadh City, the authors' knowledge of the city, and a field survey. The suitability of this classification scheme was re-evaluated twice during the training process and post-classification field survey.

Table 2. Adapted LCZs along with the description relative to Riyadh City.

LCZs	Description
LCZ-3 Compact low-rise	Closely spaced buildings 1–3 stories tall.
LCZ-5_4 Open midrise with Open high-rise	Open arrangement of buildings $\geq$ 3 stories tall.
LCZ-6 Open low-rise	Open arrangement of buildings 1–3 stories tall.
LCZ-8 Large low-rise	Large and low buildings 1–3 stories (e.g., warehousing, and shopping centers).
LCZ-B Scattered trees	Parks and agricultural lands with frequent trees.
LCZ-D Low plants	Agricultural lands and grassy parks
LCZ-E Bare rock/paved	Impervious ground (e.g., bedrock, asphalt, and concrete).
LCZ-F bare soil/sand	Pervious ground and land with construction in progress.

2.3. LCZ Classification

The LCZ Generator was used to generate the LCZ classification. One of the main advantages of the LCZ Generator is that the random forest classifier is fed with 33 Earth observation input features, whereas the default WUDAPT uses only Landsat-8 data. Studies have shown that multi-input features improve LCZ classification [20]. For instance, the WUDAPT method tends to misclassify built-type LCZ classes in arid climate cities, even when multitemporal and multispectral remotely sensed data are used [1]. Ren et al. [35] explored the proficiency of the WUDAPT method over 20 Chinese cities and demonstrated that this issue (i.e., misclassification) is largely related to a lack of building height data. Another advantage is that the LCZ Generator provides an automated quality control approach to evaluate training areas and identify suspicious areas that require further attention. More information and references about the LCZ Generator can be found in Demuzere et al. [20].

The LCZ Generator requires two inputs: training areas, and the date of the imagery used to collect the training areas. Google Earth was used for collecting training samples. The choice of training samples is a leading source of errors in classification processes [36]. Thus, a set of quality control strategies was applied during the digitizing of training areas as outlined by WUDAPT and Demuzere et al. [20], including (1) avoiding mixed pixels, feature edges, and sites experiencing changes; (2) between 5 and 15 samples for each LCZ (9–20, in this work); (3) homogeneous spectral and surface characteristics; and (4) a less complex polygon shape. In addition, to reduce subjectivity in distinguishing between low-rise and midrise zones during digitizing training areas, building height data were used and acquired from the Saudi General Authority for Survey and Geospatial Information. The data were available for 2017 at a spatial resolution of 10 m. Accordingly, the training samples and LCZ classification were performed for 2017 using midsummer imagery to maintain a higher sun elevation angle to reduce shading effects.

Once the training areas file was uploaded to the LCZ Generator online platform, a random forest classification was performed using the training samples and 33 input features at 100 m resolution (see [20] for the list of input features). The use of multiple input features was within the recommendation of Eldesoky et al. [24]. The LCZ Generator applied a three-step automated quality control on the training samples to detect samples/polygons with (1) lower area size (below 400 m^2) or complex shape, (2) abnormal average spectral value compared to other samples in the corresponding LCZ class, and (3) pixels with abnormal spectral values compared to their LCZ class's average spectral value. The second and third steps were applied simultaneously on all input features. These quality control steps can be used to modify and adjust the initial training samples to reduce possible errors related to training samples in the classification processes. The initial training samples can be resubmitted to the LCZ Generator after revision and adjustment, and a minimum of three iterations is recommended. In this work, five iterations were used, where all the training samples passed the first two steps of quality control. Concerning the third quality control step, it was difficult in most cases to identify the reasons for the flagged training samples. Demuzere et al. [20] acknowledged this issue as the quality control methods are still experimental. In our case, part of the difficulty was because the third step was applied to all 33 input features, which were not currently available for users.

2.4. Accuracy Assessment and Filtering

An automated accuracy assessment was performed for the generated LCZ classification using the LCZ Generator. The accuracy assessment was based on an automated cross-validation approach using 25 bootstraps (runs). In each bootstrap (run), 70% of the training samples (stratified random sampling) were used to train the classifier and 30% for evaluation. Five accuracy metrics were generated over the 25 bootstraps: overall accuracy (OA), OA for urban LCZ classes only (OA_u), OA of the built versus land-cover LCZ classes only (OA_{bu}), weighted accuracy (OA_w), and F1-score. The main advantage of the bootstrapping procedure is that the accuracy metrics are generated with confidence intervals. For more details on accuracy metrics and bootstrapping, refer to Demuzere et al. [37] and Demuzere et al. [20].

Although the classification was performed for features with a 100 m resolution, the derived LCZ classification did not necessarily have homogenous structures that reflected the local climate concept [10]. Thus, the initial classification was further filtered to account for heterogeneity and granularity as an LCZ class does not consist of single isolated pixels [10,37]. The LCZ Generator executes morphological Gaussian filtering using a Gaussian kernel with varying standard deviation (σ_i) values for different LCZs. The σ_i values are not fixed values and are expert-driven as they differ among LCZs and cities [37]. However, the current version of the LCZ Generator does not allow users to adjust σi values. Thus, a majority filter post-classification of 3×3 pixels (i.e., 300 m) was used to reduce granularity and increase homogeneous structures as implemented by WUDAPT and several studies (e.g., [17,19,38]).

2.5. Evaluating Thermal Characteristics of LCZs Using LSTs

To examine the seasonal variations of thermal characteristics of LCZs, 40 multi-seasonal and cloud-free Landsat-8 images (Table 3) were obtained from the United States Geological Survey (USGS). The data were collected as level-2 surface reflectance (i.e., atmospherically, and geometrically corrected), and thus, LSTs were readily available in a surface temperature band. USGS produces this band using a single band approach with several atmospheric parameters (e.g., atmospheric transmission, and upwelled/downwelled radiance), thermal radiance band, and emissivity band. For more specific details and equations, one can refer to Malakar et al. (2018) [39] and USGS (2021) [40]. The Landsat-8 sensor passes over Riyadh at around 10 am (local time) in two adjacent paths (166 and 165) with a seven-day difference (Figure 1). This was considered to be an advantage as the overlapped area (Figure 1) could be sampled twice. To increase data representation and to account for temporal variation in LSTs within seasons, multi-season LSTs were obtained for each season (Table 3). This approach of multi-sampling was implemented to minimize any possible abnormality in LSTs within seasons and to reduce potential bias related to the inclusion of

Landsat-8 thermal bands in the LCZ Generator. Since the winter season extends over two years, LSTs from winter 2017–2018 (Table 3) were selected as more images were available compared to winter 2016–2017. Data for spring were from 2018 due to the very limited number of cloud-free images over the city in 2017 (only 4 out of 12 available images in USGS). Thus, the study period for LST analysis was from June 2017 to May 2018 using 40 Landsat-8 images.

Table 3. Landsat-8 image information used for LST mapping. Cloud cover percentages are in parentheses.

Season	Acquisition Date	Path/Row	Acquisition Date	Path/Row
Summer	20170605 (0.0)	166/43	20170723 (0.0)	166/43
	20170614 (0.0)	165/43	20170801 (0.0)	165/43
	20170621 (0.0)	166/43	20170808 (0.0)	166/43
	20170630 (0.0)	165/43	20170817 (0.0)	165/43
	20170707 (2.1)	166/43	20170824 (0.5)	166/43
	20170716 (0.0)	165/43		
Fall	20170902 (0.0)	165/43	20171011 (0.0)	166/43
	20170909 (0.3)	166/43	20171020 (0.0)	165/43
	20170918 (0.0)	165/43	20171027 (0.0)	166/43
	20170925 (0.0)	166/43	20171105 (4.4)	165/43
	20171004 (0.0)	165/43	20171112 (0.04)	166/43
Winter	20171207 (0.7)	165/43	20180115 (0.3)	166/43
	20171214 (0.4)	166/43	20180124 (1.5)	165/43
	20171223 (2.5)	165/43	20180131 (1.8)	166/43
	20171230 (0.3)	166/43	20180209 (0.7)	165/43
	20180108 (0.8)	165/43	20180216 (0.3)	166/43
Spring	20180523 (0.0)	166/43	20180414 (0.0)	165/43
	20180507 (0.3)	166/43	20180329 (0.1)	165/43
	20180430 (0.0)	165/43	20180320 (1.1)	166/43
	20180421 (0.0)	166/43	20180313 (0.1)	165/43
	20180304 (0.1)	166/43		

To assess the differences in LSTs among LCZs, the departure of each LCZ class's mean LST from the mean LST of the region of interest (ROI) and the UG-II was computed. To quantify the differences in LSTs among LCZs, pairwise comparisons were applied using a two-sample Kolmogorov–Smirnov (K–S) test [41]. The K–S test is a nonparametric test and was applied to assess significant differences among different LCZs in terms of the cumulative distribution of LSTs. Although the magnitude/intensity of a UHI (SUHI) is simply defined as the air temperature (LST) difference between urban and rural, determining it is not a straightforward task. The current literature does not have a standard and clear method by which urban and rural areas are defined [32]. This issue is related to the lack of clear physical and climatic boundaries between urban and rural areas [5]. Additionally, LCZs can be found in urban and rural areas with different spatial coverages and distributions.

Consequently, this study relied and focused on the overall patterns of LSTs, not on the absolute magnitude of the SUHI. To investigate the nature of the SUHI, longitudinal and latitudinal profiles of mean and standard deviation (STD) of LSTs were computed. By analyzing the spatial patterns of both profiles, the overall spatial patterns of LSTs could be summarized across the city and SUHI could be reorganized. To minimize possible bias related to the size and spatial configuration of LCZs, all analyses were performed for both the ROI and the UG-II.

3. Results and Discussion

3.1. LCZ Classification

Figure 2 shows the obtained LCZ classification, and Figure 3 presents the distribution of the accuracy matrices over the 25 bootstraps. The average OA, OA_u, OA_{bu}, and OA_w of the LCZ classification were 87, 87, 98, and 95%, respectively. While the average F1-score for urban-type LCZs ranged from 59% for LCZ-5_4 (open midrise with open high-rise) to 90% for LCZ-8 (large low-rise), the average land-cover-type F1-score ranged from ~70% for LCZ-E (bare rock/paved) to 95% for LCZ-F (bare soil/sand). LCZ-5_4 showed the lowest average F1-score, and studies have shown that both LCZ 4 and 5 classes are challenging to map [37]. LCZ-5_4 recorded a lower average F1-score (a harmonic average of the user's and producer's accuracy, [37]) as the corresponding average user's accuracy (53%) was lower than the average producer's accuracy (94%). LCZ-5_4 was often confused with LCZ-3 (compact low-rise). The confusion among urban-type LCZs, particularly when the main distinctions among them were geometric characteristics (e.g., building heights, spacing, and sky view factor), is a typical issue in LCZ classification due to resampling input data at 100 m resolution and the limitation of high-quality geometric data [9,37]. Nevertheless, the OA_u indicated a fair overall accuracy (87%) for the urban-type LCZ classes, and all the average overall accuracy indices exceeded the suggested minimum accuracy of 50% [20]. Moreover, the obtained mean OA (87%) was within the commonly recognized target accuracy of 85% [9,42].

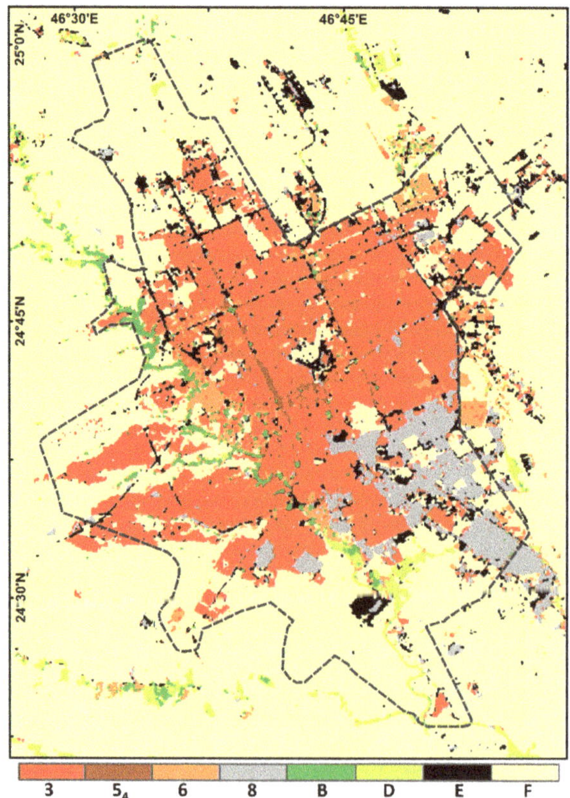

Figure 2. Developed local climate zones (LCZs) of Riyadh for 2017.

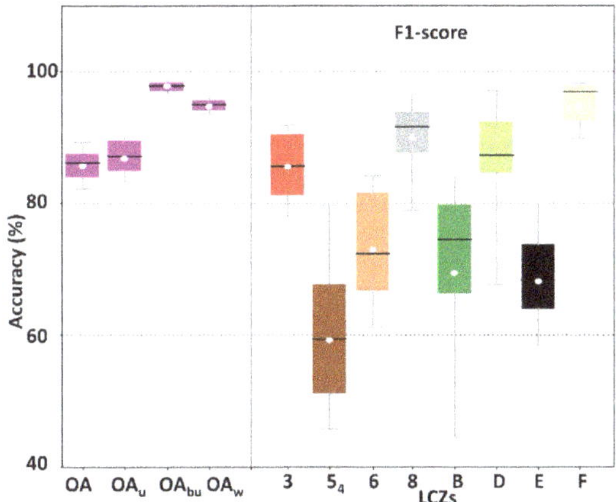

Figure 3. Distribution of the accuracy matrices over the 25 bootstraps for the developed LCZs. The plot was generated by the LCZ Generator, and the layout was modified for visualization.

Over the region of interest (ROI) and urban growth limit phase II (UG-II), bare soil/sand (LCZ-F) was the largest LCZ class, covering 68 and 41% of the total areas, respectively (Table 4). The second-largest LCZ class was compact low-rise zone (LCZ-3), and it was the most dominant urban type within both the ROI and UG-II, covering 685 (18%) and 657 km^2 (37%), respectively. LCZ-5$_4$ (open midrise with open high-rise) was confined to the UG-II (99.6%) and only constituted 1% (17 km^2) of the UG-II area. Most of this class was found within the old city center and along the main commercial roads (Figures 1 and 2). The main building types within this class were offices and tall residential apartment buildings. These results are consistent with the findings of Wang et al. (2018a) [9], who found that bare soil/sand and open low-rise zones were the predominant LCZs in Phoenix and Las Vegas, USA, as cities in desert environments tended to develop more horizontally because of abundant land for development. LCZ-6 (open low-rise) covered 114 km^2, 67% (~77 km^2) of which was within the UG-II. The zone spread mostly in the eastern and western outskirts of the city. LCZ-8 (large low-rise) covered 156 km^2, 80% (~125 km^2) of which was within the UG-II. This zone was occupied mostly by light industrial activities and was largely limited to the southeastern outskirts in compliance with city regulations.

Table 4. Area (km^2) statistic of LCZs in Riyadh for 2017.

LCZ	ROI		UG-II	
	Area	(%)	Area	(%)
3	685	18	657	37
5$_4$	17	<1	17	1
6	114	3	77	4
8	156	4	124	7
B	46	1	30	2
D	79	2	26	2
E	167	4	111	6
F	2633	68	714	41
Total	3897	100	1756	100

LCZ-E (bare rock/paved) accounted for 4% (6%) of the ROI (UG-II) area, consisted predominantly of asphalt, and was mainly found alongside transportation features (e.g., highways, airports, parking lots). Accordingly, LCZ-E could not be regarded as a natural cover in Riyadh City. LCZs B and D had the lowest area coverages of 46 (30 km^2) and

79 (26 km^2) km^2, respectively, within the ROI (UG-II). Both classes are human-made (desert city), and LCZ-B was found over large parks and farming fields with trees (commonly ornamental trees and palms), whereas agricultural lands and grassy parks constituted the LCZ-D class. Within the UG-II, both zones were spatially distributed along Wadi Hanifa (Hanifa valley, Figure 1), where agricultural sites are clustered. LCZ-D had a lower frequency across the built-up areas as low plant cover within the city was restricted to small neighborhoods parks, which are difficult to detect and classify at 100 m resolution. Thus, agricultural activities are suggested as the major contributor to those land-cover classes.

3.2. LST and LCZs

The estimated morning LST demonstrated large seasonal and spatial variability (Figure 4). Summer recorded the highest mean LST (52 °C), followed by fall (46 °C), spring (40 °C), and winter (26 °C) within the ROI (i.e., the study area as a whole). The UG-II experienced similar seasonal patterns with mean LSTs about one-degree lower (Figure 4). Alghamdi and Moore [21] reported similar findings for the city's UHI due to seasonal changes in solar inclination. In summer (winter), the sun is at higher (lower) elevation, and incoming radiation is more (less) direct and concentrated over smaller (larger) areas. As a result, LSTs are higher in summer and lower in winter. The ROI displayed higher seasonal LST variability (standard deviation) compared to the UG-II (Figure 4D). LSTs experienced increasingly higher variability during fall, summer, spring, and winter. This pattern was observed over both the ROI and UG-II, yet the seasonal differences in variability were higher over the ROI (Figure 4D). The higher mean LSTs over the ROI could be explained by its larger area and a higher presence of sand cover compared to the UG-II (Figure 1). The latter factor might be more important, as sand warms at a higher rate during the daytime and has high heat capacity [21]. In fact, maximum LSTs were detected over areas with pure sand cover in all four seasons. Pan et al. [43] reported similar results in Zhangye City, located in the northern arid region of China. These differences between the ROI and UG-II highlighted the importance of accounting for possible bias related to the size and spatial distribution of land cover (i.e., LCZs) in assessing SUHIs.

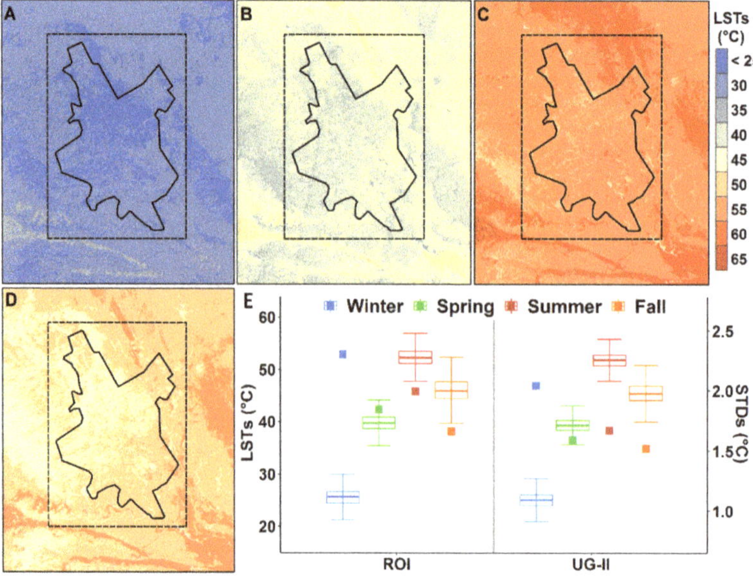

Figure 4. Mean winter (**A**), spring (**B**), summer (**C**), fall (**D**) estimated daytime LSTs (°C) and the corresponding boxplots (**E**) with standard deviations (squares).

Daytime LST exhibited considerable seasonal variability among LCZ classes over both the UG-II and the ROI (Figure 5). In summer and over the ROI, the LCZ-8, LCZ-E, and LCZ-F classes displayed the highest mean LST of ~53 °C, whereas the LCZ-B class recorded the lowest mean LST of 49 °C. Within the UG-II, LCZ-8 and LCZ-E classes displayed the highest mean LST of 53 °C, and LCZ-B class had the lowest LST of 49 °C. The similarity in summer mean LST values of LCZ-8, LCZ-E, and LCZ-B classes over both the ROI and UG-II could be due to the observation that more than two-thirds of these classes are within the UG-II. In fall, the highest (lowest) mean LST of 47 °C (43 °C) was recorded by the LCZ-8 (LCZ-B) class within both the ROI and UG-II. Over both the ROI and UG-II both LCZ 8 and D classes exhibited the highest mean winter and spring LST, whereas LCZ-B and LCZ-3 classes showed the lowest mean winter and spring LST. Overall, the large low-rise class (LCZ-8) had the highest average LST in all seasons. Other studies showed similar findings for low-rise classes in different cities (e.g., [9,44]). LCZ-B class showed the lowest mean LST across seasons. This class mainly consists of high-height vegetation (e.g., palms), and thus, lower average LSTs could be related to shadowing effects.

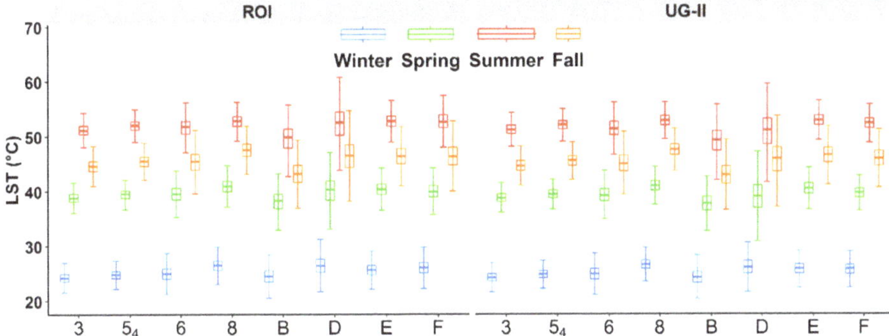

Figure 5. Boxplots of estimated seasonal daytime LSTs for mapped LCZs in the ROI (**right**) and UG-II (**left**).

To further explore differences among the LSTs of LCZs, and between the ROI and UG-II, the departure of each LCZ from the mean LST was calculated at ROI and UG-II scales (Figure 6). All LCZs showed similar anomaly directions (positive/negative) over both the ROI and UG-II and across all seasons, except the LCZ-5_4 class in summer, LCZ-D in spring and summer, and LCZ-E in winter. While urban-type LCZs tended to have large LST departure values (positive and negative) across the ROI (except LCZs 8 and E, largely distributed within the UG-II), land-cover-type LCZs showed large departure values over the UG-II, except the LCZ-B class. Thus, it was important to evaluate LSTs over the UG-II to understand the urban LST variabilities/differences (UG-II) and how urban LSTs differed from the surroundings (ROI). LCZs 3, 5_4, 6 (urban types), and B classes experienced cooler (negative departure) daytime mean LST in all seasons across the ROI. Similar results were seen across the UG-II, except for the LCZ-5_4 class during summer. The LCZ-D class exhibited a lower LST than the average in summer across both the ROI and UG-II and in spring over the UG-II. The negative departure values were larger across the ROI for LCZs 3, 5_4, and B classes compared to the UG-II. This can be attributed to the lower contribution of these classes to the ROI and the higher seasonal mean LSTs of the ROI. These classes were largely confined to the UG-II (Table 4), which had lower seasonal mean LSTs compared to the ROI (Figure 4D). Warm LST deviations (positive departures) were found for the LCZ-8 class (urban-type) and all land-cover types (except LCZ-B) during most of the seasons and across both the ROI and UG-II. For these classes, the magnitudes of LST departures were higher over the UG-II, unlike negative LST deviations.

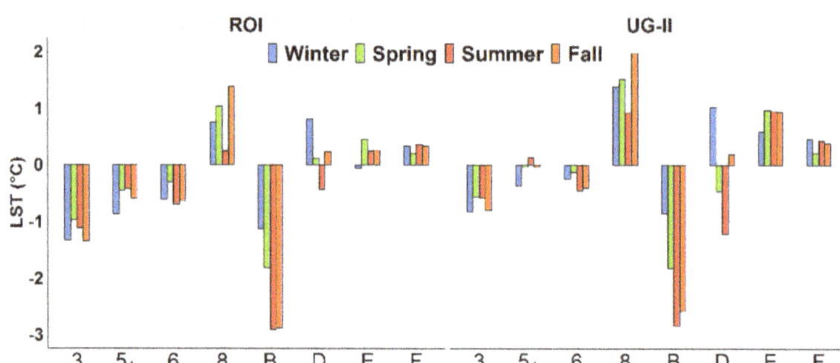

Figure 6. Departure from the ROI (**right**) and UG-II (**left**) winter (blue), spring (green), summer (red), and fall (orange) daytime mean LSTs (°C) for each LCZ class.

To quantify LST differences among LCZs and between different areas of interest/analysis (ROI vs. UG-II), the K–S test was applied. The results revealed that all LCZs differed significantly in terms of their cumulative distributions of LSTs (i.e., coming from different populations) at all seasons within both the ROI and UG-II (Figure 7, upper and mid panels). This result suggested that the developed LCZs could identify distinctive LSTs among classes [45]. That is, the developed LCZ classification is suitable for the city, as it groups distinct LSTs into different zones. The K–S test also revealed significant differences among LSTs of LCZs over the ROI and those over the UG-II across seasons (Figure 7, lower panel). This indicated that the definition of the research/analysis area is an important aspect, as different definitions can result in significantly different LST cumulative distributions. Such differences might be more related to the spatial configurations of LCZ classes, as they showed different distributions over the ROI and UG-II (Figure 2). Subsequently, different definitions would result in different LST magnitudes and, thus, different SUHI intensity. The results from Figures 6 and 7 revealed the developed LCZs can be used to investigate daytime SUHIs in Riyadh City. Accordingly, the negative LST departures of urban-type and the positive LST departures of land-cover type (excluding LCZ-B) were highly suggestive of SUHI sink presence in Riyadh City.

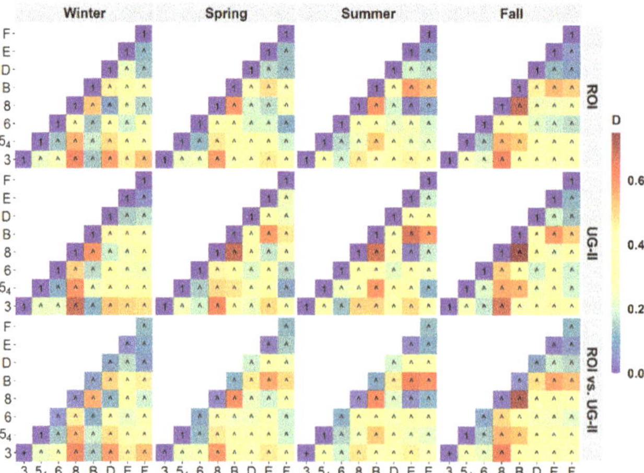

Figure 7. Results of the K–S tests to compare LCZ classes within each area of interest/analysis (upper [ROI] and mid [UG-II] panels) and between areas of interests/analysis (lower panel, ROI vs. UG-II). Cells with "^" (+) indicate significant difference at $p < 0.05$ (0.1) level. Cells with value of 1 indicate no significant differences at $p < 0.05$ or 0.1 levels.

To further explore the nature of daytime SUHIs (i.e., positive or negative) in the city, mean and standard deviation (STD) of LSTs for each LCZ class were mapped, and longitudinal and latitudinal profiles were evaluated (Figures 8 and 9). This approach allowed for a more comprehensive spatial view of the LST profiles, in which LST differences between urban and surroundings could be evaluated. To account for the observed differences in LSTs between the ROI and UG-II due to differences in LCZs distribution, the analyses were performed for both areas. Results in Figure 8 were in line with those in Figure 6 and clearly showed that urban areas experienced lower daytime mean LSTs across seasons compared to the surroundings. This indicated that daytime SUHI sink was highly evident. Longitudinal and latitudinal LST profiles (upper and left sides of each season map in Figures 8 and 9) provide west–east and north–south, respectively, views of mean LSTs and indicate the presence of SUHI sink. For instance, the latitudinal (longitudinal) mean shows the north–south (west–east) profile of LSTs and demonstrates that the mean LSTs started at higher values over northern (western) areas and decreased toward the south (east) over the urban area and then increased south (east) afterward. This observed pattern was seen across all the analyzed seasons, but with different magnitudes, as summer has higher daytime LSTs followed by fall, spring, and winter, in that order. Generally, both the ROI and UG-II areas displayed similar profile patterns, and the UG-II tended to show a detailed profile version of this over the ROI.

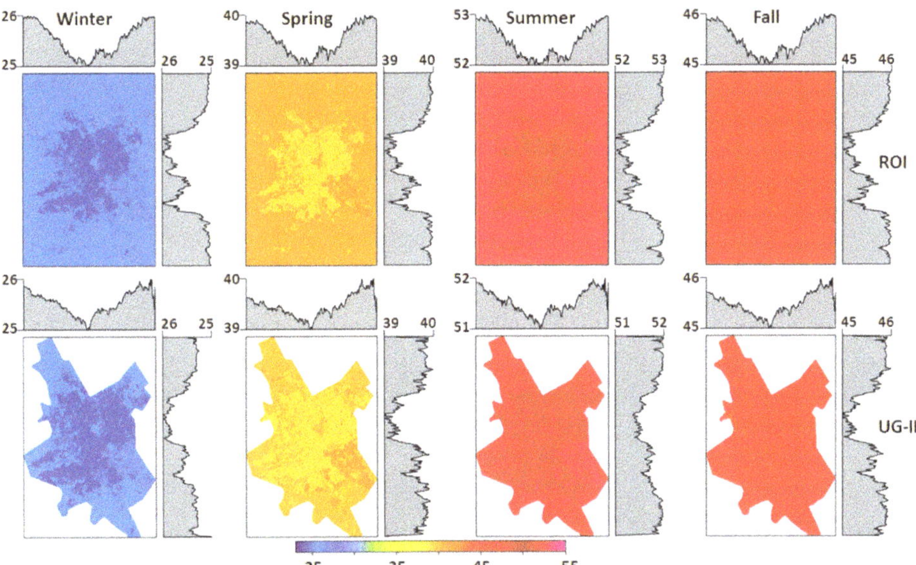

Figure 8. Seasonal mean daytime LSTs (°C) for each LCZ along with longitudinal (right-side of each map) and latitudinal (upper-side of each map) profiles over the ROI (**upper panel**) and UG-II (**lower panel**).

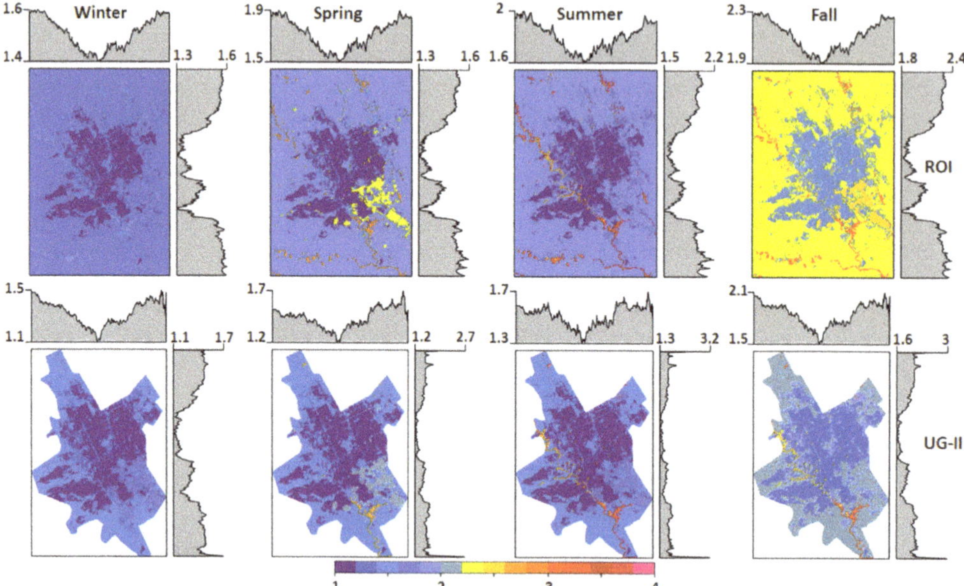

Figure 9. As in Figure 8, but for STD (°C).

The consistent profile patterns of low mean LSTs over urban areas across seasons and both the ROI and UG-II agreed with the results in Figure 6. These combined results indicated that Riyadh City had daytime SUHI sink during the studied period (June 2017–May 2018). The urban class types of LCZs 3, 5_4, and 6 were where lower mean LSTs were recorded, and these classes were what mostly constituted the urban area. The spatial distribution of low mean LSTs (Figure 8) corresponded well with the spatial destitution of these classes in Figure 2. Similarly, higher mean LSTs were recorded over the LCZ 8, E, and F classes and their spatial distribution (Figure 2) was similar to that of high mean LSTs (Figure 8). For example, a closer look at the latitudinal and longitudinal profiles showed that mean LSTs have a peak around the middle of the profiles, and this peak parallels the area of LCZ-8. This class had the largest positive departure from mean area LSTs (Figure 6). This was consistent with the findings of Aina et al. [23] and Abulibdeh [46], in which industrial lands (in our case LCZ-8) in Riyadh had the highest daytime LSTs within the urban agglomeration. The higher values at the beginning and end of LST profiles corresponded to areas where the LCZ-F class was primarily distributed.

Generally, profiles of STDs of LSTs showed patterns similar to those found for the mean profiles, where lower values are over urban areas (Figure 9). LST variabilities (STDs) were low over urban areas in all seasons over both the ROI and UG-II, and thus, the daytime SUHI sink appears to have homogenous LSTs. In fall, LSTs had higher variability, followed by summer and winter across most of the LCZs. Higher STD values were found mainly over the areas of the LCZ B and D classes across seasons, with higher magnitudes in summer and fall. The frequent small peaks along the STD profiles of LST (Figure 9) corresponded to the areas of both classes (Figure 2). In summer and fall seasons, the lowest mean LSTs were over both classes (Figure 8), more notably in summer. During warm seasons, evapotranspiration is high (higher sun elevation angle), and most of the heat is transported as latent heat [47]. Accordingly, low mean LSTs in winter were not found across LCZs B and D (Figure 8), and both showed positive LST departure values (Figure 6) as evapotranspiration is lower in this season. This is also combined with the observation that both classes were distributed largely along Wadi Hanifa, which suggested there was more heat trapped near the ground during winter. Furthermore, we noticed that there was a notable frequency of relatively lower LSTs over the LCZ-3 class coupled with the presence of small parks and sparse residential vegetation, which were omitted at the 100 m resolution. Kwarteng and Small [48] detected lower LSTs over residential areas in Kuwait City due to cooling effects of vegetation, even when the proportion of vegetation was low.

4. Summary and Conclusions

Using the LCZ Generator platform, an LCZ classification was developed for Riyadh City for 2017, based on multiple Earth observation input features. Eight LCZs were identified with an overall accuracy of 87%. The bare soil/sand (LCZ-F) class was found to be the largest LCZ class followed by the compact low-rise zone (LCZ-3), which was the largest urban-type class. Other LCZ classes were not dominant as each covered less than 8% of the total area. This is within the expected nature of an arid climate city.

The seasonal thermal characteristics of the developed LCZs were analyzed using LSTs estimated from 40 thermal bands (June 2017–May 2018). The results revealed that the urban area had lower LSTs as urban-type LCZs had low LSTs and experienced a large negative departure from the mean area LSTs across seasons. However, the large low-rise (LCZ-8) class showed the highest seasonal LSTs and tended to demonstrate the largest positive LST departure. The lowest LSTs were found over the scattered trees (LCZ-B) class, except in winter, when LCZ-3 had the lowest LSTs. All the thermal analyses suggested the presence of the SUHI sink and were consistent with the findings of previous research. Studies have indicated that the intensity of Riyadh's UHI has experienced a decreasing trend [21,31], while Aina et al. [23] reported LSTs have increased. Alghamdi and Moore [21] showed that the daytime UHI in Riyadh transitioned to a UHI sink phenomenon in the early 1990s, which was explained by increases in vegetation cover within the urban area. This study further provided seasonal analysis and showed that LSTs and, substantially SUHIs have seasonal variabilities and their magnitudes depend on how the region of study/analysis is defined. The latter is more related to the spatial distribution of LCZ classes.

The developed LCZ classification has several potential applications in different areas. For instance, one city zoning policy requires that light industrial and large storage facilities be limited to certain areas outside the city. The LCZ-8 class is an example of that policy, and it illustrates how such a strategy has helped to keep higher LSTs outside the city as that zone had the highest mean LSTs. This would also highlight the importance of a new urban design strategy for this zone to mitigate its LSTs, such as increasing spacing and vegetation cover. Another application of the developed LCZ classification is as data input for studies of population vulnerability to temperature discomfort and heat stress (e.g., [14]). Other studies have used the LCZ scheme to model and predict diseases in urban areas (e.g., [49]) as this scheme offers detailed information about urban climate. Accordingly, this developed LCZ classification can be used to provide more insights into several urban-related issues (e.g., health hazards and diseases) that have spatial patterns in the city, like those related to air pollution. Detailed knowledge about these aspects would assist the efforts of urban and health management not only to mitigate UHI/SUHI effects in the city, but also to improve the overall life quality. For instance, a better understanding of how the current urban structures influence the local climate, which contributes to heat stress during heat waves, would not only help improve heat warring and response systems, but would also help direct planning efforts for some of the mitigation strategies.

Vegetation's negative effects on UHIs/SUHIs are well established in the literature, yet the exact effects on Riyadh City's UHI/SUHI have not been evaluated, and the results presented here suggest temporal variability as in the winter case. As Collins and Dronova [50] showed, the replacement of bare soil cover (LCZ-F) with open structures and vegetation cover could lead to cooling effects. However, abundance of vegetation cover does not necessarily mitigate UHIs effects, as the low plant cover (LCZ-D) experienced relatively warm LSTs in fall and winter in this work. In a study by Collins and Dronova [50], a similar observation was found but in summer. This can be explained by the work of Li et al. [51], as they found that the spatial configuration of green space plays an important role in the association between LSTs and vegetation as higher fragmentation and density of greenspace increase LSTs. These points emphasize the need for further research as they have received limited attention in arid climate cities.

Supplementary Materials: The following are available online at https://www.mdpi.com/article/10.3390/rs13224526/s1. Figure S1: A map showing a digital elevation model of the study area. Data source: UGGS.

Author Contributions: A.I.A. and H.H.A. formulated, collected and analyzed the data, visualized the results, and drafted the manuscript. A.S.A. developed the methodology, reviewed, and revised the manuscript, and supervised the work. All authors have read and agreed to the published version of the manuscript.

Funding: This research was supported by the Deanship of Scientific Research at King Saud University through research group program (RG-1441-344). There are no conflicts of interest to declare.

Data Availability Statement: The Landsat data are available from the United States Geological Survey website at https://earthexplorer.usgs.gov (accessed on 27 July 2021).

Acknowledgments: The authors extend their appreciation to the Deanship of Scientific Research at King Saud University for funding this work through Research Group no RG-1441-344. We are grateful to four anonymous reviewers for their insightful comments.

Conflicts of Interest: The authors declare no conflict of interest.

Abbreviations

Abbreviations	Full Name
LCLU	Land cover/use
LCZ	Local climate zones
LCZs Generator	Local climate zone generator platform/tool
LST	Land surface temperature
ROI	Region of interest
SUHI	Surface urban heat island
UG-II	Urban growth boundary limit phase II
UHI	Urban heat island
WUDAPT	World Urban Database and Access Portal Tools
STD	Standard deviation
OA	Overall accuracy
OA_u	Overall accuracy for urban LCZ classes only
OA_{bu}	Overall accuracy of the built versus land-cover LCZ classes only
OA_w	Weighted accuracy

References

1. Wang, R.; Ren, C.; Xu, Y.; Lau, K.K.-L.; Shi, Y. Mapping the local climate zones of urban areas by GIS-based and WUDAPT methods: A case study of Hong Kong. *Urban Clim.* **2018**, *24*, 567–576. [CrossRef]
2. Zheng, Y.; Ren, C.; Xu, Y.; Wang, R.; Ho, J.; Lau, K.; Ng, E. GIS-based mapping of Local Climate Zone in the high-density city of Hong Kong. *Urban Clim.* **2018**, *24*, 419–448. [CrossRef]
3. Goodwin, N.R.; Coops, N.C.; Tooke, T.R.; Christen, A.; A Voogt, J. Characterizing urban surface cover and structure with airborne lidar technology. *Can. J. Remote Sens.* **2009**, *35*, 297–309. [CrossRef]
4. Stewart, I.D.; Oke, T.R. Local Climate Zones for Urban Temperature Studies. *Bull. Am. Meteorol. Soc.* **2012**, *93*, 1879–1900. [CrossRef]
5. Martin-Vide, J.; Sarricolea, P.; Moreno-García, M.C. On the definition of urban heat island intensity: The "rural" reference. *Front. Earth Sci.* **2015**, *3*, 24.
6. Ellefsen, R. Mapping and measuring buildings in the canopy boundary layer in ten U.S. cities. *Energy Build.* **1991**, *16*, 1025–1049. [CrossRef]
7. Davenport, A.G.; Grimmond, C.S.B.; Oke, T.R.; Wieringa, J. Estimating the roughness of cities and sheltered country. In Proceedings of the 12th Conference on Applied Climatology, Asheville, NC, USA, 8–11 May 2000; American Meteor Society: New York, NY, USA, 2000; Volume 96, p. 99.

8. Oke, T.R. Initial guidance to obtain representative meteorological observations at urban sites. *Instrum. Obs. Methods* **2004**, *47*, 2006.

9. Wang, C.; Middel, A.; Myint, S.W.; Kaplan, S.; Brazel, A.J.; Lukasczyk, J. Assessing local climate zones in arid cities: The case of Phoenix, Arizona and Las Vegas, Nevada. *ISPRS J. Photogramm. Remote Sens.* **2018**, *141*, 59–71. [CrossRef]

10. Bechtel, B.; Alexander, P.J.; Böhner, J.; Ching, J.; Conrad, O.; Feddema, J.; Mills, G.; See, L.; Stewart, I. Mapping Local Climate Zones for a Worldwide Database of the Form and Function of Cities. *ISPRS Int. J. Geo-Inf.* **2015**, *4*, 199–219. [CrossRef]

11. Xue, J.; You, R.; Liu, W.; Chen, C.; Lai, D. Applications of Local Climate Zone Classification Scheme to Improve Urban Sustainability: A Bibliometric Review. *Sustainability* **2020**, *12*, 8083. [CrossRef]

12. Lehnert, M.; Savić, S.; Milošević, D.; Dunjić, J.; Geletič, J. Mapping Local Climate Zones and Their Applications in European Urban Environments: A Systematic Literature Review and Future Development Trends. *ISPRS Int. J. Geo-Inf.* **2021**, *10*, 260. [CrossRef]

13. Verdonck, M.-L.; Demuzere, M.; Hooyberghs, H.; Beck, C.; Cyrys, J.; Schneider, A.; Dewulf, R.; Van Coillie, F. The potential of local climate zones maps as a heat stress assessment tool, supported by simulated air temperature data. *Landsc. Urban Plan.* **2018**, *178*, 183–197. [CrossRef]

14. Kotharkar, R.; Ghosh, A.; Kotharkar, V. Estimating summertime heat stress in a tropical Indian city using Local Climate Zone (LCZ) framework. *Urban Clim.* **2021**, *36*, 100784. [CrossRef]

15. Yang, X.; Peng, L.L.H.; Jiang, Z.; Chen, Y.; Yao, L.; He, Y.; Xu, T. Impact of urban heat island on energy demand in buildings: Local climate zones in Nanjing. *Appl. Energy* **2020**, *260*, 114279. [CrossRef]

16. Ching, J.; Mills, G.; Bechtel, B.; See, L.; Feddema, J.; Wang, X.; Ren, C.; Brousse, O.; Martilli, A.; Neophytou, M.; et al. WUDAPT: An Urban Weather, Climate, and Environmental Modeling Infrastructure for the Anthropocene. *Bull. Am. Meteorol. Soc.* **2018**, *99*, 1907–1924. [CrossRef]

17. Mushore, T.D.; Dube, T.; Manjowe, M.; Gumindoga, W.; Chemura, A.; Rousta, I.; Odindi, J.; Mutanga, O. Remotely sensed retrieval of Local Climate Zones and their linkages to land surface temperature in Harare metropolitan city, Zimbabwe. *Urban Clim.* **2019**, *27*, 259–271. [CrossRef]

18. Verdonck, M.L.; Okujeni, A.; van der Linden, S.; Demuzere, M.; De Wulf, R.; Van Coillie, F. Influence of neighbourhood information on 'Local Climate Zone' mapping in heterogeneous cities. *Int. J. Appl. Earth Obs. Geoinf.* **2017**, *62*, 102–113. [CrossRef]

19. Fricke, C.; Pongrácz, R.; Gál, T.; Savić, S.; Unger, J. Using local climate zones to compare remotely sensed surface temperatures in temperate cities and hot desert cities. *Morav. Geogr. Rep.* **2020**, *28*, 48–60. [CrossRef]

20. Demuzere, M.; Kittner, J.; Bechtel, B. LCZ Generator: A Web Application to Create Local Climate Zone Maps. *Front. Environ. Sci.* **2021**, *9*. [CrossRef]

21. Alghamdi, A.S.; Moore, T.W. Detecting temporal changes in Riyadh's urban heat island. *Appl. Geogr.* **2015**, *1*, 312–325. [CrossRef]

22. Bakarman, M.A.; Chang, J.D. The Influence of Height/width Ratio on Urban Heat Island in Hot-arid Climates. *Procedia Eng.* **2015**, *118*, 101–108. [CrossRef]

23. Aina, Y.A.; Adam, E.M.; Ahmed, F. Spatiotemporal variations in the impacts of urban land use types on urban heat island effects: The case of Riyadh, Saudi Arabia. *Int. Arch. Photogramm. Remote Sens. Spat. Inf. Sci.* **2017**, *42*, 9–14. [CrossRef]

24. Eldesoky, A.H.; Gil, J.; Pont, M.B. The suitability of the urban local climate zone classification scheme for surface temperature studies in distinct macroclimate regions. *Urban Clim.* **2021**, *37*, 100823. [CrossRef]

25. Alqurashi, A.F.; Kumar, L.; Sinha, P. Urban Land Cover Change Modelling Using Time-Series Satellite Images: A Case Study of Urban Growth in Five Cities of Saudi Arabia. *Remote Sens.* **2016**, *8*, 838. [CrossRef]

26. Rahman, M.T.; Planning, R. Land use and land cover changes and urban sprawl in Riyadh, Saudi Arabia: An analysis using multi-temporal Landsat data and Shannon's Entropy Index. *Int. Arch. Photogramm. Remote Sens. Spat. Inf. Sci.* **2016**, *41*, 1017–1021. [CrossRef]

27. Aina, Y.A.; Adam, E.; Ahmed, F.; Wafer, A.; Alshuwaikhat, H.M. Using multisource data and the V-I-S model in assessing the urban expansion of Riyadh city, Saudi Arabia. *Eur. J. Remote Sens.* **2019**, *52*, 557–571. [CrossRef]

28. Alghamdi, A.; Cummings, A.R. Assessing Riyadh's Urban Change Utilizing High-Resolution Imagery. *Land* **2019**, *8*, 193. [CrossRef]

29. AlQurashi, A.F.; Kumar, L. An assessment of the impact of urbanization and land use changes in the fast-growing cities of Saudi Arabia. *Geocarto Int.* **2017**, *34*, 78–97. [CrossRef]

30. Altuwaijri, H.A.; Alotaibi, M.H.; Almudlaj, A.M.; Almalki, F.M. Predicting urban growth of Arriyadh city, capital of the Kingdom of Saudi Arabia, using Markov cellular automata in TerrSet geospatial system. *Arab. J. Geosci.* **2019**, *12*, 135. [CrossRef]

31. Alkolibi, F.M. Possible Effects of Global Warming on Agriculture and Water Resources in Saudi Arabia: Impacts and Responses. *Clim. Chang.* **2002**, *54*, 225–245. [CrossRef]

32. Sobrino, J.A.; Irakulis, I. A Methodology for Comparing the Surface Urban Heat Island in Selected Urban Agglomerations Around the World from Sentinel-3 SLSTR Data. *Remote Sens.* **2020**, *12*, 2052. [CrossRef]

33. Nasrallah, H.A.; Brazel, A.J.; Balling, R.C., Jr. Analysis of the Kuwait City urban heat island. *International Journal of Climatology* **1990**, *10*, 401–405. [CrossRef]

34. Chen, X.-L.; Zhao, H.-M.; Li, P.-X.; Yin, Z.-Y. Remote sensing image-based analysis of the relationship between urban heat island and land use/cover changes. *Remote Sens. Environ.* **2006**, *104*, 133–146. [CrossRef]

35. Ren, C.; Cai, M.; Li, X.; Zhang, L.; Wang, R.; Xu, Y.; Ng, E.Y.Y. Assessment of Local Climate Zone Classification Maps of Cities in China and Feasible Refinements. *Sci. Rep.* **2019**, *9*, 1–11. [CrossRef]
36. Radoux, J.; Lamarche, C.; Van Bogaert, E.; Bontemps, S.; Brockmann, C.; Defourny, P. Automated Training Sample Extraction for Global Land Cover Mapping. *Remote Sens.* **2014**, *6*, 3965–3987. [CrossRef]
37. Demuzere, M.; Hankey, S.; Mills, G.; Zhang, W.; Lu, T.; Bechtel, B. Combining expert and crowd-sourced training data to map urban form and functions for the continental US. *Sci. Data* **2020**, *7*, 1–13. [CrossRef]
38. Bechtel, B.; Demuzere, M.; Mills, G.; Zhan, W.; Sismanidis, P.; Small, C.; Voogt, J. SUHI analysis using Local Climate Zones—A comparison of 50 cities. *Urban Clim.* **2019**, *28*, 100451. [CrossRef]
39. Malakar, N.K.; Hulley, G.C.; Hook, S.J.; Laraby, K.; Cook, M.; Schott, J.R. An Operational Land Surface Temperature Product for Landsat Thermal Data: Methodology and Validation. *IEEE Trans. Geosci. Remote Sens.* **2018**, *56*, 5717–5735. [CrossRef]
40. USGS. Landsat 8-9 Calibration and Validation (Cal/Val) Algorithm Description Document (ADD). 2021. Available online: https://prd-wret.s3.us-west-2.amazonaws.com/assets/palladium/production/atoms/files/LSDS-1747_Landsat8-9_CalVal_ADD-v4.pdf (accessed on 14 August 2021).
41. Massey, F.J., Jr. The Kolmogorov-Smirnov test for goodness of fit. *J. Am. Stat. Assoc.* **1951**, *46*, 68–78. [CrossRef]
42. Zhao, C.; Jensen, J.; Weng, Q.; Currit, N.; Weaver, R. Application of airborne remote sensing data on mapping local climate zones: Cases of three metropolitan areas of Texas, U.S. *Comput. Environ. Urban Syst.* **2019**, *74*, 175–193. [CrossRef]
43. Pan, X.; Zhu, X.; Yang, Y.; Cao, C.; Zhang, X.; Shan, L. Applicability of Downscaling Land Surface Temperature by Using Normalized Difference Sand Index. *Sci. Rep.* **2018**, *8*, 1–14. [CrossRef]
44. Du, P.; Chen, J.; Bai, X.; Han, W. Understanding the seasonal variations of land surface temperature in Nanjing urban area based on local climate zone. *Urban Clim.* **2020**, *33*, 100657. [CrossRef]
45. Zhao, C.; Jensen, J.L.R.; Weng, Q.; Currit, N.; Weaver, R. Use of Local Climate Zones to investigate surface urban heat islands in Texas. *GIScience Remote Sens.* **2020**, *57*, 1083–1101. [CrossRef]
46. Abulibdeh, A. Analysis of urban heat island characteristics and mitigation strategies for eight arid and semi-arid gulf region cities. *Environ. Earth Sci.* **2021**, *80*, 1–26. [CrossRef] [PubMed]
47. Li, N.; Yang, J.; Qiao, Z.; Wang, Y.; Miao, S. Urban Thermal Characteristics of Local Climate Zones and Their Mitigation Measures across Cities in Different Climate Zones of China. *Remote Sens.* **2021**, *13*, 1468. [CrossRef]
48. Kwarteng, A.Y.; Small, C.L. Comparative analysis of thermal environments in New York City and Kuwait City. In Proceedings of the Remote Sensing of Urban Areas, Tempe, AZ, USA, 14–16 March 2005.
49. Brousse, O.; Georganos, S.; Demuzere, M.; Dujardin, S.; Lennert, M.; Linard, C.; Snow, R.W.; Thiery, W.; Van Lipzig, N.P.M. Can we use local climate zones for predicting malaria prevalence across sub-Saharan African cities? *Environ. Res. Lett.* **2020**, *15*, 124051. [CrossRef]
50. Collins, J.; Dronova, I. Urban Landscape Change Analysis Using Local Climate Zones and Object-Based Classification in the Salt Lake Metro Region, Utah, USA. *Remote Sens.* **2019**, *11*, 1615. [CrossRef]
51. Li, X.; Zhou, W.; Ouyang, Z. Relationship between land surface temperature and spatial pattern of greenspace: What are the effects of spatial resolution? *Landsc. Urban Plan.* **2013**, *114*, 1–8. [CrossRef]

Article

Urban Sprawl and Changes in Land-Use Efficiency in the Beijing–Tianjin–Hebei Region, China from 2000 to 2020: A Spatiotemporal Analysis Using Earth Observation Data

Meiling Zhou [1,2], Linlin Lu [2,*], Huadong Guo [2], Qihao Weng [3], Shisong Cao [4], Shuangcheng Zhang [1] and Qingting Li [5]

[1] College of Geological Engineering and Geomatics, Chang'an University, Xi'an 710054, China; 2019126022@chd.edu.cn (M.Z.); shuangcheng369@chd.edu.cn (S.Z.)

[2] Key Laboratory of Digital Earth Science, Aerospace Information Research Institute, Chinese Academy of Sciences, Beijing 100094, China; hdguo@radi.ac.cn

[3] Center for Urban and Environmental Change, Department of Earth and Environmental Systems, Indiana State University, Terre Haute, IN 47809, USA; qihao.weng@indstate.edu

[4] School of Geomatics and Urban Spatial Informatics, Beijing University of Civil Engineering and Architecture, Beijing 102616, China; caoshisong@bucea.edu.cn

[5] Airborne Remote Sensing Center, Aerospace Information Research Institute, Chinese Academy of Sciences, Beijing 100094, China; liqt@radi.ac.cn

* Correspondence: lull@radi.ac.cn

Citation: Zhou, M.; Lu, L.; Guo, H.; Weng, Q.; Cao, S.; Zhang, S.; Li, Q. Urban Sprawl and Changes in Land-Use Efficiency in the Beijing–Tianjin–Hebei Region, China from 2000 to 2020: A Spatiotemporal Analysis Using Earth Observation Data. *Remote Sens.* **2021**, *13*, 2850. https://doi.org/10.3390/rs13152850

Academic Editors: Yuyu Zhou, Elhadi Adam, John Odindi and Elfatih Abdel-Rahman

Received: 2 June 2021
Accepted: 16 July 2021
Published: 21 July 2021

Publisher's Note: MDPI stays neutral with regard to jurisdictional claims in published maps and institutional affiliations.

Abstract: Sustainable development in urban areas is at the core of the implementation of the UN 2030 Agenda and the Sustainable Development Goals (SDG). Analysis of SDG indicator 11.3.1—Land-use efficiency based on functional urban boundaries—provides a globally harmonized avenue for tracking changes in urban settlements in different areas. In this study, a methodology was developed to map built-up areas using time-series of Landsat imagery on the Google Earth Engine cloud platform. By fusing the mapping results with four available land-cover products—GlobeLand30, GHS-Built, GAIA and GLC_FCS-2020—a new built-up area product (BTH_BU) was generated for the Beijing–Tianjin–Hebei (BTH) region, China for the time period 2000–2020. Using the BTH_BU product, functional urban boundaries were created, and changes in the size of the urban areas and their form were analyzed for the 13 cities in the BTH region from 2000 to 2020. Finally, the spatiotemporal dynamics of SDG 11.3.1 indicators were analyzed for these cities. The results showed that the urban built-up area could be extracted effectively using the BTH_BU method, giving an overall accuracy and kappa coefficient of 0.93 and 0.85, respectively. The overall ratio of the land consumption rate to population growth rate (LCRPGR) in the BTH region fluctuated from 1.142 in 2000–2005 to 0.946 in 2005–2010, 2.232 in 2010–2015 and 1.538 in 2015–2020. Diverged changing trends of LCRPGR values in cities with different population sizes in the study area. Apart from the megacities of Beijing and Tianjin, after 2010, the LCRPGR values were greater than 2 in all the cities in the region. The cities classed as either small or very small had the highest LCRPGR values; however, some of these cities, such as Chengde and Hengshui, experienced population loss in 2005–2010. To mitigate the negative impacts of low-density sprawl on environment and resources, local decision makers should optimize the utilization of land resources and improve land-use efficiency in cities, especially in the small cities in the BTH region.

Keywords: land-use efficiency; urban sprawl; Landsat; SDG 11; Google Earth Engine

1. Introduction

According to the United Nations, at present, more than 4 billion people live in urban areas worldwide, and this number continues to rise [1]. The global urbanization trend lead to excessive land use change and expansion of urban land [2]. The physical growth of urban areas is often disproportionate in relation to population growth which results in inefficient land use. This type of growth contradicts the principle of sustainability by leading to several negative impacts such as fragmentation of agro-forest landscape, increase of energy consumption and carbon emission [3]. Therefore, the efficient use of land resources is crucial for urban sustainability from ecological and socioeconomic points of view [4].

Several indicators have been developed to examine the impact of urbanization on land resources [5]. In 2015, the United Nations set out 17 social, economic and environmental sustainable development goals and 169 specific targets in the Transforming Our World: The 2030 Agenda for Sustainable Development, which will guide the global development effort during the period 2015–2030 [6]. Among them, SDG indicator 11.3.1 measures urban land-use efficiency (LUE), which is the ratio of the rate of consumption of urban land to the rate of population growth [7]. The production and dissemination of the SDG 11.3.1 indicator can characterize the evolution of urban settlements. By monitoring the spatio-temporal changes in urban land use efficiency, city authorities and decision makers can identify new areas of growth and project demand for public goods and services. They can also formulate policies that encourage optimal use of urban land and effectively protect natural and agricultural lands. Therefore, the information of LUE evolution is necessary for providing adequate infrastructure and services for improving living conditions of urban residents, as well as preserving environmentally sensitive areas from development [4].

Previous studies of urban expansion and sprawl have mainly been based on administrative boundaries [8,9]. However, urban areas experienced an outward expansion exceeding their formal administrative boundaries in a large proportion of cities due to the poor urban and regional planning and land speculation [10]. A dynamic and functional definition of urban boundaries was proposed and applied in the LUE analysis framework produced by UN-Habitat [7]. The use of functional urban boundaries makes it possible to track changes across human settlements and provides a globally harmonized avenue for tracking changes in urban settlements in different areas [11]. Unlike administrative boundaries, the functional urban area is defined as actual areas where urban growth happens. Using spatial analysis approaches from the built environment measures extracted from satellite imagery, the functional urban area can be identified [7]. Hence, the LUE indicators can be measured using built-up area and population grid data within the boundary of functional urban area. The produced results can represent the actual prevailing growth trends in the city and provide accurate aggregates to better inform decision makers at local and national levels. The use of functional urban boundaries makes it possible to track changes across human settlements and provides a globally harmonized avenue for tracking changes in urban settlements in different areas [11].

By collecting a wide range of information about the planet at large spatial scales, remote sensing can complement traditional data sources to track progress in achieving sustainable development goals and targets [12]. In particular, the development of free and open remote sensing data and high-performance cloud-computing platforms has made it possible to map urban built-up areas or impervious surfaces using Earth Observation data at the global scale [13–17]. Urban land-cover products derived from remote sensing data have been used to monitor and assess SDG 11.3.1 trends at national and global scales. The free and open Global Human Settlement Layer data were used to estimate SDG 11.3.1 in approximately 10,000 urban centers globally from 1990 to 2015 [11,18]. Based on the 1990, 2000 and 2010 CLUDS datasets, DMSP/OLS night-light data and census data, the spatial heterogeneity and dynamic trend of urban expansion and population growth in mainland China during the period 1990–2010 were analyzed in combination with SDG 11.3.1 [19]. These efforts at LUE monitoring at different spatial scales confirmed the applicability of Earth Observation data for providing consistent information on urban SDG indicators and informing the formulation of urban land-use policy.

However, the accuracy of Earth observation products can vary between different regions, which can lead to uncertainties in evaluation results. Efforts have been made to measure LUE changes using satellite data at the local city scale. Using Landsat and SPOT images from 1996, 2001–2002 and 2011, the SDG 11.3.1 indicator was evaluated for four South African cities, and it was demonstrated that the population growth rate and rate of spatial expansion of small and second-tier cities in Africa was faster than that of large cities [20]. Ghazaryan et al. developed an automated classification approach and monitored urban sprawl and densification processes in western Germany using SDG Indicator 11.3.1 [21]. Spatiotemporal changes in urban land-use efficiency were analyzed and urban land lease policy gaps were emphasized to help ensure sustainable urban growth in fast-growing cities in Ethiopia [22].

Since the reform and opening up which began in 1978, China's urbanization rate has increased from 17.9% in 1978 to 60.6% in 2019 [23]. This urbanization process is constantly accelerating and greatly promotes the nation' s social and economic development. One results of the country's industrialization and urbanization over the past four decades has been the gradual formation of large urban agglomerations or megalopolises [24,25]. The Beijing–Tianjin–Hebei (BTH) region is the largest such urban agglomeration in northern China. Monitoring and analyzing the dynamic changes in urban space is necessary to understand the urban development processes in this region. Therefore, the objectives of this study include: (1) to develop an approach to extract urban built-up areas by using Earth observation data; (2) to analyze urban expansion processes based on remote sensing products of built-up areas; and (3) to assess the spatial and temporal changes in urban land-use efficiency in the cities of the Beijing–Tianjin–Hebei urban agglomeration, China.

2. Study Area and Datasets

2.1. Beijing–Tianjin–Hebei Region

The Beijing-Tianjin-Hebei (BTH) region lies between the latitudes 36°03′ N and 42°40′ N and the longitudes 113°27′ E and 119°50′ E in northern China (Figure 1). It consists of Beijing and Tianjin municipalities, and 11 prefecture-level cities in Hebei Province (Zhangjiakou, Chengde, Qinhuangdao, Tangshan, Cangzhou, Hengshui, Langfang, Baoding, Shijiazhuang, Xingtai and Handan). Among them, Beijing and Tianjin are megacities with a population of more than 10 million. The area of this region comprises 218,000 km², accounting for 2.3% of China's land area. The total population of the study area in 2020 is 113 million, accounting for 7.8% of China's total population.

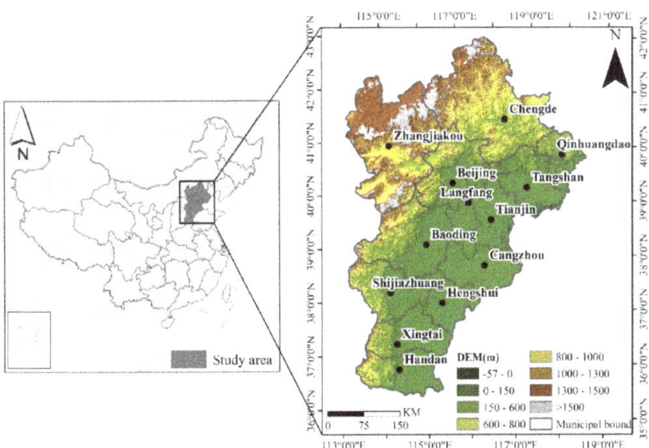

Figure 1. The geographic location of the Beijing–Tianjin–Hebei region, China.

Since 2014, the coordinated development of the BTH region has become a major national strategic policy of China. As the center of economic development in the north of the country, the BTH region plays a significant role in China's political and economic development [8]. To realize the coordinated development of the region and build a world-class urban agglomeration, the urbanization of the region should be developed in a planned and sustainable way, which requires timely monitoring of the processes and status of urban growth.

2.2. Datasets

The datasets used in this study included Landsat satellite images, topographic data, population data and administrative boundaries. Details of these datasets are described in the following sections.

2.2.1. Landsat Imagery

The Google Earth Engine (GEE), which can process satellite and other Earth observation data, is a cloud-based computing platform [26]. The platform stores nearly 40 years of publicly available global remote sensing images, including Landsat, Sentinel, MODIS and DMSP/OLS satellite imagery [27]. In this study, Landsat 5 tier-1 surface reflectance data from 2000, 2005 and 2010 and Landsat 8 data from 2015 and 2020 provided by the Google Earth Engine were used as the main data source for built-up area extraction. These datasets were atmospherically corrected using the GEE platform; this processing included cloud, shadow, water and snow masks.

2.2.2. Built-Up Area Products

Four urban built-up land or land-cover products were used in this study, including GlobeLand30, GAIA, GHS-Built and GLC_FCS-2020. These all have a spatial resolution of 30 m. The characteristics of these datasets are shown in Table 1.

The GAIA product is a global high-spatial-resolution (30 m) annual artificial surface dynamic data product (1985–2018) released in 2019. GAIA is produced using the Google Earth Engine platform together with long-time series of Landsat images (nearly 1.5 million scenes) as the main data source and auxiliary data such as night-light data and Sentinel-1 SAR data [28]. The GAIA dataset reveals the different rates of change in artificial impervious surface cover in major countries and regions of the world.

The Global Human Settlement Layer (GHSL) was developed by the Joint Research Centre (JRC) and contains spatial information about human settlements at a global scale [29]. The GHSL uses the core method of symbolic machine learning (SML) for data extraction and uses Landsat images from the past 40 years as the main data source. The GHSL is an open and free dataset. The GHS-Built product was used in this study. This product consists of multi-temporal data extracted from Landsat imagery (consisting of the GLS1975, GLS1990 and GLS2000 datasets and ad-hoc Landsat 8 data for 2013/2014). There are three types of land use classified in this product: built-up areas, non-built-up areas and water bodies.

The GLC_FCS30-2020 data is a global 30 m land-cover product with a fine classification system for the year 2020 [30]. The GLC_FCS30-2020 dataset is produced based on time-series of Landsat surface reflectance data (from 2019–2020), Sentinel-1 SAR data, DEM data, and auxiliary datasets. These data reflect the land cover types found globally (except for Antarctica) at a spatial resolution of 30 m in 2020.

The GlobeLand30 datasets from 2000, 2010 and 2020 were used in this study. These data include 10 land-cover types such as artificial land and bare land and cover the land within the range 80° S–80° N. The GlobeLand30 data were mainly classified using Landsat TM\ETM+ images and China Environmental Disaster Reduction Satellite (HJ-1) data based on a comprehensive pixel–object–knowledge classification method [31].

The data set used in this study are built-up area mapping results from these products. In the GAIA, GHS-Built, GLC_FCS30-2020, and GlobeLand30 products, the built-up areas are defined as impervious, built-up, impervious surfaces, and artificial surfaces, respectively. Hence, the four products are reclassified into two land cover types: non-built-up areas and built-up areas. The non-built-up area includes all other land cover types except built-up area, such as water, cropland, and forest. The specific reclassification schemes for each land cover product are shown in Table 2. In order to fuse theses datasets with the mapping results we derived using Google Earth Engine, the four built-up area products were re-projected to the same geo-reference system consisting of the WGS-84 coordinate system and UTM projection and resampled to a spatial resolution of 30 m.

Table 1. Characteristics of urban built-up area datasets in the Beijing-Tianjin-Hebei region from 2000 to 2020.

Product	Data Source	Classification Method	Publisher	Time Coverage	Spatial Resolution (m)	Geographic Projection
GAIA	Landsat TM, ETM+, OLI, VIIRS NTL, Sentinel-1 SAR	Exclusion–inclusion algorithm and temporal consistency check	Tsinghua University http://data.ess.tsinghua.edu.cn (10 March 2021)	1985–2018	30	WGS 1984
GHS-Built	Landsat MSS, TM, ETM+	Symbolic machine learning	Joint Research Centre, European Commission https://ghsl.jrc.ec.europa.eu/ (20 November 2020)	1975; 1990; 2000; 2015	30	WGS 1984; WGS 1984 Web Mercator Auxiliary Sphere
GLC_FCS-2020	Landsat 8 surface reflectance, Sentinel-1 SAR	Random forest classifier	Chinese Academy of Sciences https://zenodo.org/record/428092 3#.YHamCjim02w (10 October 2020)	2020	30	WGS 1984
GlobeLand30	Landsat TM, ETM+, OLI, HJ-1	POK (based on pixels, objects, and knowledge)	National Geomatics Center of China http://www.globallandcover.com (10 January 2021)	2000; 2010; 2020	30	WGS 1984; UTM

Table 2. Reclassification schemes of land-cover products in Beijing–Tianjin–Hebei region from 2000 to 2020.

DATA	1: Non-Built-Up	2: Built-Up
GAIA	2000:0 Non-impervious, 1–18 impervious 2010:0 Non-impervious, 1–8 impervious 2018:0 Non-impervious	2000:19–34 impervious 2010:9–34 impervious 2018:1–34 impervious
GHS-Built	1: Water surface, 2: Land not built-up in any epoch, 3: Built-up from 2000 to 2014	4: Built-up from 1990 to 2000, 5: Built-up from 1975 to 1990, 6: Built-up until 1975
GLC_FCS-2020	10: Rainfed cropland, 11: Herbaceous cover, 12: Tree or shrub cover (orchard), 20: Irrigated cropland, 51: Open evergreen broadleaved forest, 52: Open evergreen broadleaved forest, 61: Open deciduous broadleaved forest (0.15 < fc < 0.4), 62: Closed evergreen needle-leaved forest (fc > 0.4), 71: Open evergreen needle-leaved forest (0.15 < fc < 0.4), 72: Closed evergreen needle-leaved forest (fc > 0.14), 81: Open deciduous needle-leaved forest (0.15 < fc < 0.4), 82: Closed deciduous needle-leaved forest (fc > 0.4), 91: Open mixed-leaf forest (broadleaved and needle-leaved), 92: Closed mixed-leaf forest (broadleaved and needle-leaved), 120: Shrubland, 121: Evergreen shrubland, 122: Deciduous shrubland, 130: Grassland, 140: Lichens and mosses, 150: Sparse vegetation (fc < 0.15), 152: Sparse shrubland (fc < 0.15), 153: Sparse herbaceous (fc < 0.15), 180: Wetlands, 200: Bare areas, 201: Consolidated bare areas, 202: Unconsolidated bare areas, 210: Water body, 220: Permanent ice and snow, 250: Filled value	190: Impervious surfaces
GlobeLand30	10: Cultivated land, 20: Forest, 30: Grassland, 40: Shrubland, 50: Wetland, 60: Water bodies, 70: Tundra, 90: Bare Land, 100: Permanent snow and ice	80: Artificial surfaces

2.2.3. Population Data

The WorldPop project, which is funded by the Bill & Melinda Gates Foundation, aims to offer spatial population datasets for Central and South America, Africa and Asia to support development, disaster response and health applications. The WorldPop data constitute an open-access gridded population data set (https://www.worldpop.org/ (10 April 2021)), produced by using a combination of machine learning and asymmetric modeling [32]. In this study, the population datasets for China for 2000, 2005, 2010, 2015 and 2020, which have a spatial resolution of 100 m and are based on the WGS84 geographic coordinate system, were used. The WorldPop population totals for different countries have been adjusted to match the corresponding official United Nations population estimates obtained from the Population Division of the Department of Economic and Social Affairs of the United Nations Secretariat.

2.2.4. Ancillary Datasets

The SRTM (Shuttle Radar Topography Mission) DEM is a digital elevation terrain model published by NASA and the National Bureau of Surveying and Mapping (NIMA). It has a spatial resolution of 30 m [33] and can be accessed directly in GEE. The administrative boundaries, including municipal district boundaries, provincial boundaries and municipal boundaries, can be obtained from National Geomatics Center of China (http://www.ngcc.cn/ngcc// (6 March 2021)).

3. Methodology

3.1. Built-Up Area Extraction

Three spectral indices—the normalized difference built-up index (NDBI), normalized difference vegetation index (NDVI) and normalized difference water index (NDWI)—were calculated from time-series of Landsat images [34–36]. Annual composite images were created from the optical, near-infrared, NDVI, NDBI and NDWI bands using the 'reduce' method by provided by GEE. The reduce method reduces a collection of images to an image by performing operation such as summation, median, etc. pixel by pixel. Four reduce functions—min, max, mean and median—were applied to improve the classification accuracy [37]. Sample points were selected by visual interpretation using the GEE platform. A stratified random sampling method was implemented with mapped land cover classes as strata [38]. The numbers of the sample points were determined based on the areal proportion of each land cover type. 80% of the sample points were randomly taken as training samples and 20% as validation sample points. Elevation and slope data calculated from the SRTM DEM were used as supplementary data; image classification was performed using the random forest algorithm(RFA) [39]. The number of classification trees in the random forest classifier is 500, and the remaining parameters are default. The confusion matrix between the validation samples and the classified products was calculated and the overall accuracy (OA) and kappa coefficients were obtained [40]. For the dichotomy of built-up area products (built-up area and non-built-up area) in this study, OA is the accuracy of the overall pixel being correctly classified, which can directly reflect the proportion of the correct classification. Kappa coefficient is a statistical index that can measure the balance of the correct classification of different categories. Omission error refers to the pixels belonging to the built-up area but classified as non-built-up area. Commission error refers to the pixels that is not built-up but is classified as built-up area. We tested different strategies for splitting the training and validation samples. If the classification result meets the requirement of OA > 0.85 and kappa > 0.8 [41,42], it was exported. The RFA was run iteratively and classification result with the highest OA was taken as the final output. The resolution of classification result was 30 m, and the spatial reference system was WGS-84 coordinate system and UTM projection.

The voting method is one of the most commonly used multi-classifier combination methods [43,44] in which the category with the most votes is taken as the final classification result. When multiple categories have the same number of votes, one of them is randomly selected as the final result. Based on the existing products for classifying urban built-up land and the classification results obtained from the RFA, a multi-temporal urban built-up area dataset was generated using the majority voting method. After the voting method had been applied, a temporal filtering method was used to ensure that the amount of built-up land in the product exhibited a growing trend [45]. A majority filter with a 3 × 3 window size was used to remove isolated pixels to obtain the final built-up area products with 30 m resolution, BTH_BU 2000, BTH_BU 2005, BTH_BU 2010, BTH_BU 2015 and BTH_BU 2020 [36]. The data-processing workflow is shown in Figure 2.

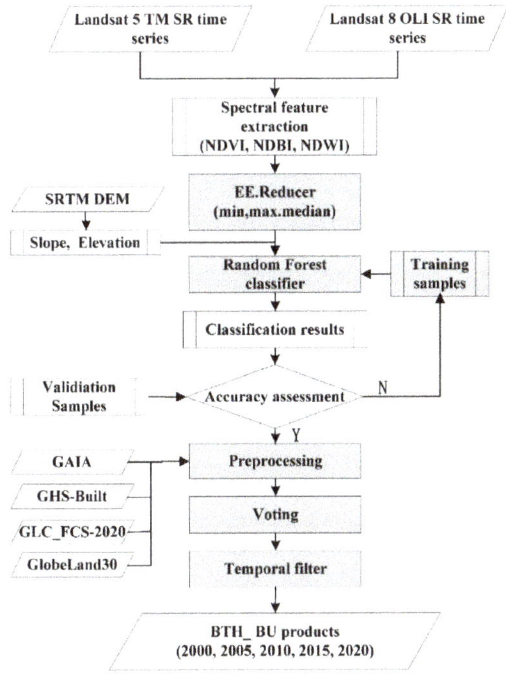

Figure 2. The workflow used for the generation of the built-up area products in Beijing–Tianjin–Hebei region from 2000 to 2020.

3.2. Accuracy Assessment

In order to confirm the feasibility of the BTH_BU product for urban expansion analysis across time, it is necessary to evaluate its thematic accuracy. In this study, confusion matrix was used to evaluate and compare the accuracy of the built-up area products [46–48]. The comparative analysis can also provide useful information for applications of these products in urban expansion and other fields. The confusion matrix compares the consistency of the land cover categories in the reference data and the data to be verified at a specific location, and then creates a two-dimensional table that compares the two (i.e., a confusion matrix). Based on this matrix, several different indices, including the overall accuracy (*OA*), kappa coefficient, omission error (*OE*) and misclassification error (*CE*), which can measure the accuracy of the product, can be calculated [49]. The equations used to calculate these indices are:

$$OA = \frac{TP + TN}{TP + FP + FN + TN} \tag{1}$$

$$p_e = \frac{(TP + FN) \cdot (TP + FP) + (FP + TN) \cdot (FN + TN)}{TP + FP + FN + TN} \tag{2}$$

$$kappa = \frac{p_o - p_e}{1 - p_e} \tag{3}$$

$$CE = \frac{FP}{TP + FP} \tag{4}$$

$$OE = \frac{FN}{TP + FN} \tag{5}$$

where TP = true positive, TN = true negative, FP = false positive, FN = false negative and P_o = OA. In this study, a true positive result indicated that the built-up area pixels had been correctly extracted. A true negative indicated that image pixels considered to be non-built-up areas had been classified as not built-up areas. A false positive meant that image pixels that did not correspond to built-up areas had mistakenly been extracted, and a false negative meant that image pixels that did correspond to built-up areas had not been extracted.

A total of 300 points including 150 classed as built-up land and 150 classed as non-built-up land were randomly created for each city using ArcGIS 10.5 software (Esri, Redlands, CA, USA). High-resolution satellite images from Google Earth obtained in the same year as the Landsat data were used to identify the land-cover types corresponding to these random points manually. An accuracy evaluation database was then built and used to calculate the error matrix and accuracy assessment indicators.

3.3. Urban Growth Analysis

3.3.1. Functional City Boundary

Determining the extent of the built-up area is a prerequisite for measuring and comparing SDG 11.3.1. Instead of the administratively defined boundaries, the dynamic and functional urban boundary proposed by UN-Habitat—the actual area within which urban growth occurs over a defined period of time—was adopted in our study [10]. The functional urban boundary was obtained using several steps (Figure 3). The BTH_BU products are reclassified based on the density of built-up area per square kilometer. When the built-up area density is greater than 50%, it is defined as urban. If the built-up area density is greater than 0.25 and less than 0.5, it is defined as suburban. If the built-up area density is less than 0.25, it is considered as rural. The urban and suburban pixels were then used to determine the fringe open spaces, which are defined as areas within 100 m of urban and suburban pixels. A polygon was created by merging urban, sub-urban and the fringe open space pixels. Finally, buffer that extended the area of the polygon by 25% was created around each feature and the maximum extent of the buffer area was taken to be the functional urban boundary [7].

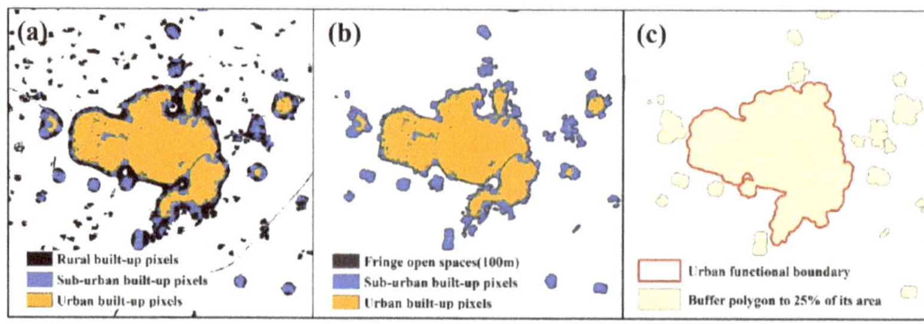

Figure 3. Urban functional boundary extraction for the city of Hengshui in 2010. (**a**) reclassification of built-up pixels, (**b**)delimitation of fringe open spaces, and (**c**) urban functional boundary creation.

3.3.2. Change in Urban Built-Up Area

After obtaining the functional urban boundaries of the 13 cities in the study area in 2000, 2005, 2010, 2015 and 2020, the annual expansion area (AEA, km²) was used to quantify the amount and rate of urban expansion in our study [17]. The AE is defined as follows:

$$AEA = \frac{A_{end} - A_{start}}{d} \tag{6}$$

where A_{end} represents the urban built-up area at the end of a certain period, A_{start} is the built-up area at the beginning of a period and d is the number of years between the start and end of the period. The *AEA* can directly measure the annual change of built-up area and reveal the difference of the expansion amount of a city in different stages.

3.3.3. Change in Urban Form

The changes in urban form were analyzed by calculating compactness and aggregation indices which are spatial metrics. The compactness (*C*) of a city's boundaries is an indicator that can reflect the form of the city's space. It is defined as:

$$C = 2\sqrt{\pi A}/P \tag{7}$$

where *C* is the compactness of the city—that is, the compactness of the urban functional boundary. *A* is the area of the city and *P* denotes the length of the city perimeter. The greater the compactness value, the more compact the city is. A circle is the most compact shape: the compactness of a space in which each part is highly compressed is 1. In contrast, less compact shapes tend to be narrow or long. Also, the more complex a shape, the smaller its compactness.

The aggregation index (*AI*) of a city is defined as:

$$AI = \left(\frac{g_{ii}}{max \rightarrow g_{ii}} \right) * 100 \tag{8}$$

where *AI* denotes the degree of aggregation and g_{ii} denotes the number of adjacent similar patches of the same landscape type. The AI is calculated based on the length of the common boundary between pixels corresponding to the same type of patch. When there is no common boundary between any of these pixels, the degree of aggregation is the lowest; when the common boundary between all pixels corresponding to the same type of patch reaches its maximum value, the aggregation index is the highest. In this study, the compactness and aggregation index for each built-up area, as defined by the functional urban boundaries, were calculated for each time period.

3.4. Derivation of LCR, PGR and LCRPGR

The SDG 11.3.1 indicators give the ratio of the land consumption rate (LCR) to the population growth rate (PGR). The PGR refers to the rate of population change in a city/urban area over a given period (usually a year). The LCR is defined as the rate of appropriation of land by urban development. The ratio of land consumption rate to population growth rate(LCRPGR) is calculated as the ratio of LCR to PGR. SDG 11.3.1 indicators are calculated using the following formulae:

$$LCR = \frac{\ln\left(Urb_{(t+n)}/Urb_t \right)}{y}, \tag{9}$$

$$PGR = \frac{\ln\left(Pop_{(t+n)}/Pop_t \right)}{y} \tag{10}$$

$$LCRPGR = \frac{LCR}{PGR} \tag{11}$$

where Pop_t is the total population of the urban area in the past/initial year, $Pop_{(t+n)}$ is the total population in the current/final year, Urb_t is the total extent of the urban area (in km^2)

for the past/initial year, $Urb_{(t+n)}$ is the sum of the urban areas in the current/final year and y is the number of years between the two measurement periods.

Ideally, the ratio of land consumption to the population growth rate would be equal to 1, implying that the amount of urban land is increasing at the same rate as the population. Based on LCRPGR values, cities can be categorized into 5 types. These are shown in Table 3.

Table 3. City classification based on the ratio of land consumption rate to population growth rate values.

LCRPGR Value	Meaning
LCRPGR < −1	the rate of population decline is greater than the rate of built-up area expansion
−1 < LCRPGR < 0	the rate of population decline is less than the rate of built-up area expansion
0 < LCRPGR < 1	the rate of population growth is greater than the rate of built-up area expansion
1 < LCRPGR < 2	the rate of built-up area expansion is 1–2 times the rate of population growth
LCRPGR > 2	the rate of built-up area expansion is greater than 2 times the rate of population growth

4. Results

4.1. Accuracy Assessment

Accuracy evaluation indices were obtained for each of these built-up area products for each study period (Table 4). Among the built-up area products, BTH_BU products had the highest average OA and Kappa coefficients and the lowest average OE and CE (Table 4). It shows that the proposed method can effectively improve the extraction accuracy of built-up area.

Table 4. Overall accuracy assessment for the built-up area products in Beijing–Tianjin–Hebei region from 2000 to 2020.

Product	Year	OA	KAPPA	OE	CE
GAIA	2000	0.82	0.64	0.07	0.31
	2010	0.84	0.67	0.05	0.29
	2018	0.86	0.72	0.05	0.24
	Average	0.84	0.67	0.06	0.28
GHS-Built	2000	0.87	0.75	0.04	0.22
GLC_FCS	2020	0.88	0.77	0.05	0.29
Globeland30	2000	0.84	0.68	0.05	0.28
	2010	0.85	0.7	0.05	0.26
	2020	0.86	0.72	0.04	0.25
	Average	0.85	0.7	0.05	0.26
BTH_BU	2000	0.91	0.83	0.05	0.13
	2005	0.91	0.83	0.04	0.14
	2010	0.94	0.87	0.05	0.08
	2015	0.93	0.86	0.04	0.10
	2020	0.94	0.89	0.04	0.07
	Average	0.93	0.85	0.04	0.11

Figure 4 shows the OA, kappa coefficient, OE and CE of the five remote sensing products for each city. The classification accuracy of BTH_BU products in 11 out of 13 cities was the highest. BTH_BU and GlobeLand30 have the smallest variability in classification accuracy, followed by GLC_FCS and GHS-Built. GAIA has the largest variability in classification accuracy between the different cities.

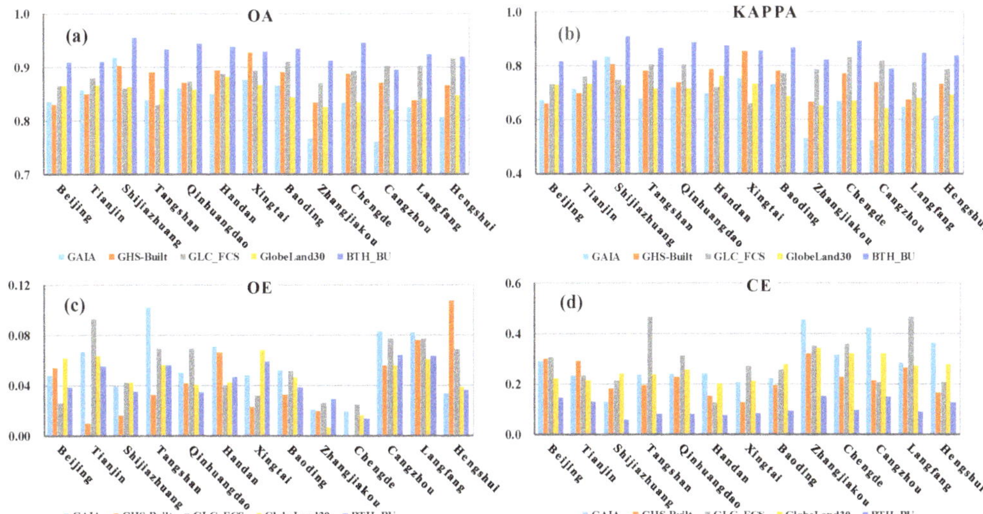

Figure 4. Accuracy assessment results for the built-up area products for each city in the Beijing–Tianjin–Hebei region. (**a**)overall accuracy, (**b**) kappa coefficient, (**c**)omission error, and (**d**) commission error.

4.2. Urban Growth Analysis

4.2.1. Changes in Urban Built-Up Area

Figure 5 shows the spatial pattern of urban expansion in the cities in the Beijing–Tianjin–Hebei region. Different types of urban spatial growth, infilling, edge-expansion, and leapfrogging, can be observed in these cities. The scattered built-up patches in suburban and rural areas was gradually connected with the central urban area, and thus forming a larger urban core in large cities.

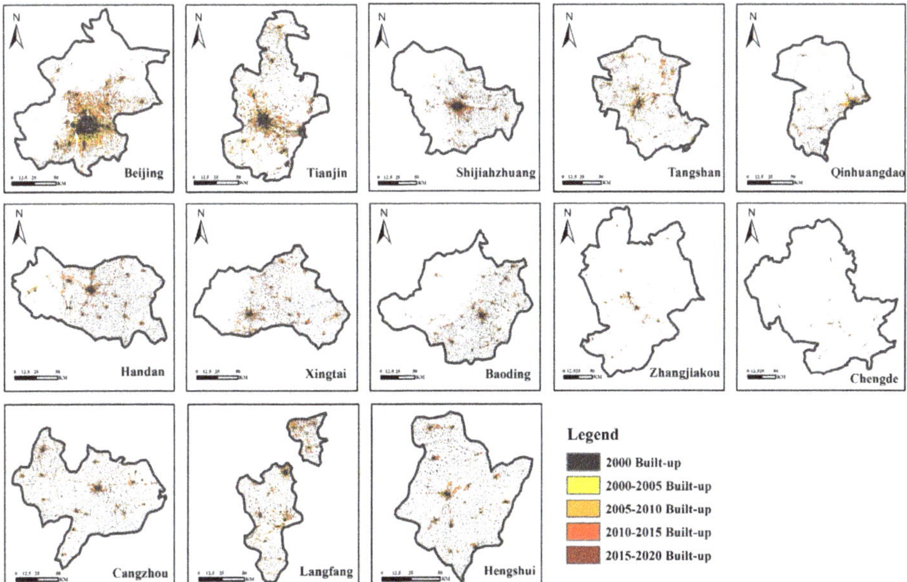

Figure 5. Urban expansion modelled by built-up product (BTH-BU) in each city in the Beijing–Tianjin–Hebei region from 2000 to 2020.

The annual expansion area (AEA) of urban built-up area showed obvious regional differences (Figure 6). From 2010 to 2015, the AEA of the Beijing–Tianjin–Hebei built-up area is much higher than that in other periods. The dramatic changes in these values may indicate that a fusion between suburban patches and the core urban district or explosive growth resulting from the sprawl of built-up areas occurred during this period.

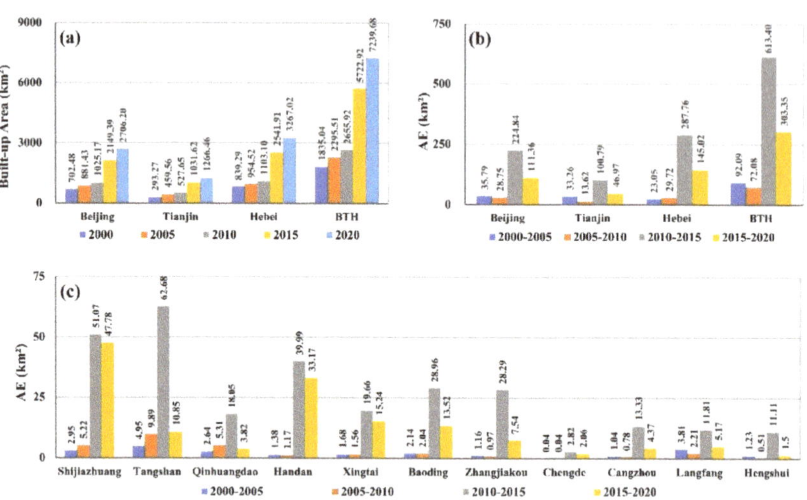

Figure 6. The area (**a**) and annual expansion area (**b**,**c**) in the built-up area for each city in the Beijing–Tianjin–Hebei region from 2000 to 2020.

4.2.2. Changes in Urban Form

The changes in the compactness (C) and aggregation index (AI) for each city are shown in Figure 7. The compactness of all the cities in the Beijing–Tianjin–Hebei region shows a downward trend, while for the aggregation index (AI) there is an overall upward trend from 2000 to 2020 (Figure 7). The increasing trend in AI and the decreasing trend in C for all of the cities indicate that the built-up areas in all of these cities are gradually amalgamating and the spatial forms of the functional urban areas are becoming increasingly complex [50].

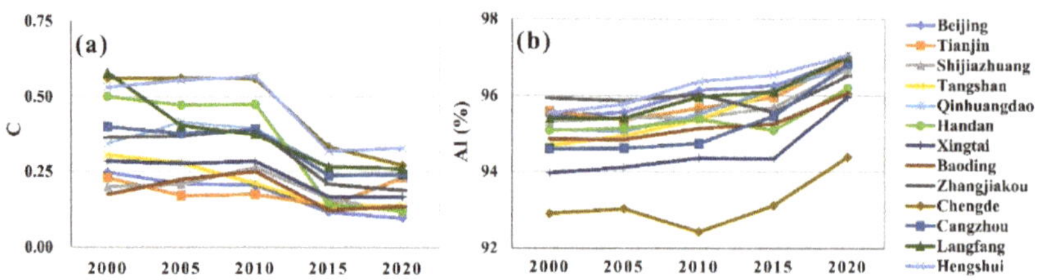

Figure 7. The temporal variation in (**a**) compactness and (**b**) aggregation index for each city in the Beijing–Tianjin–Hebei region from 2000 to 2020.

4.3. Spatiotemporal Dynamics of SDG 11.3.1

4.3.1. Variations in LCR, PGR and LCRPGR

The process of urbanization in the Beijing–Tianjin–Hebei region from 2000 to 2020 was analyzed by calculating the LCR, PGR and LCRPGR values (Table 5). The LCR, PGR and LCRPGR for the region all followed the same trend from 2000 to 2020: a decrease, followed by an increase, followed by another decrease. The decrease and increase in the LCRPGR during the period 2000–2015 agrees with the observed trend for all Chinese cities [51]. As a result of development in the early decades of the 21st century, there was also a consistent outward spread of the urban area in the Beijing–Tianjin–Hebei urban agglomeration. Before 2010, the rate of urban land growth was less than or similar to urbanization rate of the population. After Beijing hosted the Olympic Games in 2008, economic development led to rapid urban expansion in the Beijing–Tianjin–Hebei region. From 2010 to 2015, due to the acceleration in urban expansion and population growth, the LCR increased to 0.154, the PGR increased to 0.069 and the LCRPGR increased to 2.232. The urbanization rate of land was 2.232 times of the urbanization rate of population. In 2014, the *National New Urbanization Plan (2014–2020)* was issued by the national government and put forward a coordinated development strategy for the Beijing–Tianjin–Hebei region. As a guideline for China's urbanization, the plan emphasized the importance of achieving a more environmentally and sustainable form of urbanization. During the period 2015–2020, the rate of increase in urban land and population urbanization slowed down, the LCR decreased to 0.048, the PGR decreased to 0.031 and the LCRPGR decreased to 1.538.

Table 5. The temporal variation in the land consumption rate, population growth rate, and the ratio of land consumption rate to population growth rate in the Beijing–Tianjin–Hebei region from 2000 to 2020.

Time Period	LCR	PGR	LCRPGR
2000–2005	0.045	0.039	1.142
2005–2010	0.027	0.028	0.946
2010–2015	0.154	0.069	2.232
2015–2020	0.048	0.031	1.538

The temporal changes in the LCR, PGR and LCRPGR values of each city in the Beijing–Tianjin–Hebei region were analyzed. In the economically developed areas such as Beijing and Tianjin, the population growth and urban expansion were more rapid than in most cities in Hebei province. From 2000 to 2010 and from 2015 to 2020, although Beijing, as the capital of China, attracted a great inward flow of people, the rate of expansion of the built-up area was less than the population growth rate and the LCRPGR values were less than 1. Tianjin, the second largest city in the BTH region, had a higher LCRPGR value of between 1.0 and 1.5 (Figure 8). From 2010 to 2015, the built-up areas in all the cities in the Beijing–Tianjin–Hebei region expanded rapidly and their LCRPGR values were all greater than 1.5. The LCRPGR values for Langfang, Cangzhou, Baoding, Zhangjiakou and Tangshan have been increasing since 2000. By 2020, the LCRPGR values of these cities were all greater than 2, indicating that the cities had entered a phase of extensive development. In Qinhuangdao, the LCRPGR was greater than 2 during the period 2000–2020. Population growth in Hengshui has been relatively slow. The population growth here was negative from 2005 to 2010 and the absolute value of the LCRPGR from 2000 to 2020 was greater than 3. This indicates that a balance was maintained between the rates of urban expansion and population growth in Beijing and Tianjin. However, development in the cities in Hebei was unbalanced, with most of these cities underwent a phase of extensive development.

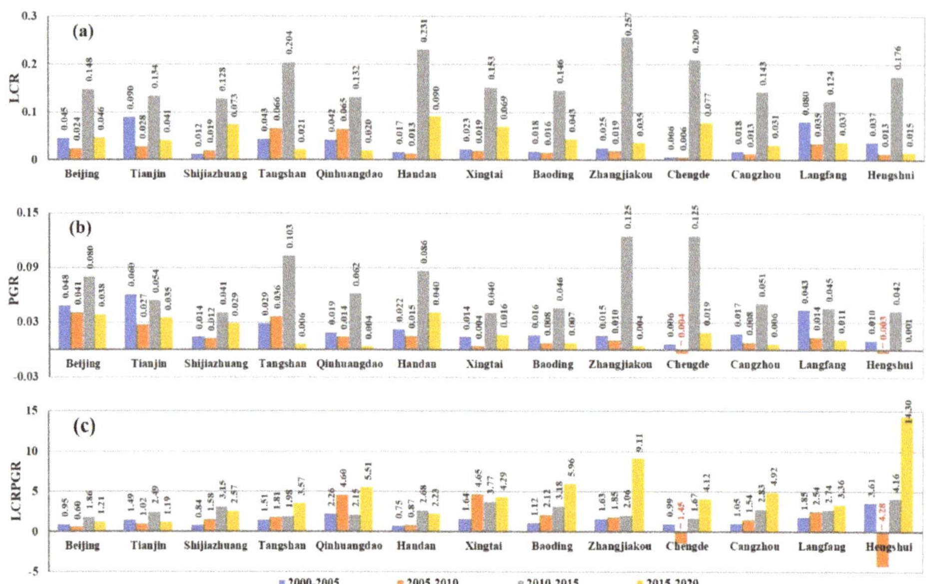

Figure 8. The temporal variation in the (**a**) land consumption rate, (**b**) population growth rate and (**c**) the ratio of land consumption rate to population growth rate in each city in the Beijing–Tianjin–Hebei region from 2000 to 2020.

4.3.2. Differences in LCRPGR by City Type

We divided the 13 cities in the study into five categories based on population: very small cities, small cities, medium cities, large cities, and megacities (Table 6). The LCRPGR values for cities with different population sizes are shown in Figure 9.

Table 6. Classification of different types of cities in the Beijing–Tianjin–Hebei region.

Type	Cities	Population
Megacities	Beijing, Tianjin	≥10,000,000
Large cities	Shijiazhuang	5,000,000–10,000,000
Medium cities	Tangshan, Qinhuangdao, Handan, Xingtai, Baoding, Zhangjiakou	1,000,000–5,000,000
Small cities	Canzhou, Langfang	500,000–1,000,000
Very small cities	Chengde, Hengshui	<500,000

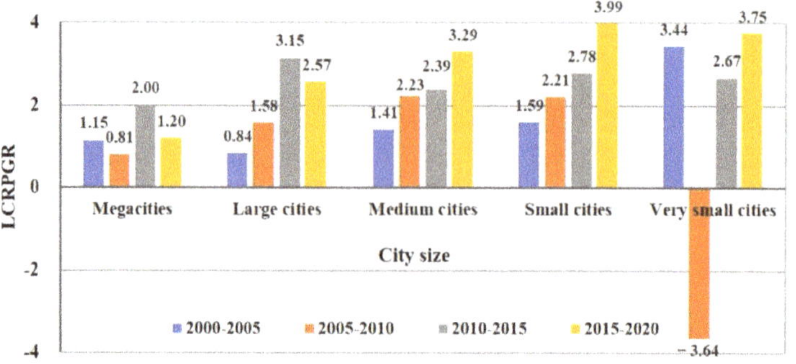

Figure 9. Temporal variation in the ratio of land consumption rate to population growth rate in cities with different populations in the Beijing–Tianjin–Hebei region from 2000 to 2020.

Cities of different sizes undergone diverged urbanization processes during the study period (Figure 9). The LCRPGR values of megacities and large cities in 2015–2020 were lower than that before 2015, while the LCRPGR value of medium-sized cities, small cities and very small cities in 2015–2020 is higher than that before 2015. The slowed down sprawling development of megacities and large cities since 2015 might be attributed to the implementation of the new urbanization policy. However, the land use efficiency remains very low in medium, small and very small cities after 2015.

5. Discussion

5.1. Urban Growth Analysis

In this study, a new built-up area product (BTH_BU) was generated for the Beijing–Tianjin–Hebei (BTH) region by fusing the mapping results with four available land-cover products—GlobeLand30, GHS-Built, GAIA and GLC_FCS-2020. The accuracy of different built-up area or land cover products varies across cities. The accuracy of built-up area mapping mainly depends on the data sources, sample samples and classification methods employed. BTH_BU and GAIA rely mostly on the spectral features of Landsat data and a pixel-based classifier, whereas GHS-Built and GlobeLand30 also employ textural and contextual information about settlements that is obtained from remote sensing images. The GAIA and GLC_FCS-2020 uses both optical and Sentinel-1 SAR data. Our results showed that the integrative use of optical and SAR data improved the identification of buildings in less-dense urban areas. The sensitivity of SAR backscatter to building structures can help distinguish bare lands or sparse vegetated lands from built-up areas which are often confused in multispectral images [52]. By fusing the urban mapping results with four available products with a majority voting method, the BTH_BU takes advantage of different data products, and thus improved the accuracy of the final classification results.

Using the BTH_BU product and urban land use efficiency indicator, our results revealed diverged changing trends of LCRPGR values in cities with different population sizes in the study area. The extensive sprawling growth patterns in medium and small sized cities of this region suggest that economical and efficient use of land resources was neglected in urban planning and management strategies in these cities [17]. As the third largest urban agglomeration in China, the urban sprawl trajectory of cities with different sizes in Beijing–Tianjin–Hebei region can provide implications for other urban agglomerations [9]. Moreover, the urban mapping method and spatio-temporal urban sprawl analysis workflow using Earth observation data and geospatial technology can be applied to other cities and urban agglomerations.

5.2. Uncertainties and Limitations

Although the urban sprawl was characterized with the LCRPGR indicator effectively in our study, there are some limitations which should be considered. Multi-temporal Landsat imagery with 30 m spatial resolution was used as the main data source to extract urban land in this study, which limits the accuracy of built-up area extraction. The application of high resolution satellite data such as Sentinel-1/2, SPOT-5/6/7 and WorldView-2/3 may provide more detailed and accurate information on urban built-up areas [53,54]. For built-up area data fusion, different methods such as entropy weighting and performance weighting and the Dempster-Shafer theory can be applied in future studies [43]. When the LCRPGR value less than 0 is obtained, it should be interpreted with cautions. If PGR is less than 0 and LCR is greater than 0, the LCRPGR represents urban expansion and population loss. If PGR is greater than 0 and LCR is less than 0, it represents the decrease of built-up area and the increase of population in cities. Moreover, the indicators of SDG11.3.1 only use land and population as indicators, which may not fully characterize the urbanization process. Other environmental and economic indicators in SDGs can be used together with LCRPGR to perform a more comprehensive assessment of the urban development trajectories and trends.

6. Conclusions

In this study, a land-cover classification method was developed using a random forest classifier and Google Earth Engine cloud platform. The built-up area was extracted using time-series of Landsat imagery, a DEM and other ancillary data of the Beijing–Tianjin–Hebei region for the years 2000, 2005, 2010, 2015 and 2020. The data were also fused with four existing land-classification products—GlobeLand30, GHS-Built, GAIA and GLC_FCS-2020 to generate the built-up products BTH_BU 2000, BTH_BU 2005, BTH_BU 2010, BTH_BU 2015 and BTH_BU 2020. An accuracy assessment produced an overall accuracy of 0.93 and a kappa coefficient of 0.85 for the BTH_BU products. The built-up areas in all of these cities are aggregating and the spatial forms of the functional urban areas are becoming more complex. The spatiotemporal dynamics of the SDG 11.3.1 land-use efficiency indicators were monitored using BTH_BU and population data. The results showed that, for the BTH region overall, the LCRPGR values were close to 1 from 2000 to 2010 but rose to 2 or higher after 2010. Diverged changing trends of LCRPGR values in cities with different population sizes in the study area. The rates of land and population urbanization were found to be relatively balanced in the megacities of Beijing and Tianjin. Except for the megacities, the LCRPGR values were greater than 2 after 2010. The small and very small cities had the highest LCRPGR values after 2015; however, some of these cities, such as Chengde and Hengshui, experienced population loss. To mitigate the negative impacts of low-density sprawl on the environment and land resources, local decision makers should optimize the utilization of land resources and improve land-use efficiency in cities, especially in the small cities in the Beijing–Tianjin–Hebei region.

Author Contributions: Conceptualization: H.G. and L.L.; methodology: L.L.; software: L.L.; validation: M.Z. and S.Z.; formal analysis: M.Z.; investigation: S.C.; resources: Q.L.; data curation: M.Z.; original draft preparation: M.Z.; review and editing of draft: Q.W. and L.L.; visualization: M.Z.; supervision: L.L.; project administration: Q.L.; funding acquisition: Q.L. All authors have read and agreed to the published version of the manuscript.

Funding: This research was funded by the National Key Research and Development Program of China, grant number 2019YFE0127700; the National Natural Science Foundation of China, grant number 42071321; and the Strategic Priority Research Program of the Chinese Academy of Sciences, grant number XDA19030502.

Institutional Review Board Statement: Not applicable.

Informed Consent Statement: Not applicable.

Data Availability Statement: The data presented in this study are available on request from the corresponding author.

Acknowledgments: The authors would like to thank Yifang Ban from the KTH Royal Institute of Technology for her help regarding the SDG11.3.1 calculations. The authors are grateful for the constructive comments from anonymous reviewers and the editors.

Conflicts of Interest: The authors declare no conflict of interest.

References

1. United Nations Department of Economic and Social Affairs. World Urbanization Prospects: The 2018 Revision (ST/ESA/SER.A/420). In *Population Division*; United Nations: New York, NY, USA, 2019.
2. Grimm, N.B.; Faeth, S.H.; Golubiewski, N.E.; Redman, C.L.; Wu, J.; Bai, X.; Briggs, J.M. Global Change and the Ecology of Cities. *Science* **2008**, *319*, 756. [CrossRef]
3. Jaeger, J.A.G.; Bertiller, R.; Schwick, C.; Cavens, D.; Kienast, F. Urban permeation of landscapes and sprawl per capita: New measures of urban sprawl. *Ecol. Indic.* **2010**, *10*, 427–441. [CrossRef]
4. Zitti, M.; Ferrara, C.; Perini, L.; Carlucci, M.; Salvati, L. Long-Term Urban Growth and Land Use Efficiency in Southern Europe: Implications for Sustainable Land Management. *Sustainability* **2015**, *7*, 3359. [CrossRef]
5. Hasse, J.E.; Lathrop, R.G. Land resource impact indicators of urban sprawl. *Appl. Geogr.* **2003**, *23*, 159–175. [CrossRef]
6. Assembly, U.N.G. *Transforming Our World: The 2030 Agenda for Sustainable Development A/RES/70/1*; United Nations: New York, NY, USA, 2015.
7. UN-Habitat. *SDG Indicator 11.3.1 Training Module: Land Use Efficiency*; UN-Habitat: Nairobi, Kenya, 2018.

8. Haas, J.; Ban, Y. Urban growth and environmental impacts in Jing-Jin-Ji, the Yangtze, River Delta and the Pearl River Delta. *Int. J. Appl. Earth Obs. Geoinf.* **2014**, *30*, 42–55. [CrossRef]

9. Sun, Y.; Zhao, S. Spatiotemporal dynamics of urban expansion in 13 cities across the Jing-Jin-Ji Urban Agglomeration from 1978 to 2015. *Ecol. Indic.* **2018**, *87*, 302–313. [CrossRef]

10. UN-Habitat. *Metadata on SDGs Indicator 11.3.1*; UN-Habitat: Nairobi, Kenya, 2019.

11. Melchiorri, M.; Pesaresi, M.; Florczyk, A.J.; Corbane, C.; Kemper, T. Principles and Applications of the Global Human Settlement Layer as Baseline for the Land Use Efficiency Indicator—SDG 11.3.1. *ISPRS Int. J. Geo-Inf.* **2019**, *8*, 96. [CrossRef]

12. Anderson, K.; Ryan, B.; Sonntag, W.; Kavvada, A.; Friedl, L. Earth observation in service of the 2030 Agenda for Sustainable Development. *Geo Spat. Inf. Sci.* **2017**, *20*, 77–96. [CrossRef]

13. Wu, B.; Tian, F.; Zhang, M.; Zeng, H.; Zeng, Y. Cloud services with big data provide a solution for monitoring and tracking sustainable development goals. *Geogr. Sustain.* **2020**, *1*, 25–32. [CrossRef]

14. Li, W.; El-Askary, H.; Lakshmi, V.; Piechota, T.; Struppa, D. Earth Observation and Cloud Computing in Support of Two Sustainable Development Goals for the River Nile Watershed Countries. *Remote Sens.* **2020**, *12*, 1391. [CrossRef]

15. Weng, Q. Remote sensing of impervious surfaces in the urban areas: Requirements, methods, and trends. *Remote Sens. Environ.* **2012**, *117*, 34–49. [CrossRef]

16. Li, Q.; Lu, L.; Weng, Q.; Xie, Y.; Guo, H. Monitoring Urban Dynamics in the Southeast U.S.A. Using Time-Series DMSP/OLS Nightlight Imagery. *Remote Sens.* **2016**, *8*, 578. [CrossRef]

17. Lu, L.; Guo, H.; Corbane, C.; Li, Q. Urban sprawl in provincial capital cities in China: Evidence from multi-temporal urban land products using Landsat data. *Sci. Bull.* **2019**, *64*, 955–957. [CrossRef]

18. Pesaresi, M.; Huadong, G.; Blaes, X.; Ehrlich, D.; Ferri, S.; Gueguen, L.; Halkia, M.; Kauffmann, M.; Kemper, T.; Lu, L.; et al. A Global Human Settlement Layer From Optical HR/VHR RS Data: Concept and First Results. *IEEE J. Sel. Top. Appl. Earth Observ. Remote Sens.* **2013**, *6*, 2102–2131. [CrossRef]

19. Wang, Y.C.; Huang, C.L.; Feng, Y.Y.; Zhao, M.Y.; Gu, J. Using Earth Observation for Monitoring SDG 11.3.1-Ratio of Land Consumption Rate to Population Growth Rate in Mainland China. *Remote Sens.* **2020**, *12*, 357. [CrossRef]

20. Mudau, N.; Mwaniki, D.; Tsoeleng, L.; Mashalane, M.; Beguy, D.; Ndugwa, R. Assessment of SDG Indicator 11.3.1 and Urban Growth Trends of Major and Small Cities in South Africa. *Sustainability* **2020**, *12*, 7063. [CrossRef]

21. Ghazaryan, G.; Rienow, A.; Oldenburg, C.; Thonfeld, F.; Trampnau, B.; Sticksel, S.; Jürgens, C. Monitoring of Urban Sprawl and Densification Processes in Western Germany in the Light of SDG Indicator 11.3.1 Based on an Automated Retrospective Classification Approach. *Remote Sens.* **2021**, *13*, 1694. [CrossRef]

22. Koroso, N.H.; Zevenbergen, J.A.; Lengoiboni, M. Urban land use efficiency in Ethiopia: An assessment of urban land use sustainability in Addis Ababa. *Land Use Policy* **2020**, *99*, 105081. [CrossRef]

23. National Bureau of Statistics of China. *Statistical Communiqué of the People's Republic of China on the 2019 National Economic and Social Development*; National Bureau of Statistics of China: Beijing, China, 2020.

24. Fang, C. Important progress and future direction of studies on China's urban agglomerations. *J. Geogr. Sci.* **2015**, *25*, 1003–1024. [CrossRef]

25. Fang, C.; Yu, D. Urban agglomeration: An evolving concept of an emerging phenomenon. *Landsc. Urban Plan.* **2017**, *162*, 126–136. [CrossRef]

26. Moore, R.; Hansen, M. Google Earth Engine: A new cloud-computing platform for global-scale earth observation data and analysis. *AGU Fall Meet. Abstr.* **2011**, *2011*, IN43C-02.

27. Gorelick, N.; Hancher, M.; Dixon, M.; Ilyushchenko, S.; Thau, D.; Moore, R. Google Earth Engine: Planetary-scale geospatial analysis for everyone. *Remote Sens. Environ.* **2017**, *202*, 18–27. [CrossRef]

28. Gong, P.; Li, X.; Wang, J.; Bai, Y.; Chen, B.; Hu, T.; Liu, X.; Xu, B.; Yang, J.; Zhang, W.; et al. Annual maps of global artificial impervious area (GAIA) between 1985 and 2018. *Remote Sens. Environ.* **2020**, *236*. [CrossRef]

29. Martino, P.; Daniele, E.; Stefano, F.; Aneta, F.; Vasileios, S. Operating procedure for the production of the Global Human Settlement Layer from Landsat data of the epochs 1975, 1990, 2000, and 2014. *Publ. Off. Eur. Union* **2016**, 1–62.

30. Zhang, X.; Liu, L.; Chen, X.; Gao, Y.; Xie, S.; Mi, J. GLC_FCS30: Global land-cover productwith fine classification system at 30 m using time-series Landsat imagery. *Earth Syst. Sci. Data* **2021**, *13*, 2753–2756. [CrossRef]

31. Chen, J.; Chen, J.; Liao, A.; Cao, X.; Chen, L.; Chen, X.; He, C.; Han, G.; Peng, S.; Lu, M.; et al. Global land cover mapping at 30 m resolution: A POK-based operational approach. *ISPRS J. Photogramm. Remote Sens.* **2015**, *103*, 7–27. [CrossRef]

32. Lloyd, C.T.; Sorichetta, A.; Tatem, A.J. High resolution global gridded data for use in population studies. *Sci. Data* **2017**, *4*, 170001. [CrossRef] [PubMed]

33. Farr, T.G.; Kobrick, M. Shuttle radar topography mission produces a wealth of data. *Eos Trans. Am. Geophys. Union* **2000**, *81*. [CrossRef]

34. Li, Q.; Lu, L.; Wang, C.; Li, Y.; Sui, Y.; Guo, H. MODIS-Derived Spatiotemporal Changes of Major Lake Surface Areas in Arid Xinjiang, China, 2000–2014. *Water* **2015**, *7*, 5731. [CrossRef]

35. Lu, L.; Kuenzer, C.; Wang, C.; Guo, H.; Li, Q. Evaluation of Three MODIS-Derived Vegetation Index Time Series for Dryland Vegetation Dynamics Monitoring. *Remote Sens.* **2015**, *7*, 7597–7614. [CrossRef]

36. Lu, L.; Weng, Q.; Xiao, D.; Guo, H.; Li, Q.; Hui, W. Spatiotemporal Variation of Surface Urban Heat Islands in Relation to Land Cover Composition and Configuration: A Multi-Scale Case Study of Xi'an, China. *Remote Sens.* **2020**, *12*, 2713. [CrossRef]

37. Zhang, Z.; Wei, M.; Pu, D.; He, G.; Wang, G.; Long, T. Assessment of Annual Composite Images Obtained by Google Earth Engine for Urban Areas Mapping Using Random Forest. *Remote Sens.* **2021**, *13*, 748. [CrossRef]
38. Olofsson, P.; Foody, G.M.; Herold, M.; Stehman, S.V.; Woodcock, C.E.; Wulder, M.A. Good practices for estimating area and assessing accuracy of land change. *Remote Sens. Environ.* **2014**, *148*, 42–57. [CrossRef]
39. Breiman, L. Random Forests. *Mach. Learn.* **2001**, *45*, 5–32. [CrossRef]
40. Foody, G.M. Status of land cover classification accuracy assessment. *Remote Sens. Environ.* **2002**, *80*, 185–201. [CrossRef]
41. Kundel, H.L.; Polansky, M. Measurement of Observer Agreement. *Radiology* **2003**, *228*, 303–308. [CrossRef]
42. Wickham, J.; Stehman, S.V.; Gass, L.; Dewitz, J.A.; Sorenson, D.G.; Granneman, B.J.; Poss, R.V.; Baer, L.A. Thematic accuracy assessment of the 2011 National Land Cover Database (NLCD). *Remote Sens. Environ.* **2017**, *191*, 328–341. [CrossRef]
43. Zhang, H.; Xu, R. Exploring the optimal integration levels between SAR and optical data for better urban land cover mapping in the Pearl River Delta. *Int. J. Appl. Earth Obs. Geoinf.* **2018**, *64*, 87–95. [CrossRef]
44. Smits, P.C. Multiple classifier systems for supervised remote sensing image classification based on dynamic classifier selection. *IEEE Trans. Geosci. Remote Sens.* **2002**, *40*, 801–813. [CrossRef]
45. Zhang, L.; Weng, Q. Annual dynamics of impervious surface in the Pearl River Delta, China, from 1988 to 2013, using time series Landsat imagery. *ISPRS J. Photogramm. Remote Sens.* **2016**, *113*, 86–96. [CrossRef]
46. Kang, J.; Wang, Z.; Sui, L.; Yang, X.; Ma, Y.; Wang, J. Consistency Analysis of Remote Sensing Land Cover Products in the Tropical Rainforest Climate Region: A Case Study of Indonesia. *Remote Sens.* **2020**, *12*, 1410. [CrossRef]
47. Yang, Y.; Xiao, P.; Feng, X.; Li, H. Accuracy assessment of seven global land cover datasets over China. *ISPRS J. Photogramm. Remote Sens.* **2017**, *125*, 156–173. [CrossRef]
48. Gao, Y.; Liu, L.; Zhang, X.; Chen, X.; Mi, J.; Xie, S. Consistency Analysis and Accuracy Assessment of Three Global 30-m Land-Cover Products over the European Union using the LUCAS Dataset. *Remote Sens.* **2020**, *12*, 3479. [CrossRef]
49. Sabo, F.; Corbane, C.; Florczyk, A.J.; Ferri, S.; Pesaresi, M.; Kemper, T. Comparison of built-up area maps produced within the global human settlement framework. *Trans. GIS* **2018**, *22*, 1406–1436. [CrossRef]
50. Liu, F.; Zhang, Z.X.; Shi, L.F.; Zhao, X.L.; Xu, J.Y.; Yi, L.; Liu, B.; Wen, Q.K.; Hu, S.G.; Wang, X.; et al. Urban expansion in China and its spatial-temporal differences over the past four decades. *J. Geogr. Sci.* **2016**, *26*, 1477–1496. [CrossRef]
51. Deng, Y.; Qi, W.; Fu, B.; Wang, K. Geographical transformations of urban sprawl: Exploring the spatial heterogeneity across cities in China 1992–2015. *Cities* **2020**, *105*, 102415. [CrossRef]
52. Cao, H.; Zhang, H.; Wang, C.; Zhang, B. Operational Built-Up Areas Extraction for Cities in China Using Sentinel-1 SAR Data. *Remote Sens.* **2018**, *10*, 874. [CrossRef]
53. Liu, C.; Huang, X.; Zhu, Z.; Chen, H.; Tang, X.; Gong, J. Automatic extraction of built-up area from ZY3 multi-view satellite imagery: Analysis of 45 global cities. *Remote Sens. Environ.* **2019**, *226*, 51–73. [CrossRef]
54. Geiß, C.; Schrade, H.; Aravena Pelizari, P.; Taubenböck, H. Multistrategy ensemble regression for mapping of built-up density and height with Sentinel-2 data. *ISPRS J. Photogramm. Remote Sens.* **2020**, *170*, 57–71. [CrossRef]

 *remote sensing*

Article

Comparative Analysis of Variations and Patterns between Surface Urban Heat Island Intensity and Frequency across 305 Chinese Cities

Kangning Li, Yunhao Chen * and Shengjun Gao

State Key Laboratory of Remote Sensing Science, Faculty of Geographical Science, Beijing Normal University, Beijing 100875, China; 201631190001@mail.bnu.edu.cn (K.L.); 201931051039@mail.bnu.edu.cn (S.G.)
* Correspondence: cyh@bnu.edu.cn; Tel.: +86-010-5880-4056

Abstract: Urban heat island (UHI), referring to higher temperatures in urban extents than its surrounding rural regions, is widely reported in terms of negative effects to both the ecological environment and human health. To propose effective mitigation measurements, spatiotemporal variations and control machines of surface UHI (SUHI) have been widely investigated, in particular based on the indicator of SUHI intensity (SUHII). However, studies on SUHI frequency (SUHIF), an important temporal indicator, are challenged by a large number of missing data in daily land surface temperature (LST). Whether there is any city with strong SUHII and low SUHIF remains unclear. Thanks to the publication of daily seamless all-weather LST, this paper is proposed to investigate spatiotemporal variations of SUHIF, to compare SUHII and SUHIF, to conduct a pattern classification, and to further explore their driving factors across 305 Chinese cities. Four main findings are summarized below: (1) SUHIF is found to be higher in the south during the day, while it is higher in the north at night. Cities within the latitude from 20° N and 40° N indicate strong intensity and high frequency at day. Climate zone-based variations of SUHII and SUHIF are different, in particular at nighttime. (2) SUHIF are observed in great diurnal and seasonal variations. Summer daytime with 3.01 K of SUHII and 80 of SUHIF, possibly coupling with heat waves, increases the risk of heat-related diseases. (3) K-means clustering is employed to conduct pattern classification of the selected cities. SUHIF is found possibly to be consistent to its SUHII in the same city, while they provide quantitative and temporal characters respectively. (4) Controls for SUHIF and SUHII are found in significant variations among temporal scales and different patterns. This paper first conducts a comparison between SUHII and SUHIF, and provides pattern classification for further research and practice on mitigation measurements.

Keywords: surface urban heat island frequency; surface urban heat island intensity; all-weather land surface temperature; spatiotemporal variations; factor analysis

Citation: Li, K.; Chen, Y.; Gao, S. Comparative Analysis of Variations and Patterns between Surface Urban Heat Island Intensity and Frequency across 305 Chinese Cities. *Remote Sens.* **2021**, *13*, 3505. https://doi.org/10.3390/rs13173505

Academic Editors: Elhadi Adam, John Odindi, Elfatih Abdel-Rahman and Yuyu Zhou

Received: 9 August 2021
Accepted: 1 September 2021
Published: 3 September 2021

1. Introduction

Over half of the world's population have aggregated in urban areas [1], in particular with the rapid development of urbanization in recent years. Urbanization brings not only demographic and socioeconomic advancement, but also some urban environmental issues. The transformation, from natural surface to impervious areas, introduces evident perturbations to the Earth's energy balance [2]. These perturbations can possibly be attributed to low albedo and high thermal capacity of impervious areas compared to evaporative vegetated cover [3,4]. Coupling with a mass of anthropogenic heat from rapid urbanization, urban heat island (UHI), with a higher temperature in urban extents than their surrounding areas, are reported in a large number of cities throughout the world [5–16]. The UHI is reported to be closely associated with a wide range of environmental issues, including air pollution [17], biodiversity reduction [18], and increased energy consumption [19,20]. Moreover, The UHI poses a potential threat to resident comfort and even human health, with increasing heat-related mortality [21,22], especially when heat waves strike simultaneously [23]. Thus, to propose effective mitigation measurements in future urban planning, extensive exploration of the variations and controls of UHI is necessary.

According to the way and height of observation, the UHI is generally grouped into two major categories: namely atmospheric (AUHI) and surface UHI (SUHI) [24,25]. The AUHI includes canopy and boundary UHI [5], which are detected by ground-based and high-altitude equipment, respectively. The sparse distribution and expensive installation of these observation measurements pose great challenges to AUHI studies, especially on a large scale. On the contrary, the SUHI takes advantage of thermal remote sensing with the capability of large-scale periodic observations [26]. Therefore, SUHI is reported as the main part of the UHI research from local to global scales [5].

Since Rao [27] first employed satellite-based data in the study of SUHI, a large number of SUHI studies have been published. Voogt and Oke [25] conducted a literature review on the application of thermal remote sensing in urban climates, and demonstrated their advantages in UHI studies. Streutker [28] explored the growth of SUHI in Houston, Texas from 1989 to 1999 based on an Advanced Very High Resolution Radiometer (AVHRR). Weng et al. [29] reported that interactions between land surface temperature (LST) and vegetation largely contribute to the spatial patterns of UHI, based on Landsat ETM+ thermal observation. Weng et al. [30] summarized the studies of urban climates based on remotely sensed thermal infrared (TIR) and pointed out that less attention has been paid to the estimation of UHI parameters. Imhoff et al. [7] analyzed the relationship between impervious surface area (ISA) and LST, and found great effects from the ecological context on the amplitude of SUHII during the summer daytime. Schwarz et al. [31] reported a large discrepancy and a low correlation among eleven different SUHI indicators, and suggested a combination of several indicators for comprehensive SUHI studies. Peng et al. [6] investigated SUHI across 419 global big cities and reported different driving mechanisms between daytime and nighttime. Quan et al. [32] employed the Gaussian volume model to explore the trajectory of SUHI in Beijing, and found temporal variations of the UHI centroid. Zhao et al. [33] reported that the efficiency of thermal convection make a strong contribution to the geographic variations of daytime UHI, and proposed albedo managements as promising mitigation measurements on a large scale. Zhou et al. [34] analyzed spatial–temporal patterns of SUHI across 32 major cities in China, and found that vegetation and anthropogenic heat were closely related to the SUHI during summer daytime. Lai et al. [35] proposed a four-parameter diurnal temperature cycle (DTC) model, and characterized 354 Chinese cities into five typical temporal patterns. Manoli et al. [36] introduced a coarse-grained model based on population and precipitation, and reported their strong contribution to magnitude of SUHI.

Taking advantage of remote sensing, SUHI intensity (SUHII) has been widely investigated. Despite considerable applications of satellite-based TIR data, TIR measurements are largely limited by their low tolerance to cloud cover [37–41], resulting in over half of the missing data [42]. Since there is a hard availability of daily seamless LST directly from the TIR products, the spatial–temporal variations of SUHI frequency (SUHIF) and its controls remain largely unknown. To compensate for missing data in the TIR products, several methods were proposed to reconstruct seamless LST. These methods can generally be grouped into two broad classes, namely spatial–temporal interpolation [37,41–44] and model simulation [38–40,45]. The spatial–temporal interpolation refers to two major ways, including gap-filling based on spatial neighboring or temporal adjacent pixels [37], and correlation between LST and other data sources. The former way is limited by a deficiency in adjacent pixels, while the latter one is challenged by the quality of auxiliary data [38]. In contrast, LST reconstruction based on model simulation benefits from the stability of its physical or semi-physical model. Liu et al. [38] proposed a hybrid ATC model on the base of TIR LST, taking into account both the estimation accuracy and efficiency. To compensate for the limitation of TIR, passive microwave (PMW), with the capability of penetrating cloud, was employed to estimate all-weather LST. Zhang et al. [40] proposed a temporal decomposition-based method including ATC, diurnal temperature cycle (DTC) and a weather temperature component by combining TIR and PMW to reconstruct all-weather LST. To compensate for the swath gap of PMW, Zhang et al. [39] modified this method on the base of reanalysis data at the low–mid latitudes. Thanks to their great efforts, a daily all-weather LST dataset proposed by Zhang et al. [39,40] for China and its surrounding areas has recently been released.

Great efforts have been paid to SUHI studies from local to global scales. In particular, SUHII rather than urban temperature was widely applied as an evaluation indicator in large-scale studies [46] due to its comparability among different cities. Compared to quantitative information from SUHII, SUHIF is a frequency indicator to show the number of occurrences of SUHI effects during a period of time. As the annual average of SUHII is the most popular temporal scale, the occurrences of SUHI effects during a year were taken as the major concentration here. However, the spatiotemporal variations of SUHIF remain largely unknown as a result of the absence of daily seamless LST. Fortunately, the recent publication of a daily seamless LST dataset across China has provided the data basis for the investigation on SUHIF. There are three issues about SUHIF requiring further exploration. Frist, the spatial and temporal variations of SUHIF are not clear. Second, the relationship between SUHII and SUHIF remains unknown. There are several questions on their comparisons. Does high SUHII necessarily accompany high SUHIF? Is there any city with a strong SUHII but weak SUHIF (vice versa)? Can the pattern classification of these cities based on SUHII and SUHIF generalize their characteristics? Third, there are some questions on their driving mechanisms which remain unclear. Are controls of SUHII and SUHIF coincident or different? Do their driving factors vary with different SUHI patterns?

To address the aforementioned issues, this paper proposed to compare spatial–temporal variations between SUHII and SUHIF, to identify SUHI patterns, and to explore their driving factors across 305 Chinese cities. These comparisons can bring insights into the quantitative and temporal characteristics of SUHI. Furthermore, it can provide a foundation for further exploration, which is closely correlated to heat-related diseases and poses stronger negative effects on human health, SUHII or SUHIF.

2. Study Area and Data

2.1. Study Area

China is located on the eastern Eurasian continent and on the western Pacific coast (from 72° E to 135° E, from 19° N to 55° N). Thanks to its vast territory, there are great gradients in terms of temperature and precipitation across China. Additionally, there is significant economic and social development in China with the implementation of the reform and opening-up policy. Considerable urbanization brings not only economic boosts, but also great perturbations to the urban environment. Zhou et al. [5] reported that the UHI possibly became serious issues under conditions of global warming and rapid urbanization, especially in China. Accordingly, various natural environments and rapid urbanization make China an ideal place to conduct studies on SUHI. A total of 305 Chinese cities (Figure 1) with population over than 500,000 in 2018 were selected. A population of 500,000 was utilized as the threshold for city selection due to its wide applications in the difference of small cities and large towns.

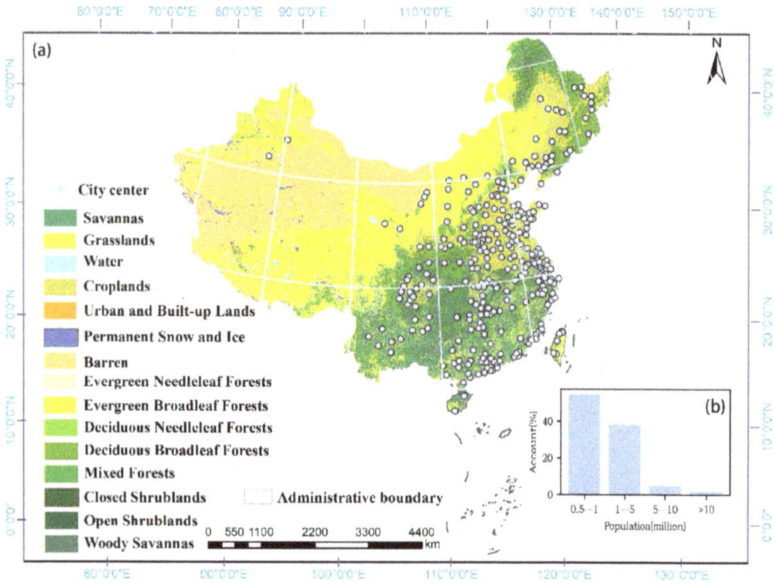

Figure 1. Spatial distribution of the 305 selected cities across China. The background of the map shows Land use and land cover from MODIS MCD12Q1. The bar plot (**b**) shows the percentage of these cities according to five population classifications.

2.2. Data

LST was collected from the daily 1-km all-weather LST dataset for the Chinese landmass and its surrounding areas [39,40,47,48], also named TRIMS LST (Thermal and Reanalysis Integrating Moderate-resolution Spatial-seamless LST), in 2018. This dataset is available at the National Qinghai-Tibet Plateau Science Data Center (http://data.tpdc.ac. cn/zh-hans/, accessed on 1 July 2021). Good image quality and accuracy of TRIMS LST were illustrated according to data validations. The consistency between TRIMS LST and Aqua MODIS LST was validated. Compared to MODIS LST, the MAE (mean absolute error) of TRIMS LST is 0.08 K and 0.16 K at day and night. Thanks to the advantages of the high accuracy and seamless images, TRIMS LST provides a data foundation for the study on SUHIF.

Land cover, land use (LULC), and Enhanced Vegetation Index (EVI) were collected from MODIS products in 2018. LULC was collected from the annual 500-m MCD12Q1, which was produced based on supervised classification and further post-processing. EVI was collected from 16-Day 500-m MYD13A1 V6 (Version 6), which utilized the blue band to mitigate effects from atmosphere contamination. These two MODIS data are available at the Level-1 and Atmosphere Archive and Distribution System (LAADS) Distributed Active Archive Center (DAAC) (https://ladsweb.modaps.eosdis.nasa.gov/search/, accessed on 1 July 2021).

Total precipitation (TP) was collected from ERA5-Land hourly 0.1° × 0.1° climate reanalysis dataset at the European Centre for Medium-Range Weather Forecasts (ECWMF). A digital elevation model (DEM) was collected from GTOPO30, a global 1-km DEM product. GTOPO30 is available at the United States Geological Survey (https://earthexplorer.usgs. gov/, accessed on 1 July 2021). Monthly 1-km nighttime light (NTL) was collected from the Suomi National Polar-orbiting Partnership (SNPP) Visible Infrared Imaging Radiometer Suite (VIIRS) Day/Night Band (DNB), available at the NOAA National Geophysical Data Center (https://ngdc.noaa.gov/eog/download.html, accessed on 1 July 2021). Administrative boundaries for the 305 Chinese cities were collected from the Global Administrative

Unit Layers (GAUL). The population of the selected cities was collected from the yearbook of Chinese cities.

3. Methods

The flowchart was generally grouped into four steps, namely SUHII calculation, SUHIF estimation, pattern classification and factor analysis (Figure 2).

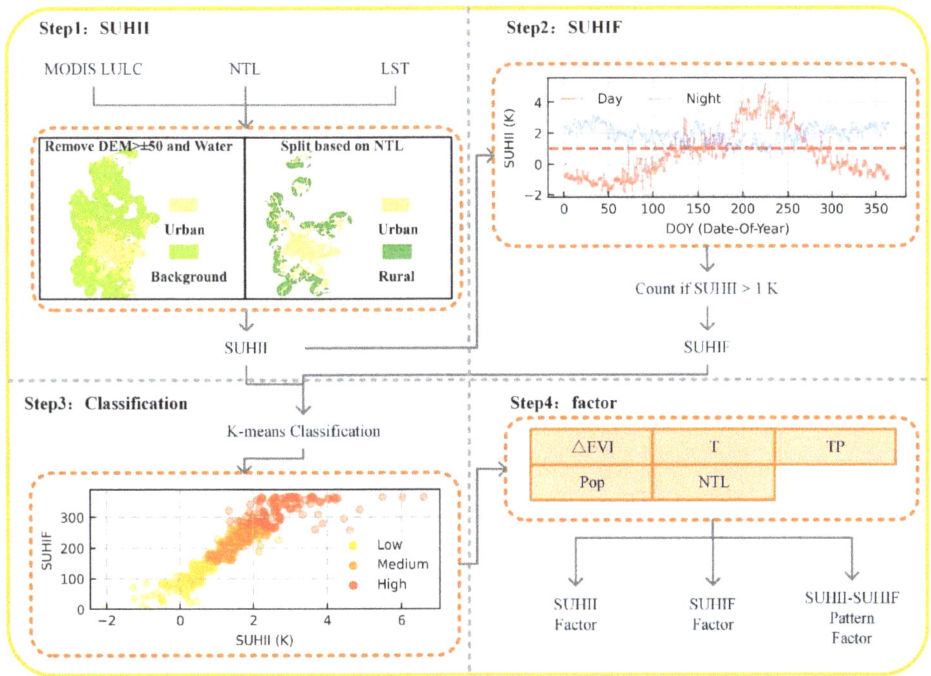

Figure 2. Flowchart of this study. There are four steps, including SUHII calculation, SUHIF estimation, pattern classification and factor analysis. To make it clear, Tianjin is taken as an example in the flowchart.

3.1. SUHII Calculation

SUHII is commonly defined as a temperature difference between urban and rural areas (Equation (1)). Before SUHII calculation, two basic preparations were the definition of these two regions and elimination of the effects from water and DEM. First, urban extents were defined as urban land cover from MODIS LULC data within the administrative boundaries of the selected cities. There were two major methods for defining rural extents for SUHII, namely buffer-based and size-based methods. The buffer-based rural definition were reported to be limited by the difficulties of buffer selection [46] and the in-adaptability of one fixed buffer for various cities [35]. Thus, the size-based method was employed to define rural reference with an equal size of urban extents. Before generating equal-size rural surroundings, water and ice cover were removed to mitigate their effects on temperature. Besides, pixels with DEM over ±50 m of the urban average were eliminated. To mitigate the impacts from human activities, the rural reference was defined as an equal-size ring distant to urban cores. Nighttime light (NTL) benefiting from a representation of human impacts was applied to distinguish rural extents. As a twice-size ring was first prepared, the median NTL of it was used to extract two equal-sized areas with a higher and lower

NTL respectively. To keep temporal consistency, SUHII was first calculated at a daily scale, and then temporally aggregated to seasonal and annual scales.

$$SUHII_{i,d} = LST_{Urban,i,d} - LST_{Rural,i,d} \, , \tag{1}$$

where i represents the ith city, and d represents the DOY (Date-Of-Year). $LST_{Urban,i,d}$ represents average urban LST of the ith city in the dth day.

3.2. SUHIF Estimation

To our knowledge, few studies on SUHIF were conducted due to a lack of daily seamless LST. Therefore, the selection of the threshold for SUHI was first discussed. 1 K was selected as the threshold to estimate SUHIF. There were three major reasons for the selection of 1 K rather than 0 K. First, a large number of large-scale studies on SUHII have reported that the average of SUHII was approximately 1 K [6–9,34,46,49]. Second, the selection of 0 K as the threshold was possibly more sensitive to the accuracy of LST, compared to 1 K. Third, the selection of the threshold was determined on the basis of the application requirements. After determination of the threshold, SUHIF was estimated based on daily SUHII from seamless LST under all-weather conditions. According to the difference between temporal periods, SUHIF was estimated at annual (Equation (2)) and seasonal (Equation (3)) scales, respectively.

$$SUHIF_{annual,i} = count(if\ SUHII_{i,d} \geq 1\ K), \quad d \in (0, d_{annual}) \tag{2}$$

$$SUHIF_{season.i} = count(if\ SUHII_{i,d} \geq 1\ K), \quad d \in (d_{season,start}, d_{season,end}) \tag{3}$$

where i represents the ith city, and d represents the DOY (Date-Of-Year). $SUHIF_{annual,i}$ represents the annual $SUHIF$ of the ith city, $SUHIF_{season,i}$ represents $SUHIF$ at a certain season (spring, summer, autumn and winter) of the ith city. d_{annual} represents the number of days in the year. $d_{season,start}$ and $d_{season,end}$ represent the DOY of the start and end date of a certain season.

3.3. Pattern Classification

This paper was proposed to discuss the consistency and difference between SUHII and SUHIF, and then to conduct a pattern classification for better characterization of urban heat island in China. As SUHII and SUHIF represent quantity and temporal information respectively, they were taken as features for the pattern classification. The K-means clustering algorithm, aiming at the partition of a dataset into K clusters [50], was employed to determine the pattern classification. There were two major steps of this iterative algorithm. First, K objects from the dataset were randomly selected as cluster centers, and each object was divided into clusters based on its distance to the center. Second, the cluster center was recalculated when a new object was divided into it. The second step was iterated until no more new object was grouped into different clusters. Since the number of classes is important in the K-means clustering algorithm, there are three reasons for the number selection. (1) According to the Gap statistics, a method to validate the efficiency of the number, three clusters can distinguish the characteristics of the dataset. (2) Three classes, namely the low, medium and high patterns, can not only characterize their features but also provide meaningful information for related researchers, policymakers and even residents who concern for the surrounding thermal environment and its heat-related illness. (3) Great differences among these three patterns was found across 305 Chinese cities.

3.4. Factor Analysis

There were five features being selected in the factor analysis, namely EVI difference (ΔEVI), LST, total precipitation (TP), population (Pop) and NTL. Since these factors have been widely reported as important controls for UHI [6,7,29,33,34,36,51], a factor analysis was conducted to explore whether they also largely contributed to the spatiotemporal variations of SUHIF. ΔEVI was the absolute value of EVI difference between urban and rural extents. Two indicators for background climate, LST and TP, were the average of urban and rural regions. NTL was estimated based on its average within urban areas. Despite various resolutions of data sources for these factors, the spatial average within certain boundaries makes them comparable. According to the literature review, the ordinary lest squares (OLS) with Pearson's correlation were reported as the dominant methods for investigating controls for SUHI [52]. The Pearson's correlation was utilized to explore the relationship between these selected factors and SUHIF. The factor analysis was conducted in terms of SUHII, SUHIF, and their comparisons. Despite great efforts on SUHII drivers, limited by the lack of daily data, the correlation between SUHII and these factors was first conducted at daily scales. Factor analysis for SUHIF was conducted at annual and seasonal scales. Additionally, to determine whether the controls for SUHII or SUHIF varied among different patterns, factor analysis was investigated in terms of pattern classification.

4. Results and Discussion

4.1. Comparison of Spatial Distribution between SUHII and SUHIF

To explore the similarities and differences of geographic variations between SUHII and SUHIF, two major steps were conducted. They are a qualitative analysis of the overall distribution (Figure 3) and further quantitative estimation in terms of latitude and climate zone-based variations (Figure 4). Figure 3 shows the overall spatial distribution of SUHII and SUHIF across 305 Chinese cities. At daytime, SUHII is observed in an evident north–south contrast, with the higher intensity in the south and the weaker one in the northern regions. According to the previous studies on the driving mechanism, vegetation is widely reported as an important indicator for daytime SUHII [6,8,53]. Referring to LULC in Figure 1, a north–south contrast of vegetation distribution is found to be similar to that of SUHII. This similar distribution possibly suggests that the difference of vegetation distribution between northern and southern China is related to the north–south contrast of SUHII. Conversely, an opposite distribution pattern with a weaker intensity in the south, and a higher one in the northern areas, is found from nighttime SUHII. Despite different city selections across China, these spatial patterns and diurnal contrasts of SUHII are consistent with Zhou's findings [34]. Additionally, the daytime SUHIF shows an increasing gradient from the north to the southern regions. At night, an inverse gradient of SUHIF is observed. SUHIF is also found to be higher in the southern regions at day, while they are higher in the north at night. On the whole, SUHIF is observed to have similar spatial patterns to SUHII at both day and night.

Furthermore, Figure 4 shows a quantitative analysis based on the latitudinal and climatic variations of SUHII and SUHIF. Considering the geographic distribution of the selected cities, latitudinal variations were conducted from 20° to 50° N with an interval of 5°. At daytime, there are three primary changing points of SUHII. Daytime SUHII decreases from 1.14 K at the 50° N zone to 0.43 K at the 40° N zone, and rises to 3.65 K at the 20° N zone. This decrease from 50° N to 40° N possibly results from different vegetation distributions between them. According to Figure 1, vegetation cover of rural areas in Hegang and Jiamusi (at 50° N zone), with high daytime SUHII, are deciduous broadleaf forests (Figure 1), while cities at 40° N zone are mainly surrounded by croplands. At night, SUHII decreases from 1.31 K at 45° N to 0.64 K at 25° N, and then increases to 1.33 K at 20° N. Nighttime SUHII is observed to have a less significant north–south contrast. Latitudinal variations of SUHIF indicate similar patterns to that of SUHII, especially at daytime. Daytime SUHIF decreases from 167.50 at 50° N to 121.73 at 40° N, and rises to 364 at 20° N. Accordingly, due to the latitude between 35° N and 20° N, more than 200 days in a year are affected by SUHII over 1 K. Despite general consistency to the low north and high south patterns of daytime SUHIF (Figure 3), latitudinal analysis provides more details about the north–south contrast. At night, SUHIF shows greater variations from 25° N to 20° N compared to nighttime SUHII.

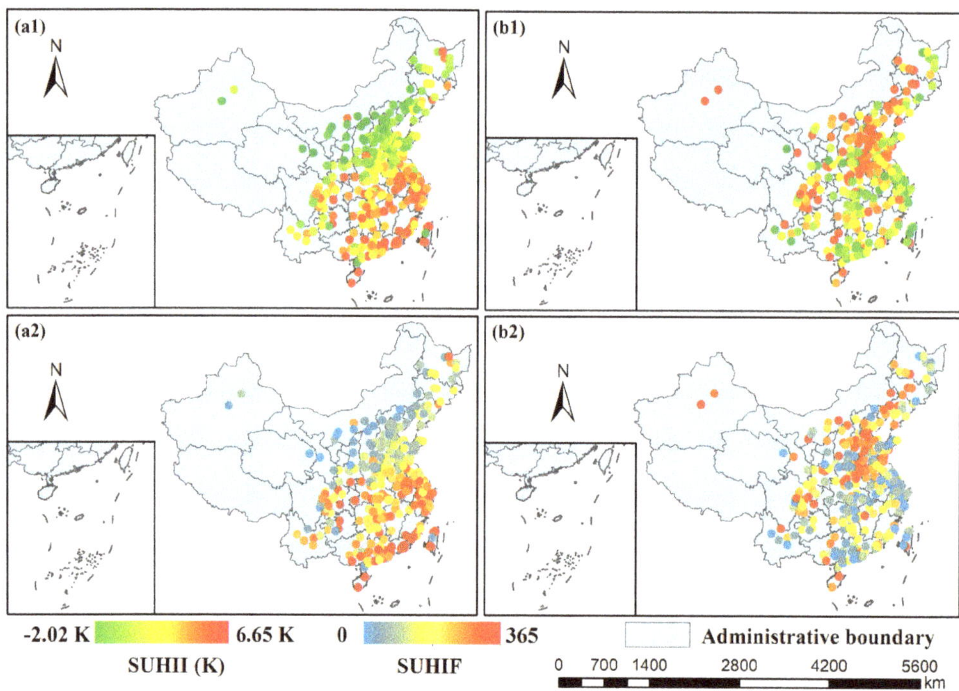

Figure 3. Spatial distribution of SUHII and SUHIF at daytime (**a**) and nighttime (**b**) in China. The spatial maps in the first row represent SUHII (**a1,b1**). The second row maps represent SUHIF (**a2,b2**).

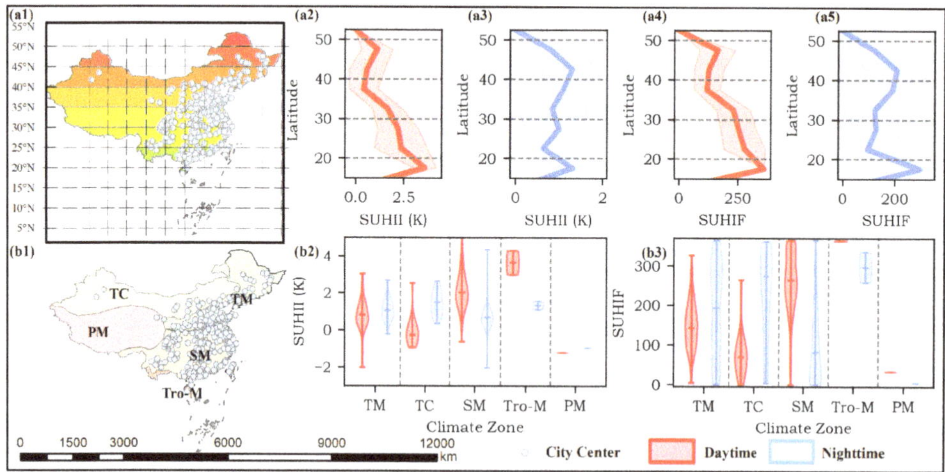

Figure 4. Latitudinal and climate zone-based variations of SUHII and SUHIF at daytime and nighttime. The two maps (**a1,b1**) show the latitudinal and climatic distributions of the selected cities, respectively. The plots in the first row show latitudinal variations of SUHII (**a2,a3**) and SUHIF (**a4,a5**) based on the combination of their average and variance. The violin plots in the second row show climate zone-based variations of SUHII (**b2**) and SUHIF (**b3**). The daytime and nighttime patterns are represented based on red and blue colors, respectively.

The climate zone-based variations were conducted in terms of five major climate types, namely Temperate Monsoon (TM), Temperate Continental (TC), Subtropical Monsoon (SM), Tropical Monsoon (Tro-M) and Plateau Mountain (PM) climate. At daytime, the average of the climate zones from the highest to the lowest are listed as Tro-M, SM, TM, TC and PM, with values of 3.64 K, 2.17 K, 0.81 K, −0.07 K and −1.21 K, respectively. Stronger daytime SUHII in the monsoon climate is observed compared to that in TC and PM. In addition, previous studies reported a larger daytime SUHII in humid-hot cities than cold-drier ones [6,7,10]. Despite the similar climate variations on average, daytime SUHIF indicates larger value ranges than that of SUHII. During the day, over 200 of SUHIF in the SM and Tro-M climate means that a large number of cities in these two climate zones are influenced by SUHI for more than 200 days in a year. Additionally, the order of average SUHII at night is listed as TC, Tro-M, TM, SM, and PM, from 1.44 K to −0.94 K. The highest average of nighttime SUHII is found in the TC climate, whereas that of SUHIF is found in Tro-M. Compared to daytime SUHIF, the nighttime one indicates larger internal variations in each climate zone.

4.2. Comparison of Temporal Variations between SUHII and SUHIF

To compare temporal patterns between SUHII and SUHIF, annual and seasonal variations were conducted. Figure 5 shows the Probability Distribution Function (PDF) and Cumulative Distribution Function (CDF) of SUHII and SUHIF across the selected cities. Daytime SUHII is primarily found between 0 and 4 K with an average of 1.53 K. As Zhou et al. [5] reported that UHI largely challenged the ecological environment and human health in China, we found that more than 66.87% of Chinese cities suffer from SUHII over 1 K. Nighttime SUHII, mainly varying from 0 to 2 K with an average of 0.95 K, is less evident than the daytime one. This diurnal contrast was also reported in some previous large-scale studies [6,8,9]. This contrast is possibly attributed to more driving factors for daytime SUHI than for the nighttime one [8]. Additionally, rather than an approximate Gaussian distribution of SUHII, the CDF and PDF of SUHIF indicate different patterns. There is an aggregation in the highest value range (from 354 to 365) accounting for 8.50% of daytime SUHIF. Conversely, nighttime SUHIF shows an aggregation within the lowest value range, accounting for 15.41%. The annual average of SUHIF is 210.88 and 145.27 at day and night respectively, indicating a strong diurnal contrast. Greater SUHIF is found at day than at night. To explore more detail about SUHIF across China, the extremum is excluded from zooming graphs. PDF of SUHIF in the zooming graphs is found to have an approximately homogeneous distribution, which further illustrates the difference from SUHII.

Figure 6 shows the seasonal variations of SHUII and SUHIF on the basis of both seasonal average and value distribution, to provide more information and details. At daytime, SUHII indicates evident seasonal variations. Over four seasons, SUHII over 1 K accounts for 63.93%, 92.78%, 58.36%, and 25.90%, respectively. The strongest season of SUHII is found in summer, with an average of 3.01 K, while the weakest one is found in winter with 0.39 K. Summer daytime is also widely reported to be the strongest season of SUHII in a large number of related studies [5], which is possibly attributed to a strong and negative correlation between ΔEVI and SUHII in the summer [34]. Nevertheless, less evident seasonal variations are observed from nighttime SUHII. Summer, with an average of 1.05 K, is found to be the strongest season, while winter with 0.83 K is found to be the weakest one. This seasonal contrast is also found in some SUHII studies at a global scale [6,8]. Additionally, daytime SUHIF indicates significant seasonal variations, also referred to as being strongest in summer and weakest in winter. SUHIF over 80 days in a season (about 90 days) accounts for about 70.16% of the selected cities in summer, while a SUHIF less than 20 accounts for 60.32% in winter. For both daytime SUHII and SUHIF, the extremums are found in summer and winter, while transition seasons (spring and autumn) are found to have similar variations to each other. Tan et al. [54] reported that daytime SUHI in summer was responsible for the exacerbation of heat waves and was closely associated with heat-related mortality. Thus, the summer SUHI at day, with a strong intensity (3.01 K) and high frequency (80), should be further explored in terms of driving factors and mitigation measurements. At night, a similar average of SUHIF during four seasons (36, 37, 38 and 31) indicates less evident seasonal variations.

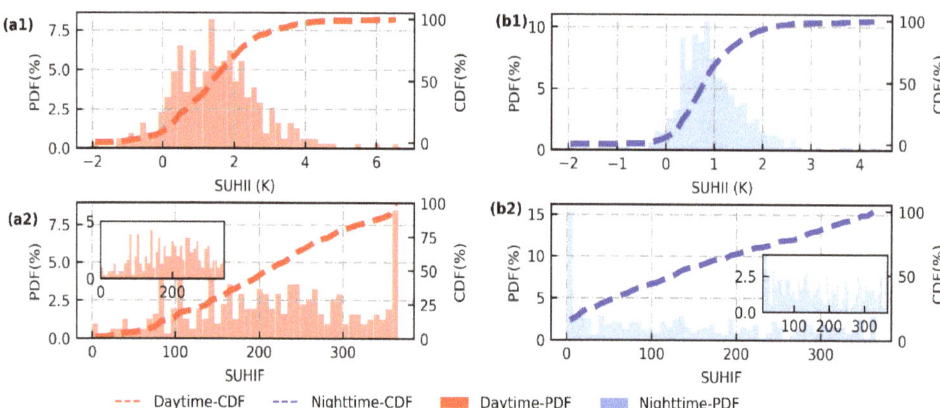

Figure 5. Annual variations of SUHII and SUHIF at daytime (**a**) and nighttime (**b**). The plots in the first row show the value distribution of annual SUHII (**a1,b1**), and the second-row plots show that of SUHIF (**a2,b2**). PDF and CDF are employed to show the value distribution across the selected cities. To clearly show their variations, zoom-in images without the extremum are added for SUHIF.

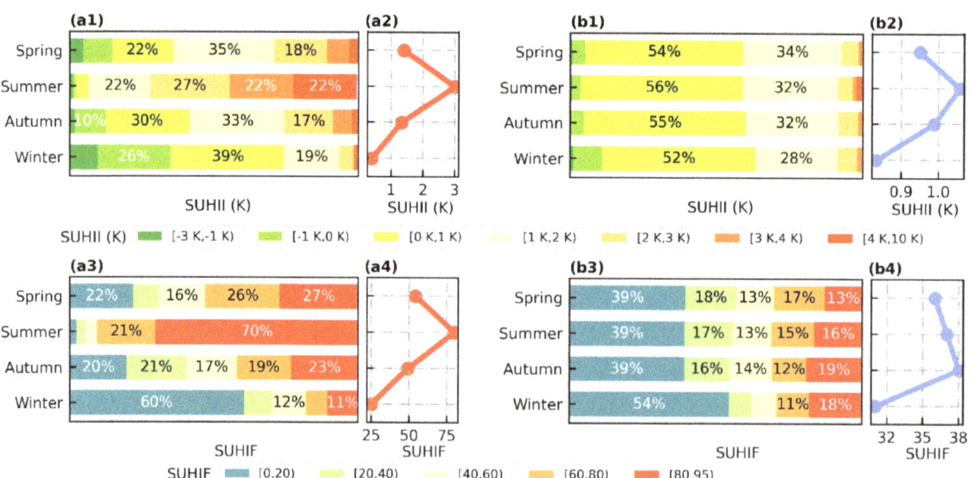

Figure 6. Seasonal variations of SUHII and SUHIF at daytime (**a**) and nighttime (**b**). The discrete distribution as horizontal bar charts in the first row shows seasonal variations of SUHII (**a1,b1**) and the second row charts show that of SUHIF (**a3,b3**). Percentage over 10% is labelled in the graph. The line plots in the first row show the average of SUHII (**a2,b2**) and plots in the second row show that of SUHIF (**a4,b4**) among the different seasons.

4.3. Pattern Classification Based on SUHII and SUHIF

After comparing their spatial and temporal variations between SUHII and SUHIF across 305 Chinese cities, further exploration was conducted to determine their relationship and to conduct a pattern classification (Figure 7). Scatter plots based on SUHII and SUHIF show that there is no city with a high intensity and a low frequency of UHI (vice versa).

Despite different value distributions based on their PDF, SUHII and SUHIF of the same city are observed to have a relative consistency, especially at night. Accordingly, a city with high SUHI intensity possibly accompanies high frequency. This closer correlation at night results from the less significant variations of nighttime SUHI. In spite of the consistency between SUHII and SUHIF, it is noted that they provide two different aspects of information from quantity and time. Furthermore, the K-means clustering was employed in the attempts of SUHI pattern classification based on these two important features. This pattern classification allows easy characterization and further exploration to determine whether driving factors vary among these patterns. Three categories, namely the low, medium, and high patterns, are not only proven by the gap statistics as a rational number of classification, but are also meaningful to related researchers and decision-makers. The amount of the selected cities in these three classes is 89, 123, and 93 at day, and 124, 94, and 87 at night. To further explore the discrepancy of these patterns, grouped statistics were conducted. At day, the average of SUHII among these three classes is 0.18 K, 1.49 K, and 2.89 K, and that of SUHIF is 94.06, 210.62, and 323.01. At night, SUHII of three patterns is 0.35 K, 0.93 K, and 1.82 K, and SUHIF is 28.84, 154.31, and 301.46. Cities labelled into the high pattern, of which intensity and frequency are found to be much higher than the average of them all, should attract more attention in both research and practice.

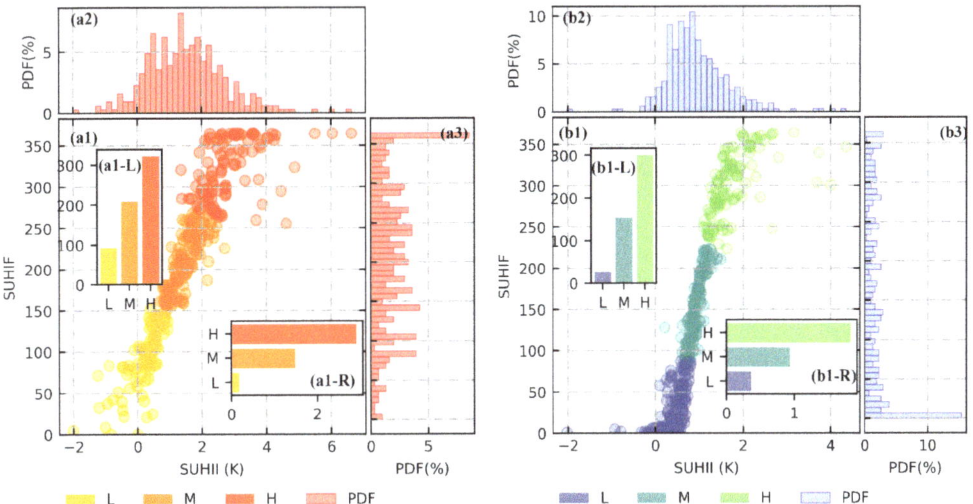

Figure 7. Relationship between SUHII and SUHIF at daytime (**a**) and nighttime (**b**). The scatter plots show their relationship, with color representation of three categories, namely the low (L), medium (M), and high (H) patterns. To clearly show the characteristics of these three categories, the bar charts show the average of SUHII (**a1-R,b1-R**) and SUHIF (**a1-L,b1-L**). The histograms represent the PDF of SUHII (**a2,b2**) and SUHIF (**a3,b3**).

Figure 8 shows the spatial distribution of these SUHI patterns in three aspects, namely the overall, latitudinal, and climate zone-based variations. On the whole, cities labelled as the high pattern at daytime are densely located in the south of China, while those of the low one are primarily located in the northern regions. At night, cities of the high pattern are distributed in the north, while the low ones are mainly in the southern regions. Although a distribution of the two extreme patterns differ to each other, cities of the medium pattern are scattered across China at both day and night. Only 86 cities are found to have consistency patterns during the whole day, and most of the cities accounting for about 71% are found to have a pattern transformation from day to night. This significant diurnal contrast can be possibly attributed to different driving mechanisms of UHI at day and night, which has been widely reported in previous studies [6,17,33,35]. Furthermore, latitudinal variations of patterns show that cities of the low pattern are mainly located in the 35–50° N zone at day, while those of the high ones are densely distributed from 25° to 35°. Apart from scatter distribution, cities of the medium pattern are further observed to have an aggregation of around 35° N based on the quantitative analysis of their latitudinal variations. At night, these three patterns are densely distributed in 45° N, 35° N, and 35° N, respectively. Moreover, almost all of cities grouped into the daytime high pattern are aggregated in the SM climate, according to the climate zone-based variations. Therefore, the climate characteristics of SM are closely associated with high intensity and frequency of SUHI. This close relation between SM and the high pattern of SUHI is possibly due to daytime SUHI being reported to be stronger in humid-hot regions than in cold-drier ones [6,7,10]. Conversely, there is no complete aggregation of any patterns in one climate zone at night.

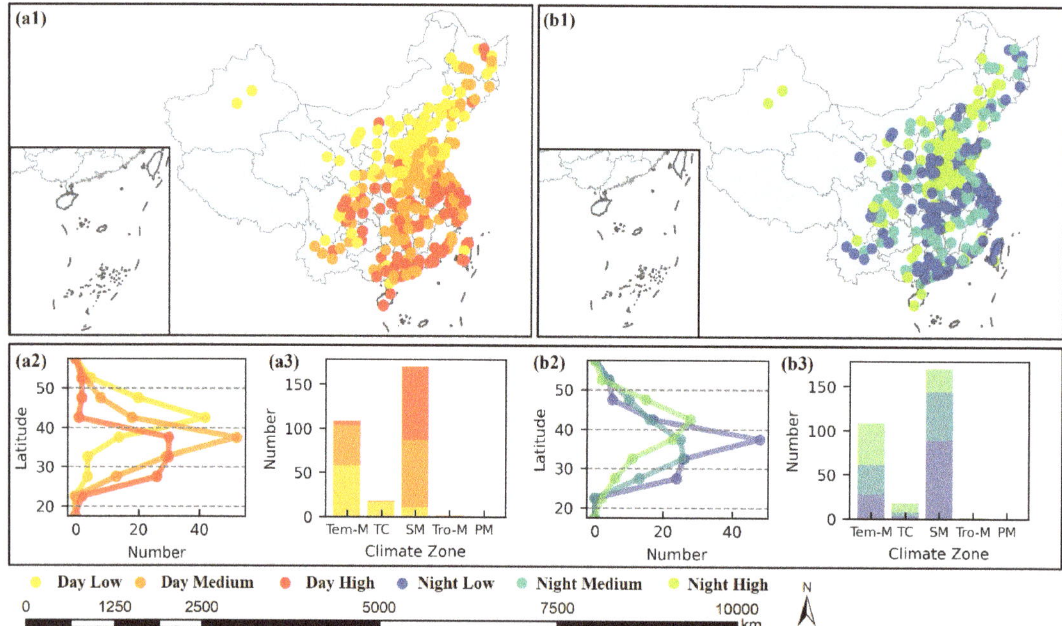

Figure 8. Spatial distribution of the aforementioned three categories at daytime (**a**) and nighttime (**b**). The maps show the spatial distribution of three pattern classifications (**a1,a2**). The line plots show their latitudinal variations (**a2,b2**). The bar plots show their climate zone-based variations (**a3,b3**). Different patterns are represented by colors.

4.4. Factor Analysis of SUHII and SUHIF

To compare the controls between SUHII and SUHIF, and to further explore pattern effects on controls, the factor analysis was conducted based on SUHII (Figure 9), SUHIF (Figure 10) and pattern classifications (Figure 11), respectively.

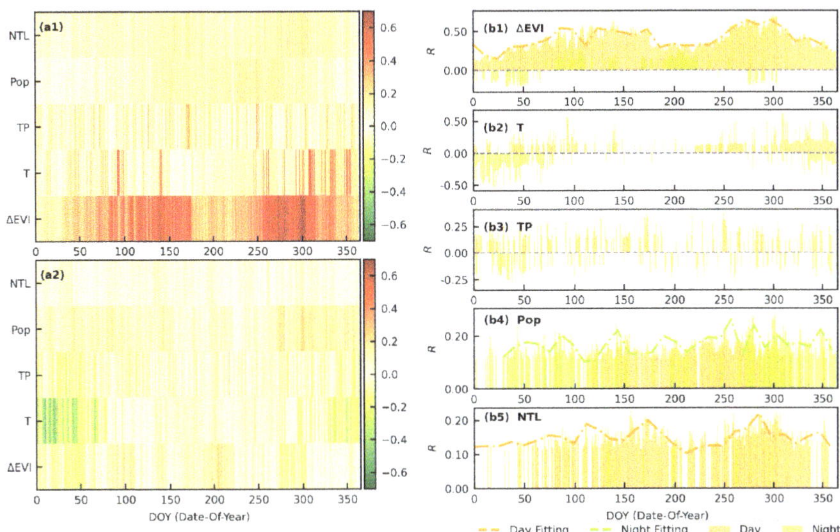

Figure 9. Relationships between SUHII and potential factors at a daily scale. The heat maps represent their correlation at daytime (**a1**) and nighttime (**a2**). The bar charts (**b**) show a diurnal contrast of their relationships in terms of each factor ((**b1**) for ΔEVI, (**b2**) for T, (**b3**) for TP, (**b4**) for Pop, (**b5**) for NTL). To make the results clear, the correlation with a *p*-value less than 0.05 is shown in the bar charts. The fitting lines are employed to clearly show the annual variations in their correlation, where days with *p*-values less than 0.05 are found to be over 270 during a year.

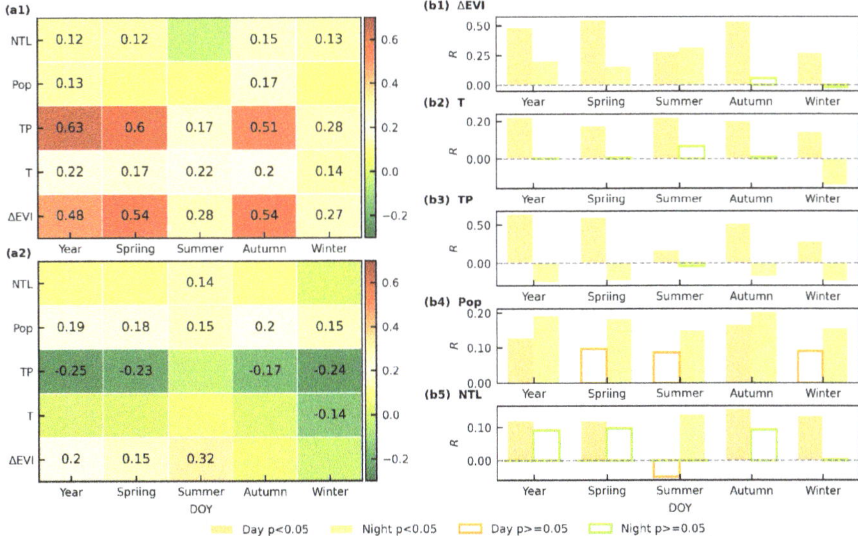

Figure 10. Relationship between SUHIF and potential factors in terms of annual and seasonal scales. The heat maps represent their correlation at daytime (**a1**) and nighttime (**a2**). The bar charts (**b**) show a diurnal contrast of their relationship in terms of each factor ((**b1**) for ΔEVI, (**b2**) for T, (**b3**) for TP, (**b4**) for Pop, (**b5**) for NTL). To make the results clear, the correlation with *p*-values less than 0.05 are labeled in the heat maps.

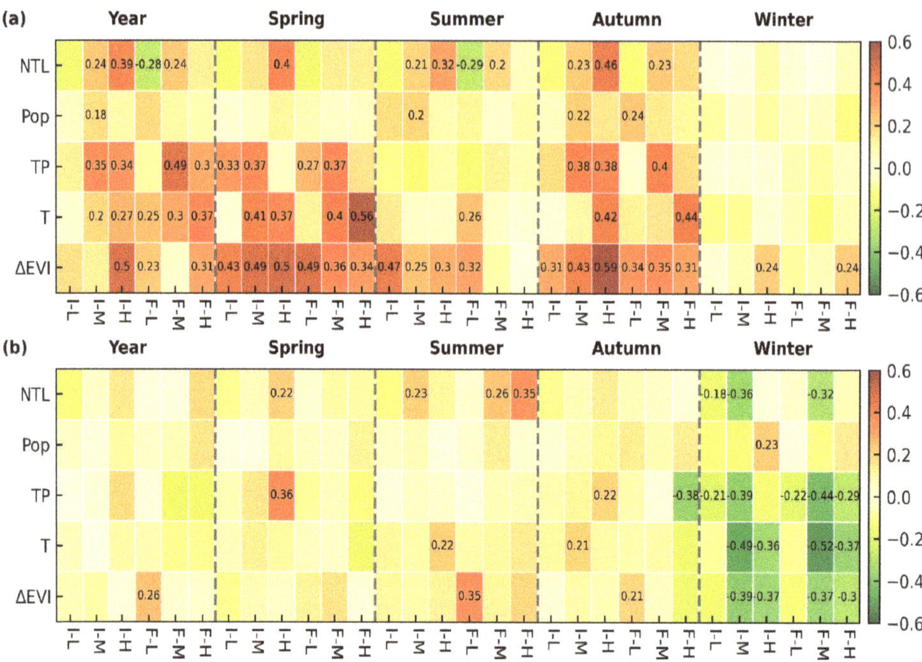

Figure 11. Comparisons of factor analysis based on SUHII and SUHIF in terms of pattern classifications at daytime (**a**) and nighttime (**b**). There are two temporal scales, namely annual and seasonal scales. I is the abbreviation of SUHII and F is the abbreviation of SUHIF. To make the results clear, the correlations with *p*-value less than 0.05 are labeled in the heat maps.

Figure 9 shows the correlation between SUHII and potential factors in terms of a daily scale. As great challenges on the reconstruction of daily seamless LST, few factor analyses are conducted on a daily scale. Thanks to the publication of daily all-weather LSTs, this paper can provide insights into daily-scale analysis of their relationship. At day, ΔEVI is found to have a significant correlation with daily SUHII. To make the EVI difference between urban and rural clear, ΔEVI is taken as its absolute value. Accordingly, a positive correlation in this paper is consistent with a negative one in previous studies [8,9]. Vegetation has been widely recognized to have an important cooling factor [6,8,34,54,55], owing to its evapotranspiration. Despite its significance on almost every day of the year (342 days), their correlation coefficients indicate a fluctuation with higher relations in the two transition seasons. This annual fluctuation of the effects from ΔEVI is also found in the studies of driving factors across highly populated cities across eastern China, as proposed by Zhou et al. [9] Temperature and total precipitation are taken as indicators for background climates across the selected cities, which have been reported as important controls for UHI [33,36]. A closer correlation between TP and daytime SUHII throughout the year compared to that in previous studies possibly results from all-weather rather than clear-sky LST employed in this paper. At night, populations in the selected cities indicates the continuous effects on SUHII. Chakraborty and Lee [8] reported that heat storage and anthropogenic heat emission (AHE) were closely related to nighttime UHI. A close association between population and AHE [56] explains the impacts from population on nighttime SUHII. Besides, an evident diurnal contrast from factor correlation is consistent with previous studies of UHI controls [6,9,51].

Figure 10 shows the correlation between SUHIF and potential factors at annual and seasonal scales. At daytime, ΔEVI and TP are found to be significantly correlated to SUHIF, in particular in spring and autumn. Since ΔEVI is an associated indicator for energy transfer between urban and rural environments, it influences not only SUHII, but also its frequency.

SUHIF is found to be closely correlated to precipitation at an annual scale. In previous studies, the temporal aggregation of clear-sky data was biased sampling with an ignorance of weather conditions. Here, SUHIF, on the basis of all-weather rather than clear-sky LST, makes it possible to analyze the relationship between it and precipitation. Manoli et al. [36] reported that precipitation influences the UHI based on changes in evapotranspiration and convection efficiency. This close relationship suggests that precipitation makes a strong contribution not only to the magnitude, but also to the frequency. There are two reasons why the effects from ΔEVI and precipitation are more significant in the transition seasons than in summer and winter. First, saturation of daytime SUHIF at summer across most of the selected cities results in a weaker correlation with continuous variables like ΔEVI and precipitation. In winter, a dense aggregation of around zero also influences the factor analysis. Secondly, there are other potential factors that contribute to variations in SUHIF. At night, the relationships between SUHIF and these factors is observed to have less seasonal variations than daytime ones. This is possibly as a result of the non-significant seasonal variations of SUHIF at nighttime. Population is related to daytime SUHIF throughout the year, whereas NTL is correlated to the nighttime one.

After exploring controls for SUHII and SUHIF respectively, comparisons on factor analysis were conducted on the basis of pattern classification at annual and seasonal scales (Figure 11). Two objectives are a comparisons on the controls of SUHII and SUHIF, exploration on temporal variations and pattern variations of factor analysis. At daytime, ΔEVI and TP indicate a significant correlation to both intensity and frequency at an annual scale, while population is found to be presumably irrelevant to their variations. ΔEVI and TP are widely reported as important factors for explaining spatial and temporal variations of SUHI [6,9,36]. In spite of the significant correlation at the annual scale, their impacts are observed to have variations between SUHII and SUHIF among different patterns. On an annual scale, ΔEVI indicates a closer relationship to SUHII, while temperature and precipitation indicate a closer relationship to SUHIF. The frequency is estimated on the basis of daily all-weather LST, which receives considerable impacts from weather conditions. In the cities classified into the high pattern, vegetation is found to be an important factor for daytime SUHII, whereas a much lesser correlation is found in the low and medium patterns. At night, the weak explanation of the selected factors suggest the necessity of a further exploration into the driving factors for nighttime SUHII and SUHIF. Additionally, the factor analysis shows evident seasonal variations. At daytime, there are greater impacts from three important factors (ΔEVI, T and TP) on both SUHII and SUHIF in spring and autumn, rather than during the extreme seasons. In particular, the cities labelled as the high patterns in summer, possibly coupled with heat waves to pose great threats on residential health [57], needs a more comprehensive factor analysis for further mitigations.

5. Conclusions

Great efforts have been devoted to the study of spatiotemporal variations and driving mechanisms of SUHI from the local to the global scale, due to its close association with environmental issues and human health. However, studies on SUHIF are largely challenged by a great number of missing LSTs on a daily scale. There are three issues concerning SUHIF which remain largely unknown, namely its spatiotemporal variations, pattern classifications, and driving factors. Additionally, it is unclear whether SUHII and SUHIF are consistent or different to each other in terms of their variations and controls. Thanks to the release of daily all-weather LSTs across China, this paper is allowed to conduct comparisons between SUHII and SUHIF, to conduct a pattern classification and to further explore their controls.

There are four major findings, which are summarized as follows. First, SUHIF indicates a different north–south contrast between day and night. Cities within the latitude of 20° N to 40° N should be have more attention paid to them, since they are found to have a strong intensity and a high frequency during the day. Despite the overall consistency of spatial distribution, discrepancy between SUHII and SUHIF is observed from the climate zone-based variations, especially at night. Second, SUHIF shows a significant diurnal and seasonal contrast. The difference in SUHIF between day and night is 65.61, while for SUHII it is 0.58 K. Summer daytime is found to have the highest intensity and frequency. Third, the selected 305 Chinese cities are grouped into three patterns based on K-means clustering. After comparative analysis, SUHIF is possibly consistent with SUHII in the same city. Despite the generally similar spatiotemporal patterns between SUHII and SUHIF, they provide quantitative and temporal characteristics, respectively. Fourth, impacts from driving factors on SUHII and SUHIF are found to have significant variations among different times and patterns. Daytime SUHIF is found to be closely associated with precipitation in spring and autumn.

In summer, strong urban heat island and heat waves are reported as great contributors for high heat-related mortality. Therefore, cities classified into the high pattern with both strong intensity and high frequency during summer should be taken as a research emphasis in future exploration, to propose targeting mitigation measurements.

Author Contributions: Conceptualization, K.L. and Y.C.; Methodology, K.L.; Formal Analysis, K.L.; Investigation, K.L.; Data K.L. and S.G.; Writing—Original Draft Preparation, K.L.; Writing—Review and Editing, K.L.; Visualization, K.L.; Supervision, Y.C.; Project Administration, Y.C.; Funding Acquisition, Y.C. All authors have read and agreed to the published version of the manuscript.

Funding: This work was supported by Beijing Natural Science Foundation (8192025), National Key R&D Program of China (2017YFC1502406), Projects of Beijing Advanced Innovation Center for Future Urban Design, Beijing University of Civil Engineering and Architecture (UDC2019031321), and in part by the Beijing Laboratory of Water Resources Security and Beijing Key Laboratory of Environmental Remote Sensing and Digital Cities.

Data Availability Statement: Not applicable.

Acknowledgments: The authors would like to thank Center for Geodata and Analysis, Faculty of Geographical Science, and Beijing Normal University for the high-performance computing support.

Conflicts of Interest: The authors declare that they have no known competing financial interests or personal relationships that could have appeared to influence the work reported in this paper.

References

1. United Nations Population Division. *The World's Cities in 2018 Data Booklet*; United Nations Population Division: New York, NY, USA, 2018.
2. Grimm, N.B.; Faeth, S.H.; Golubiewski, N.E.; Redman, C.L.; Wu, J.; Bai, X.; Briggs, J.M. Global Change and the Ecology of Cities. *Science* **2008**, *319*, 756–760. [CrossRef]
3. Li, Y.; Schubert, S.; Kropp, J.P.; Rybski, D. On the influence of density and morphology on the Urban Heat Island intensity. *Nat. Commun.* **2020**, *11*, 2647. [CrossRef]
4. Oke, T.R. The urban energy balance. *Prog. Phys. Geogr. Earth Environ.* **1988**, *12*, 471–508. [CrossRef]
5. Zhou, D.; Xiao, J.; Bonafoni, S.; Berger, C.; Deilami, K.; Zhou, Y.; Frolking, S.; Yao, R.; Qiao, Z.; Sobrino, J.A. Satellite Remote Sensing of Surface Urban Heat Islands: Progress, Challenges, and Perspectives. *Remote Sens.* **2018**, *11*, 48. [CrossRef]
6. Peng, S.; Piao, S.; Ciais, P.; Friedlingstein, P.; Ottle, C.; Bréon, F.-M.; Nan, H.; Zhou, L.; Myneni, R. Surface Urban Heat Island Across 419 Global Big Cities. *Environ. Sci. Technol.* **2011**, *46*, 696–703. [CrossRef] [PubMed]
7. Imhoff, M.L.; Zhang, P.; Wolfe, R.; Bounoua, L. Remote sensing of the urban heat island effect across biomes in the continental USA. *Remote Sens. Environ.* **2010**, *114*, 504–513. [CrossRef]
8. Chakraborty, T.; Lee, X. A simplified urban-extent algorithm to characterize surface urban heat islands on a global scale and examine vegetation control on their spatiotemporal variability. *Int. J. Appl. Earth Obs. Geoinf.* **2019**, *74*, 269–280. [CrossRef]
9. Zhou, D.; Bonafoni, S.; Zhang, L.; Wang, R. Remote sensing of the urban heat island effect in a highly populated urban agglomeration area in East China. *Sci. Total Environ.* **2018**, *628–629*, 415–429. [CrossRef] [PubMed]
10. Peng, J.; Ma, J.; Liu, Q.; Liu, Y.; Hu, Y.; Li, Y.; Yue, Y. Spatial-temporal change of land surface temperature across 285 cities in China: An urban-rural contrast perspective. *Sci. Total Environ.* **2018**, *635*, 487–497. [CrossRef]
11. Yang, Q.; Huang, X.; Tang, Q. The footprint of urban heat island effect in 302 Chinese cities: Temporal trends and associated factors. *Sci. Total Environ.* **2019**, *655*, 652–662. [CrossRef]
12. Santamouris, M. Analyzing the heat island magnitude and characteristics in one hundred Asian and Australian cities and regions. *Sci. Total Environ.* **2015**, *512–513*, 582–598. [CrossRef]
13. Zhou, D.; Zhang, L.; Hao, L.; Sun, G.; Liu, Y.; Zhu, C. Spatiotemporal trends of urban heat island effect along the urban development intensity gradient in China. *Sci. Total Environ.* **2016**, *544*, 617–626. [CrossRef]

14. Azevedo, J.A.; Chapman, L.; Muller, C.L. Quantifying the Daytime and Night-Time Urban Heat Island in Birmingham, UK: A Comparison of Satellite Derived Land Surface Temperature and High Resolution Air Temperature Observations. *Remote Sens.* **2016**, *8*, 153. [CrossRef]

15. Fabrizi, R.; Bonafoni, S.; Biondi, R. Satellite and Ground-Based Sensors for the Urban Heat Island Analysis in the City of Rome. *Remote Sens.* **2010**, *2*, 1400–1415. [CrossRef]

16. Huang, W.; Li, J.; Guo, Q.; Mansaray, L.R.; Li, X.; Huang, J. A Satellite-Derived Climatological Analysis of Urban Heat Island over Shanghai during 2000–2013. *Remote Sens.* **2017**, *9*, 641. [CrossRef]

17. Cao, C.; Lee, X.; Liu, S.; Schultz, N.; Xiao, W.; Zhang, M.; Zhao, L. Urban heat islands in China enhanced by haze pollution. *Nat. Commun.* **2016**, *7*, 12509. [CrossRef]

18. Čeplová, N.; Kalusová, V.; Lososová, Z. Effects of settlement size, urban heat island and habitat type on urban plant biodiversity. *Landsc. Urban Plan.* **2017**, *159*, 15–22. [CrossRef]

19. Santamouris, M. Cooling the cities—A review of reflective and green roof mitigation technologies to fight heat island and improve comfort in urban environments. *Sol. Energy* **2014**, *103*, 682–703. [CrossRef]

20. Li, X.; Zhou, Y.; Yu, S.; Jia, G.; Li, H.; Li, W. Urban heat island impacts on building energy consumption: A review of approaches and findings. *Energy* **2019**, *174*, 407–419. [CrossRef]

21. Wong, K.; Paddon, A.; Jimenez, A. Review of World Urban Heat Islands: Many Linked to Increased Mortality. *J. Energy Resour. Technol.* **2013**, *135*, 022101. [CrossRef]

22. Van Hove, L.; Jacobs, C.; Heusinkveld, B.; Elbers, J.; van Driel, B.; Holtslag, B. Temporal and spatial variability of urban heat island and thermal comfort within the Rotterdam agglomeration. *Build. Environ.* **2015**, *83*, 91–103. [CrossRef]

23. Dong, J.; Peng, J.; He, X.; Corcoran, J.; Qiu, S.; Wang, X. Heatwave-induced human health risk assessment in megacities based on heat stress-social vulnerabil-ity-human exposure framework. *Landsc. Urban Plan.* **2020**, *203*, 103907. [CrossRef]

24. Oke, T.R. The energetic basis of the urban heat island. *Q. J. R. Meteorol. Soc.* **1982**, *108*, 1–24. [CrossRef]

25. Voogt, J.A.; Oke, T.R. Thermal remote sensing of urban climates. *Remote Sens. Environ.* **2003**, *86*, 370–384. [CrossRef]

26. Li, X.; Zhou, X.; Asrar, G.R.; Imhoff, M.; Li, X. The surface urban heat island response to urban expansion: A panel analysis for the conterminous United States. *Sci. Total Environ.* **2017**, *605–606*, 426–435. [CrossRef] [PubMed]

27. Rao, P.K. Remote sensing of urban "heat islands" from an environmental satellite. *B. Am. Meteorol. Soc.* **1972**, *53*, 647–648.

28. Streutker, D. Satellite-measured growth of the urban heat island of Houston, Texas. *Remote Sens. Environ.* **2003**, *85*, 282–289. [CrossRef]

29. Weng, Q.; Lu, D.; Schubring, J. Estimation of land surface temperature–vegetation abundance relationship for urban heat island studies. *Remote Sens. Environ.* **2004**, *89*, 467–483. [CrossRef]

30. Weng, Q. Thermal infrared remote sensing for urban climate and environmental studies: Methods, applications, and trends. *ISPRS J. Photogramm. Remote Sens.* **2009**, *64*, 335–344. [CrossRef]

31. Schwarz, N.; Lautenbach, S.; Seppelt, R. Exploring indicators for quantifying surface urban heat islands of European cities with MODIS land surface temperatures. *Remote Sens. Environ.* **2011**, *115*, 3175–3186. [CrossRef]

32. Quan, J.; Chen, Y.; Zhan, W.; Wang, J.; Voogt, J.; Wang, M. Multi-temporal trajectory of the urban heat island centroid in Beijing, China based on a Gaussian volume model. *Remote Sens. Environ.* **2014**, *149*, 33–46. [CrossRef]

33. Zhao, L.; Lee, X.; Smith, R.B.; Oleson, K. Strong contributions of local background climate to urban heat islands. *Nat. Cell Biol.* **2014**, *511*, 216–219. [CrossRef]

34. Zhou, D.; Zhao, S.; Liu, S.; Zhang, L.; Zhu, C. Surface urban heat island in China's 32 major cities: Spatial patterns and drivers. *Remote Sens. Environ.* **2014**, *152*, 51–61. [CrossRef]

35. Lai, J.; Zhan, W.; Huang, F.; Voogt, J.; Bechtel, B.; Allen, M.; Peng, S.; Hong, F.; Liu, Y.; Du, P. Identification of typical diurnal patterns for clear-sky climatology of surface urban heat islands. *Remote Sens. Environ.* **2018**, *217*, 203–220. [CrossRef]

36. Manoli, G.; Fatichi, S.; Schläpfer, M.; Yu, K.; Crowther, T.W.; Meili, N.; Burlando, P.; Katul, G.; Bou-Zeid, E. Magnitude of urban heat islands largely explained by climate and population. *Nat. Cell Biol.* **2019**, *573*, 55–60. [CrossRef] [PubMed]

37. Li, X.; Zhou, Y.; Asrar, G.R.; Zhu, Z. Creating a seamless 1 km resolution daily land surface temperature dataset for urban and surrounding areas in the conterminous United States. *Remote Sens. Environ.* **2018**, *206*, 84–97. [CrossRef]

38. Liu, Z.; Zhan, W.; Lai, J.; Hong, F.; Quan, J.; Bechtel, B.; Huang, F.; Zou, Z. Balancing prediction accuracy and generalization ability: A hybrid framework for modelling the annual dynamics of satellite-derived land surface temperatures. *ISPRS J. Photogramm. Remote Sens.* **2019**, *151*, 189–206. [CrossRef]

39. Zhang, X.; Zhou, J.; Liang, S.; Wang, D. A practical reanalysis data and thermal infrared remote sensing data merging (RTM) method for reconstruction of a 1-km all-weather land surface temperature. *Remote Sens. Environ.* **2021**, *260*, 112437. [CrossRef]

40. Zhang, X.; Zhou, J.; Gottsche, F.-M.; Zhan, W.; Liu, S.; Cao, R. A Method Based on Temporal Component Decomposition for Estimating 1-km All-Weather Land Surface Temperature by Merging Satellite Thermal Infrared and Passive Microwave Observations. *IEEE Trans. Geosci. Remote Sens.* **2019**, *57*, 4670–4691. [CrossRef]

41. Li, K.; Chen, Y.; Xia, H.; Gong, A.; Guo, Z. Adjustment From Temperature Annual Dynamics for Reconstructing Land Surface Temperature Based on Downscaled Microwave Observations. *IEEE J. Sel. Top. Appl. Earth Obs. Remote Sens.* **2020**, *13*, 5272–5283. [CrossRef]

42. Duan, S.-B.; Li, Z.-L.; Leng, P. A framework for the retrieval of all-weather land surface temperature at a high spatial resolution from polar-orbiting thermal infrared and passive microwave data. *Remote Sens. Environ.* **2017**, *195*, 107–117. [CrossRef]

43. Kou, X.; Jiang, L.; Bo, Y.; Yan, S.; Chai, L. Estimation of Land Surface Temperature through Blending MODIS and AMSR-E Data with the Bayesian Maximum Entropy Method. *Remote Sens.* **2016**, *8*, 105. [CrossRef]
44. Sun, L.; Chen, Z.; Gao, F.; Anderson, M.; Song, L.; Wang, L.; Hu, B.; Yang, Y. Reconstructing daily clear-sky land surface temperature for cloudy regions from MODIS data. *Comput. Geosci.* **2017**, *105*, 10–20. [CrossRef]
45. Xia, H.; Chen, Y.; Gong, A.; Li, K.; Liang, L.; Guo, Z. Modeling Daily Temperatures Via a Phenology-Based Annual Temperature Cycle Model. *IEEE J. Sel. Top. Appl. Earth Obs. Remote Sens.* **2021**, *14*, 6219–6229. [CrossRef]
46. Li, K.; Chen, Y.; Wang, M.; Gong, A. Spatial-temporal variations of surface urban heat island intensity induced by different definitions of rural extents in China. *Sci. Total Environ.* **2019**, *669*, 229–247. [CrossRef]
47. Wenbin, T.; Zhou, J.; Zhang, X.; Zhang, X.; Ma, J.; Ding, L. *Daily 1-km All-Weather Land Surface Temperature Dataset for the Chinese Landmass and Its Surrounding Areas (TRIMS LST; 2000-2020)*; National Tibetan Plateau Data Center: Beijing, China, 2021.
48. Zhou, J.; Zhang, X.; Zhan, W.; Goettsche, F.-M.; Liu, S.; Olesen, F.-S.; Hu, W.; Dai, F. A Thermal Sampling Depth Correction Method for Land Surface Temperature Estimation from Satellite Passive Microwave Observation Over Barren Land. *IEEE Trans. Geosci. Remote Sens.* **2017**, *55*, 4743–4756. [CrossRef]
49. Lai, J.; Zhan, W.; Huang, F.; Quan, J.; Hu, L.; Gao, L.; Ju, W. Does quality control matter? Surface urban heat island intensity variations estimated by satellite-derived land surface temperature products. *ISPRS J. Photogramm. Remote Sens.* **2018**, *139*, 212–227. [CrossRef]
50. Likas, A.; Vlassis, N.; Verbeek, J.J. The global k-means clustering algorithm. *Pattern Recognit.* **2003**, *36*, 451–461. [CrossRef]
51. Li, L.; Zha, Y.; Zhang, J. Spatially non-stationary effect of underlying driving factors on surface urban heat islands in global major cities. *Int. J. Appl. Earth Obs. Geoinf.* **2020**, *90*, 102131. [CrossRef]
52. Deilami, K.; Kamruzzaman, M.; Liu, Y. Urban heat island effect: A systematic review of spatio-temporal factors, data, methods, and mitigation measures. *Int. J. Appl. Earth Obs. Geoinf.* **2018**, *67*, 30–42. [CrossRef]
53. Tan, J.; Zheng, Y.; Tang, X.; Guo, C.; Li, L.; Song, G.; Zhen, X.; Yuan, D.; Kalkstein, A.J.; Li, F.; et al. The urban heat island and its impact on heat waves and human health in Shanghai. *Int. J. Biometeorol.* **2009**, *54*, 75–84. [CrossRef] [PubMed]
54. Clinton, N.; Gong, P. MODIS detected surface urban heat islands and sinks: Global locations and controls. *Remote Sens. Environ.* **2013**, *134*, 294–304. [CrossRef]
55. Zhou, W.; Wang, J.; Cadenasso, M.L. Effects of the spatial configuration of trees on urban heat mitigation: A comparative study. *Remote Sens. Environ.* **2017**, *195*, 1–12. [CrossRef]
56. Flanner, M.G. Integrating anthropogenic heat flux with global climate models. *Geophys. Res. Lett.* **2009**, *36*, 36. [CrossRef]
57. Founda, D.; Santamouris, M. Synergies between Urban Heat Island and Heat Waves in Athens (Greece), during an extremely hot summer (2012). *Sci. Rep.* **2017**, *7*, 10973. [CrossRef]

remote sensing

MDPI

Article

Quantitatively Assessing the Impact of Driving Factors on Vegetation Cover Change in China's 32 Major Cities

Baohui Mu [1,2], Xiang Zhao [1,2,*], Jiacheng Zhao [1,2], Naijing Liu [1,2], Longping Si [1,2], Qian Wang [1,2], Na Sun [1,2], Mengmeng Sun [1,2], Yinkun Guo [1,2] and Siqing Zhao [1,2]

[1] State Key Laboratory of Remote Sensing Science, Jointly Sponsored by Beijing Normal University and Aerospace Information Research Institute of Chinese Academy of Sciences, Faculty of Geographical Science, Beijing Normal University, Beijing 100875, China; mubaohui@mail.bnu.edu.cn (B.M.); zhaojiacheng@mail.bnu.edu.cn (J.Z.); liunj@mail.bnu.edu.cn (N.L.); silongp@bnu.edu.cn (L.S.); qianwang@bnu.edu.cn (Q.W.); sunna_bnu@mail.bnu.edu.cn (N.S.); smm2021@mail.bnu.edu.cn (M.S.); guoyinkun_bnu@mail.bnu.edu.cn (Y.G.); zhaosiqing@mail.bnu.edu.cn (S.Z.)

[2] Beijing Engineering Research Center for Global Land Remote Sensing Products, Institute of Remote Sensing, Science and Engineering, Faculty of Geographical Science, Beijing Normal University, Beijing 100875, China

* Correspondence: zhaoxiang@bnu.edu.cn; Tel.: +86-010-5880-0181

Citation: Mu, B.; Zhao, X.; Zhao, J.; Liu, N.; Si, L.; Wang, Q.; Sun, N.; Sun, M.; Guo, Y.; Zhao, S. Quantitatively Assessing the Impact of Driving Factors on Vegetation Cover Change in China's 32 Major Cities. *Remote Sens.* **2022**, *14*, 839. https://doi.org/10.3390/rs14040839

Academic Editor: Elhadi Adam

Received: 16 December 2021
Accepted: 7 February 2022
Published: 10 February 2022

Publisher's Note: MDPI stays neutral with regard to jurisdictional claims in published maps and institutional affiliations.

Abstract: After 2000, China's vegetation underwent great changes associated with climate change and urbanization. Although many studies have been conducted to quantify the contributions of climate and human activities to vegetation, few studies have quantitatively examined the comprehensive contributions of climate, urbanization, and CO_2 to vegetation in China's 32 major cities. In this study, using Global Land Surface Satellite (GLASS) fractional vegetation cover (FVC) between 2001 and 2018, we investigated the trend of FVC in China's 32 major cities and quantified the effects of CO_2, urbanization, and climate by using generalized linear models (GLMs). We found the following: (1) From 2001 to 2018, the FVC in China generally illustrated an increasing trend, although it decreased in 23 and 21 cities in the core area and expansion area, respectively. (2) Night light data showed that the urban expansion increased to varying degrees, with an average increasing ratio of approximately 168%. The artificial surface area increased significantly, mainly from cropland, forest, grassland, and tundra. (3) Climate factors and CO_2 were the major factors that affected FVC change. The average contributions of climate factors, CO_2, and urbanization were 40.6%, 39.2%, and 10.6%, respectively. This study enriched the understanding of vegetation cover change and its influencing factors, helped to explain the complex biophysical mechanism between vegetation and environment, and guided sustainable urban development.

Keywords: fractional vegetation cover; urbanization; climate change; vegetation change

1. Introduction

Urban vegetation plays an important role in human life and environmental regulation in cities [1–4]. As an important part of the urban ecosystem, urban vegetation is the main producer of the city, participating in regulation of climate change, altering energy and matter exchange between the surface and atmosphere, and promoting complex biogeochemical cycles [5–8]. In addition, the effect of urban vegetation on the beautification and purification of the urban environment has also been concerned [9–11]. Therefore, studying the long-term dynamic change of urban vegetation and its driving factors can provide theoretical support for the protection of urban ecological environment.

Climate factors are the main drivers of vegetation change and thus have been a popular research topic. At present, many scholars have studied the effect of climate on vegetation and obtained some similar conclusions. Increased precipitation promotes photosynthesis and improves the absorption and transport of soil nutrients. However, excessive precipitation inhibits vegetation transpiration. Decreased precipitation indirectly affects vegetation activities by regulating hydrothermal conditions, thus leading to the weakening of photosynthesis and the reduction of organic yield [12–14]. Temperature change promotes photosynthesis and accelerates the release of soil nutrients in the region suppressed by temperature. At the same time, accelerated soil water loss, weakened photosynthesis, and enhanced respiration result in dry matter consumption [15–17]. As the energy source of vegetation photosynthesis, solar radiation is also an important factor affecting vegetation growth [18].

In addition to climate factors, the effect of CO_2 on vegetation growth cannot be ignored. Plants can use CO_2 to produce organic material through photosynthesis to build plant tissue [19]. Therefore, increasing CO_2 concentration will affect plant growth [20]. CO_2 has a different effect on vegetation outside and inside the city. Outside the city, increased CO_2 concentrations affect vegetation growth by speeding up carboxylation in photosynthesis [21,22]. Inside the city, increased CO_2 concentrations cause the greenhouse effect, which leads to increased temperature and accelerated soil water evaporation. Under such conditions, the growth of vegetation is mainly affected by soil water [23,24].

It was worth mentioning that urbanization has also gradually become an important factor affecting vegetation growth [25–27]. Urbanization is a phenomenon that involves simultaneous changes in the population, economy, and land use patterns [28], and can often be measured using land cover data [29] and nighttime light data [30]. Land cover data express urbanization by calculating the change area of impervious surface [29,31]. However, the spatial resolution of land cover products with a long time series time resolution of one year is coarse and the information expressed is limited. Nighttime light data represent urbanization by measuring the night light of cities, towns, and other continuously lit areas, and can be an explanatory indicator for estimating urbanization dynamics [32]. From 2001 to 2018, China experienced intense urban development and rapid land consumption, which put great pressure on urban ecosystem functions [33,34]. The effects of urbanization on vegetation growth are complex. Urbanization can not only directly affect vegetation growth by promoting land cover change [35], but also indirectly affect vegetation growth by increasing the impervious layer area. The principle of the latter is that increases in impervious surfaces reduce the latent heat flux and increase the sensible heat flux, thus leading to a change in temperature and evapotranspiration processes. Such changes indirectly promote or inhibit vegetation growth [36,37]. Therefore, accurate knowledge of the complex nonlinear relationship between urbanization and vegetation can help enhance the understanding of vegetation changes under urbanization and could be essential for formulating environmental protection strategies in cities.

Although many studies have focused on the response relationship between urban vegetation and the environment, the influence of the drivers of long-term vegetation change in multi-urban areas has been limited [38,39]. At present, scholars have carried out research on the change in vegetation coverage in some cities and the driving factors [38–40]. Nevertheless, because few cities have been investigated, such work can only reflect the vegetation driving forces of individual cities; thus, a macroscopic analysis of China as a whole has not been performed [41,42]. In addition, we also noticed that vegetation growth is comprehensively affected by climate and human factors; however, few studies have considered the comprehensive impact of climate factors, CO_2, and urbanization on vegetation [43–45]. Equally important, coarse-resolution land cover data have difficulty expressing detailed urbanization information because the land cover type data in cities have remained unchanged for many years [36].

In this study, we analyzed the effect of climate, urbanization, and CO_2 on vegetation in 32 major cities of China using Global Land Surface Satellite (GLASS) fractional vegetation cover (FVC) in conjunction with climate data from the National Tibetan Plateau Data Center, CO_2 from the National Cryosphere Desert Data Center, and nighttime light (NTL) data between 2001 and 2018. Our main objectives were to investigate the following: (1) spatiotemporal variation in FVC in China's 32 major cities from 2001 to 2018; (2) differences in FVC variations between urban core areas and urban expanded areas; and (3) relative contributions of climate-related factors, CO_2, and urbanization to FVC dynamics.

2. Materials and Methods

2.1. Study Area

China has a vast territory and abundant resources. The terrain is high in the East and low in the West. China is divided into three steps according to altitude (Figure 1). The first ladder is mainly distributed in the vicinity of the Qinghai-Tibet Plateau at an altitude of more than 4000 m. Under the influence of the southwest monsoon, the water content decreases sharply, and the precipitation is generally less than 150 mm, which decreases spatially from southeast to northwest. The average annual temperature is below zero. The average altitude of the second step is 1000–2000 m, and the local terrain highly fluctuates. The precipitation is mainly between 400 and 1000 mm. The cities in the region include Hohhot, Yinchuan, Xining, Lanzhou, Lhasa, Guiyang, and Kunming. The temperatures range between 4 °C and 15 °C, and the radiation ranges between 170 W·m^{-2} and 190 W·m^{-2}. The third step is the lowest step at an elevation of less than 500 m, and it has annual precipitation of more than 1000 mm, although it can reach more than 6000 mm in some areas. The overall trend decreases from southeast to northwest. The cities in the region include 25 major cities, such as Beijing and Tianjin. The city with the lowest temperature and radiation is Harbin, and the city with the highest temperature is Haikou (Table 1).

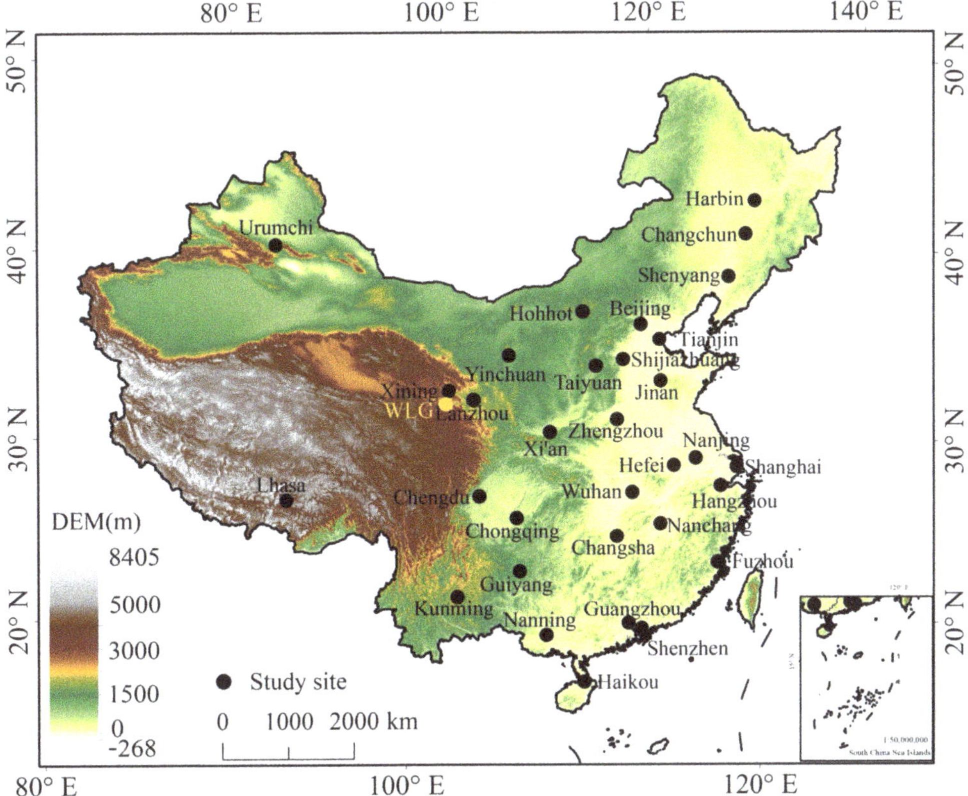

Figure 1. Positions of the 32 major cities in China, with black dots representing the locations of cities and yellow dot representing the locations of Waliguan (WLG) stations.

Table 1. Average temperature, altitude, precipitation, and radiation of 32 major cities in China.

Types / City	DEM (m)		Precipitation (mm)		Temperature (°C)		Radiation (W·m^{-2})	
	C	E	C	E	C	E	C	E
Harbin	143.2	132.6	563.3	551.6	4.2	4.7	139.7	140.1
Changchun	214.3	211.8	668.9	656.6	5.	5.9	149.5	149.5
Urumchi	823.7	800.3	236.9	239.6	6.2	6.5	177.6	177.4
Shenyang	45.6	47.4	425.9	431.8	7.9	7.9	156.8	157.2
Hohhot	1053.8	1057.3	591.3	591.2	6.9	6.9	169.3	169.6
Beijing	46.6	46.1	504.2	514.1	12.6	12.2	161.0	160.9
Tianjin	5.5	3.5	562.7	580.6	12.9	12.9	162.0	163.9
Yinchuan	1113.5	1110.3	270.2	274.0	9.9	9.9	186.1	185.1
Shijiazhuang	76.5	80.7	431.6	439.3	13.9	13.7	152.2	152.5
Taiyuan	799.3	802.3	363.3	358.5	10.3	10.1	170.1	169.4
Jinan	46.2	77.8	767.9	800.9	14.9	14.7	166.9	166.3
Xining	2261.0	2297.5	485.8	484.9	4.2	4.4	185.0	185.5
Lanzhou	1541.9	1569.8	632.1	640.5	7.5	6.7	181.2	180.8
Zhengzhou	104.2	108.1	598.8	593.4	15.5	15.5	156.4	156.3
Xi'an	411.8	404.6	474.9	473.4	14.9	14.8	159.7	159.7
Nanjing	23.0	18.3	1281.9	1292.1	15.9	15.9	160.1	159.3
Hefei	26.0	27.4	1366.2	1384.6	15.9	15.9	150.1	150.2
Shanghai	5.6	4.6	1214.5	1347.3	15.6	17.5	154.4	155.0
Chengdu	498.8	505.7	1117.2	1136.7	15.9	15.9	134.9	134.2
Wuhan	27.6	27.8	1029.9	1055.3	17.6	17.4	148.8	148.5
Hangzhou	15.0	14.9	1735.1	1732.2	17.8	17.4	154.2	154.1
Lhasa	3655.5	3653.2	569.4	566.8	8.6	8.5	227.0	226.9
Chongqing	265.7	275.8	1069.2	1068.6	18.4	18.5	129.9	129.9
Nanchang	23.9	26.9	1564.7	1566.6	18.6	18.5	155.7	155.4
Changsha	53.2	55.5	1275.2	1275.1	17.0	17.1	148.6	148.9
Guiyang	1111.2	1196.4	1180.1	1173.4	14.8	14.4	129.4	129.7
Fuzhou	14.2	24.6	1334.3	1325.9	20.9	20.3	153.3	153.5
Kunming	1897.8	1916.6	1169.8	1246.5	14.9	14.8	188.6	187.1
Guizhou	14.1	13.9	1764.5	1783.4	21.9	21.8	154.8	153.8
Nanning	83.6	97.9	1244.9	1245.1	20.9	20.9	160.0	159.8
Shenzhen	57.0	68.8	1929.3	1869.5	22.4	22.0	170.8	169.5
Haikou	16.5	17.9	2166.7	2036.3	23.9	23.9	184.7	182.1

2.2. Data Sources

GLASS FVC products: In this study, we used GLASS FVC from 2001 to 2018 produced by Beijing Normal University. The dataset processing combined Tang et al.'s MODIS reflectance data preprocessing method and machine learning method [46]. The dataset has a time resolution of 8 days and a spatial resolution of 500 m [47–49]. Due to the high stability of annual mean FVC, it is suitable for large-scale data research, and this paper synthesized the annual mean FVC using the data of 46 scenes throughout the year [20].

NTL products: In this study, we used the harmonized global NTL time series data from 2001 to 2018. The dataset includes the stepwise calibrated stable DMSP NTL observations from 2001 to 2013, and the simulated DMSP-like DNs from the VIIRS radiance data (2014–2018). The temporal resolution of the dataset was one year, and the spatial resolution was 1000 m [50]. It measures lights from cities, towns, and other continuous lighting areas at night, and can be an explanatory indicator for estimating urbanization dynamics [32].

Climate products: The first high-resolution meteorological forcing dataset for land process studies over China, which was produced by the Yangkun team from 2001 to 2018, was used in our research. The data were subjected to rigorous data quality control. A meteorological dataset covering the China was constructed using station data, satellite data, and means of convergence in the analysis of the data. These data had a spatial resolution of 0.1° and a temporal resolution of 3 h. Currently, this Chinese regional high-resolution meteorologically driven dataset has been released at the National Tibetan Plateau

Science Center. The data provider also freely provides synthesized annual average climate data [51,52].

CO$_2$ products: CO$_2$ data obtained from the National Cryosphere Desert Data Center and China Greenhouse Gas Bulletin were used in this study. The data recorded monthly averages of CO$_2$ at the China Global Atmosphere Watch Baseline Observatory Mount Waliguan (WLG) over many years. As one of 31 ground stations around the world, the atmospheric samples collected by WLG can well represent the average atmospheric conditions in China. The data values of the WLG site are monthly mean values. In order to match the time resolution of other products, we calculated the annual mean value of CO$_2$ by using the recorded values of 12 months each year, and then processed it into a grid dataset covering China with a time resolution of 1 year and a spatial resolution of 1 km.

Globe30 products: The 30 m land cover datasets in 2000 and 2020 that we used were from the National Geomatics Center of China. The images used for classification were 30 m multispectral images, including Landsat satellite and China Environmental Disaster Reduction Satellite multispectral images, which were generated after synthesizing a considerable amount of auxiliary data and reference materials. The data include 10 land cover types, including water, wetland, artificial, tundra, ice, grass, bareland, cropland, shrub, and forest. The overall accuracy of the third-party evaluation is 83.50% [53].

2.3. Data Preprocessing and Trend Analysis

The time resolution of FVC data was 8 days. In this study, the annual average FVC data were synthesized by calculating the average of 46 scenes of FVC data every year. To match the spatial resolution of nighttime light data, the nearest neighbor sampling method was used to resample vegetation coverage data, climate data, and CO$_2$ data. After the above processing, all data had the same time resolution (1 year) and spatial resolution (1 km).

To explore the vegetation and NTL data changes, the trends of vegetation and NTL were calculated using the linear regression method. The following formula expresses the relationship between them:

$$Y = k \times Year + b \tag{1}$$

where k is the trend and b is the intercept term. A positive k value represents an increasing Y trend, while a negative k value represents a decreasing Y trend [54].

2.4. Urban Boundary Extraction

Urban boundary extraction was carried out using automated and manual interventions. The automatic extraction of the urban boundary was divided into three steps. Step 1: calculate the histogram of the NTL, calculate the position of the point with the most drastic change in the histogram, and then calculate the NTL value of the corresponding position as the first threshold for edge extraction. Step two: calculate the gradient of the NTL, repeat the operation of step one for the gradient data, and calculate the second threshold of edge extraction. Step 3: Using the thresholds extracted by the first two cloths, take the NTL between the first and second thresholds as the city extraction result. Finally, the extraction results were compared with the land cover data and adjusted manually. The city boundaries extracted from the NTL in 2001 were used as the core area boundaries. The urban boundaries extracted from the NTL in 2010 were used as the extension area boundaries. Beijing is taken as an example (Figure 2), with the Figure 2a showing the boundary of the Beijing core area and Figure 2b showing the boundary of the Beijing expanded area.

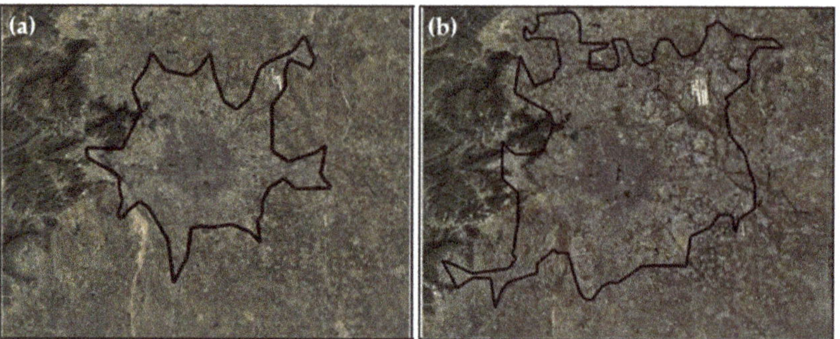

Figure 2. True color Landsat image of Beijing. (**a**,**b**) are images taken from Google Earth in 2001 and 2010, respectively. The polygons on the (**a**,**b**) are the core and extension regions, respectively.

2.5. Attribution Analysis

This study used the GLM as the method of attribution analysis. FVC was used as the dependent variable, and CO_2, climate factors, and NTL were independent variables. The GLM provided a flexible framework that described the relationship between response variables and explanatory variables well. In addition, the model was not only suitable for describing linear relationships, but also had a strong ability to describe nonlinear relationships. The link function can be applied to data with normal, Poisson, gamma, binomial, and other distributions [55–59]. The response variable in the study is FVC, which generally follows a normal distribution, so the family set in the model is Gaussian model, and the link is identity. The corresponding mathematical expression of GLM is as follows:

$$Y = g(b_0 + b_1 \times x_1 + \cdots b_m \times x_m) \tag{2}$$

where Y is the response variable, x is the explanatory variable, b is the regression coefficient, and g (.) is a link function. The corresponding expression in R language is as follows:

$$glm(FVC \sim C + U + P + T + R, \text{ family } = gaussian(link = 'identity')) \tag{3}$$

where C, U, P, T, and R represent CO_2, urbanization, precipitation, temperature, and radiation, respectively.

The calculation method of specific contribution is mainly divided into three steps:

Step 1: The mean square (MS) of each explanatory variable were obtained using GLM. The MS calculation formula is as follows:

$$MS = \frac{SS}{Df} \tag{4}$$

where Df is the degree of freedom, SS is the sum of squares, and its value is equal to the explained sum of squares (ESS) increased by addition of an independent variable to the model. The mathematical expression for the sum of squares is as follows:

$$ESS = \sum (\hat{y} - \bar{y})^2 \tag{5}$$

$$SS = ESS(model2) - ESS(model1) \tag{6}$$

According to the order in which independent variables enter the model, the MS expressions of each independent variable are as follows, taking CO_2 and urbanization as examples:

$$SS_C = ESS(glm(FVC \sim C) \tag{7}$$

$$SS_U = ESS(glm(FVC \sim C + U)) - ESS(glm(FVC \sim C)) \tag{8}$$

The SS of other independent variables was also calculated according to the above method. Finally, MS corresponding to all independent variables was calculated according to Formula (3).

Step 2: The regression coefficients of each explanatory variable were obtained using GLM. Firstly, the annual mean values of FVC, CO_2, urbanization, precipitation, temperature, and radiation of 32 cities were calculated and then input into the model. The regression coefficients of explanatory variables in the corresponding models of different cities were obtained. The positive and negative effects of explanatory variables were judged by the positive and negative values of the regression coefficient. When the regression coefficient is greater than 0, it indicates that there is a positive effect between the variable and the response variable; when the regression coefficient is less than 0, it indicates that there is a negative influence between the explanatory variable and the response variable.

Step 3: Use MS to calculate the contribution, the method refers to Tao's [55], taking CO_2 for example:

$$\text{Contribution}_C = \frac{MS_C}{MS_C + MS_U + MS_P + MS_T + MS_R + MS_{\text{other}}} \tag{9}$$

All analyses were carried out in R Version 3.6.1.

3. Results

3.1. Vegetation Cover Change in China's 32 Major Cities

From 2001 to 2018, China's vegetation coverage showed a trend of large-scale growth overall, although some areas showed a decreasing trend. The areas showing increases were mainly concentrated in central China, such as Shaanxi Province, Shanxi Province, Guizhou Province, and Guangxi Province. The areas showing a decreasing trend included the Yangtze River Delta. The fastest growth rate of vegetation coverage was 0.05/year, which showed that the greening trend in China was very fast (Figure 3).

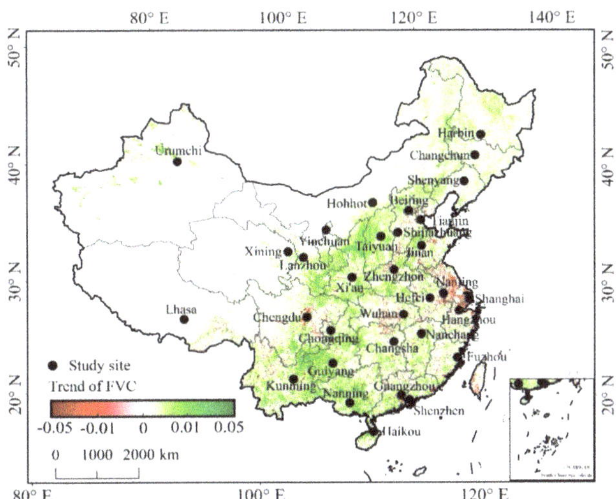

Figure 3. Spatial distribution of vegetation coverage trends from 2001 to 2018.

For different cities, the FVC changes can be divided into two categories: FVC changes that differed greatly between the core area and the expansion area, and FVC changes that showed limited differences between the core area and the expansion area. Among the 32 cities studied, similar FVC changes between the core area and expansion area were observed in Urumqi, Beijing, Lanzhou in Northern China, Shenzhen in Southern China, and other cities, and they were mainly manifested by a similar slope of FVC change in spatial distribution; whereas great differences in FVC changes between the core area and expansion area were observed in Southern China, Shanghai, Hangzhou, Chengdu, Fuzhou, Changsha, Chongqing, Nanning, Haikou, and other cities, and they manifested as unchanged growth, an increasing trend in the core area vegetation or a decreasing trend in the expansion area vegetation coverage (Figure 4).

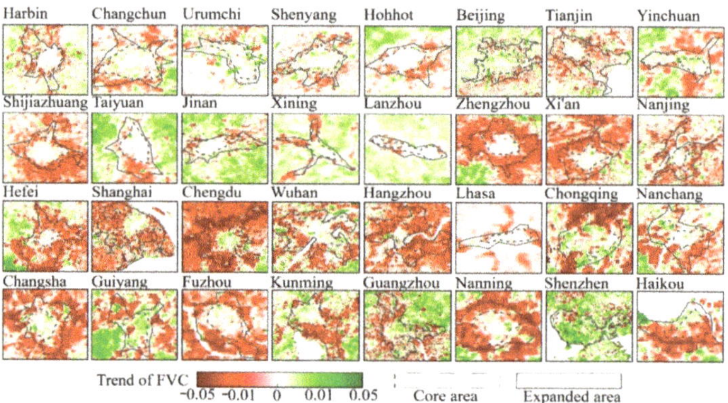

Figure 4. Spatial distribution of FVC trends in 32 major cities in China from 2001 to 2018, with dotted lines representing the boundaries of core areas and solid lines representing the boundaries of extended areas.

In order to further explore the long-term trends of FVC change in 32 major cities in China, this study quantitatively analyzed the FVC changes in these urban cores and expansions. FVC changes come in four different ways. The first was the form in which FVC of both core and expansion areas decreased, which includes 16 cities, namely Harbin, Tianjin, Shijiazhuang, Zhengzhou, Xi'an, Nanjing, Hefei, Shanghai, Chengdu, Wuhan, Hangzhou, Lhasa, Chongqing, Nanchang, Fuzhou, and Guizhou. The second was a form in which both the core area and the expansion FVC increased, and this form includes five cities, namely Changchun, Urumchi, Beijing, Yinchuan, and Shenzhen. The third was the form of FVC increased in the core area and decreased in the expansion area, which includes four cities, namely Shenyang, Xining, Changsha, and Guiyang. The fourth was the form of FVC decreased in the core area and increased in the expansion area, which includes six cities, namely Hohhot, Taiyuan, Jinan, Nanning, Haikou, and Kunming (Figure 5).

3.2. Urban Expansion Model

Figure 6 shows the changing trend of NTL in China from 2001 to 2018, and it reflects the drastic urban expansion process in China from 2001 to 2018. Figure 7 shows the spatial distribution of NTL trends in 32 major cities in China from 2001 to 2018, and it reflects the urban expansion patterns of different cities and the differences in urban development between core and expansion areas. Cities with drastic urban development in China are mainly concentrated in Eastern China, among which Shanghai, Nanjing, Yinchuan, Guangzhou, Shenzhen, and other regions had the most drastic urban expansion. In addition, we can also see that China's urbanization process was generally fast and widely distributed in Eastern China. Major urban agglomerations gradually formed in China, such as the Yangtze River Delta urban agglomerations, Pearl River Delta urban agglomerations, and the Beijing–Tianjin–Hebei region.

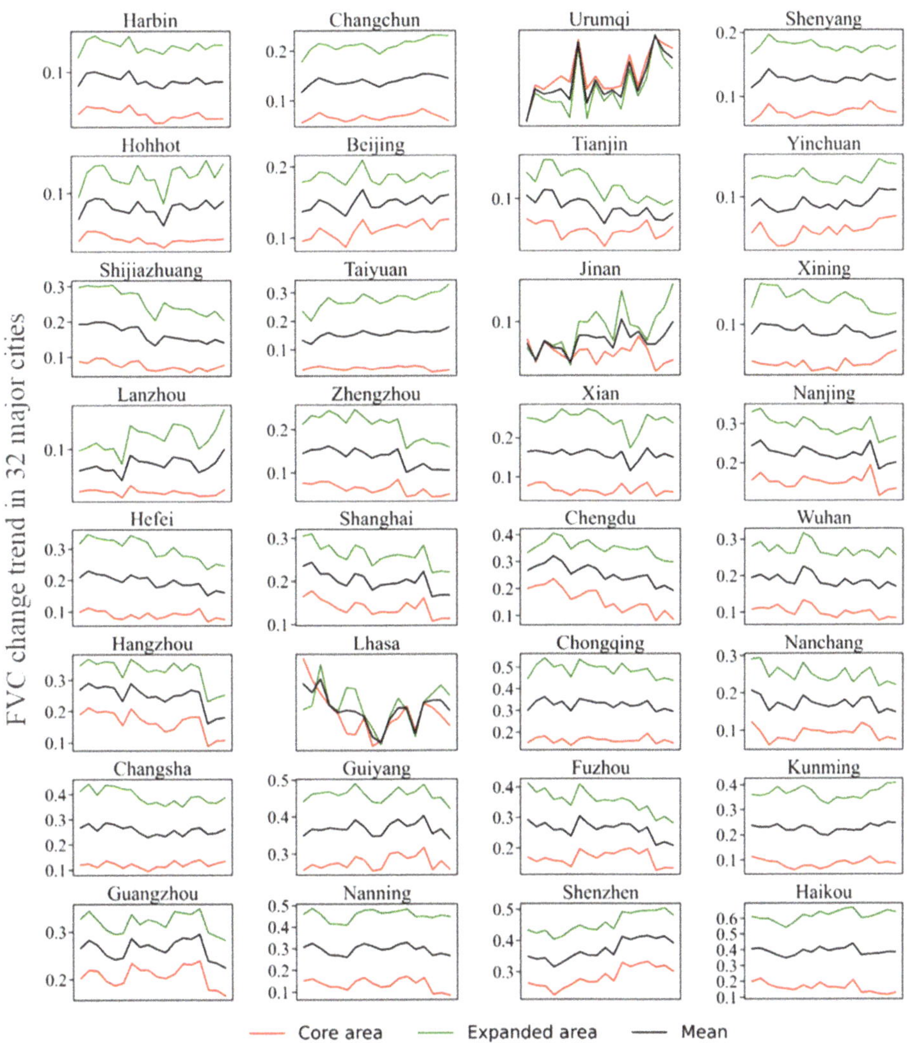

Figure 5. Temporal series changes in vegetation coverage in 32 Chinese cities from 2001 to 2018.

Urban core areas and expansion areas had great differences in the urbanization process. The study found that the NTL of 32 major urban core areas in China remained unchanged, thus reflecting urban core area stability. Some cities' urban expansion was rapid, such as Urumqi, Changchun, Yinchuan, Hefei, Chengdu, Wuhan, Nanchang, Changsha, Guiyang, and others. There was a significant trend of increasing NTL in the expansion area, thus reflecting the rapid urban expansion pattern of these cities. At the same time, there were some cities with small differences between the core areas and expansion areas, such as Beijing, Taiyuan, Lanzhou, Chongqing, Shenzhen, etc. Since urbanization, the area of each city has changed greatly. The expanded area of Shenzhen increased by 50.7%, and that of the Hefei expanded area increased by 614.2% (Figure 7).

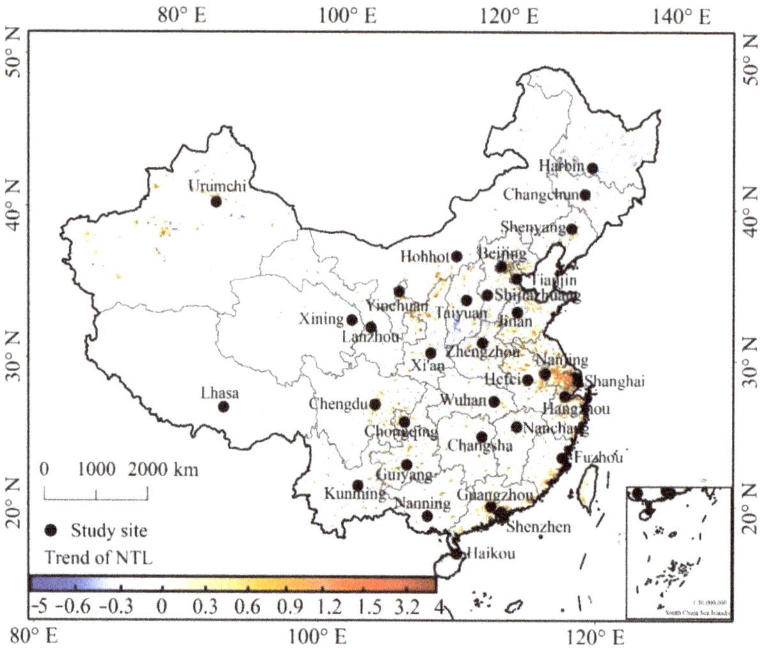

Figure 6. Spatial distribution map of the NTL change trend from 2001 to 2018.

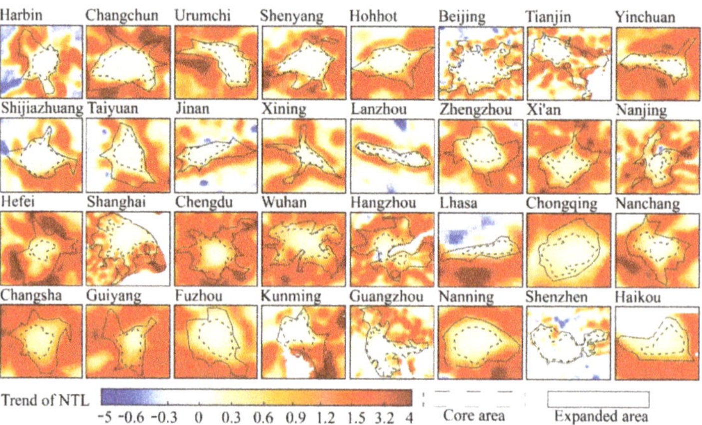

Figure 7. Spatial distribution of NTL trends in 32 major cities in China from 2001 to 2018, with dotted lines representing the boundaries of core areas and solid lines representing the boundaries of extended areas.

To further explore the relative process of urbanization in 32 cities in China, this paper used the land cover data in 2000 and 2020 to explore urban changes from the perspective of land cover transfer in terms of the method and extent. Then, we calculated table statistics on the area and proportion of the land cover types converted to artificial surfaces in 32 major cities in China between 2000 and 2020. The results on the area and proportion of land cover types transferred from artificial surfaces showed that all 32 cities had different degrees of expansion, and the area of artificial surface gain was much higher than that of artificial surface loss. Among them, Shanghai had the largest area converted to artificial surface, followed by Beijing, and the results showed that the scope of urbanization of Shanghai was enormous between 2000 and 2020. Harbin had the largest proportion of the area converted to artificial surface, followed by Zhengzhou, which reflected that the degree of urbanization was intense. The area converted from artificial land to other land cover types was far less than that from artificial land, and the smallest area was observed in Lhasa. The results showed that the area converted from artificial land was only 0.6 km^2, which accounted for approximately 1.8% of the Lhasa study area. Kunming had the smallest proportion of artificially transferred area at approximately 0.1% (Table 2).

Table 2. Statistical table of artificial surface gains and losses in China's 32 major cities from 2000 to 2020 (Unit: km^2).

Types / City	Gain		Loss		Types / City	Gain		Loss	
	Area	Ratio	Area	Ratio		Area	Ratio	Area	Ratio
Harbin	354.5	74.9	8.8	1.9	Hefei	281.7	45.5	14.2	2.3
Changchun	136.6	26.6	5.4	1.0	Shanghai	1211.4	30.6	114.4	2.9
Urumchi	34.7	14.8	4.8	2.0	Chengdu	438.2	45.3	10.3	1.1
Shenyang	86.5	12.7	10.4	1.5	Wuhan	280.3	33.2	7.4	0.9
Hohhot	49.9	23.1	5.8	2.7	Hangzhou	400.0	29.9	23.8	1.8
Beijing	1022.8	37.0	39.0	1.4	Lhasa	0.6	1.8	0.1	0.4
Tianjin	547.9	32.6	37.4	2.2	Chongqing	82.1	33.7	3.2	1.3
Yinchuan	59.6	36.5	4.3	2.6	Nanchang	87.8	22.1	3.3	0.8
Shijiazhuang	80.7	21.7	4.6	1.2	Changsha	140.9	38.3	6.0	1.6
Taiyuan	110.2	33.4	2.7	0.8	Guiyang	36.4	18.6	4.5	2.3
Jinan	157.0	32.7	7.5	1.6	Fuzhou	53.4	22.6	2.5	1.1
Xining	25.6	22.5	2.4	2.1	Kunming	180.5	45.7	0.6	0.1
Lanzhou	39.7	21.0	0.9	0.5	Guizhou	508.2	28.6	23.8	1.3
Zhengzhou	272.3	50.7	5.8	1.1	Nanning	75.3	29.2	3.5	1.4
Xi'an	232.9	33.7	9.1	1.3	Shenzhen	316.4	21.7	28.8	2.0
Nanjing	357.1	36.9	10.7	1.1	Haikou	9.1	10.1	0.2	0.2

3.3. Contribution Analysis of FVC Changes

3.3.1. Spatial Distribution of the Main Driving Factors

Two different methods of statistical analysis were applied to determine the contribution of drivers to vegetation in China's 32 major cities. The methods of analysis used in Figure 8a,b were different from those in Figure 8c,d. The main difference was that Figure 8c,d combined the contributions of precipitation, temperature, and radiation, which were named climate and represented by the green color in the pie chart. Two different methods were used to show the contributions of different factors to the change in FVC.

Certain regular trends were observed among the dominant factors of vegetation in space. The contribution of CO_2 increased slightly from north to south at close altitudes and exceeded 50% in Haikou and Shenzhen. Precipitation dominated vegetation growth in urban areas in Northern China, such as Inner Mongolia and Yinchuan in arid and semiarid regions. The contribution of temperature decreased gradually with increasing dimensionality and steadily increased with increasing shoreline distance. The contribution of radiation generally increased with decreasing dimensions and increasing coastline distance. The difference in urbanization contribution among the 32 cities was not obvious, and the contribution of coastal cities was slightly higher than that of inland cities. Nevertheless, the overall contribution was similar, which also reflected the good situation of synchronous development.

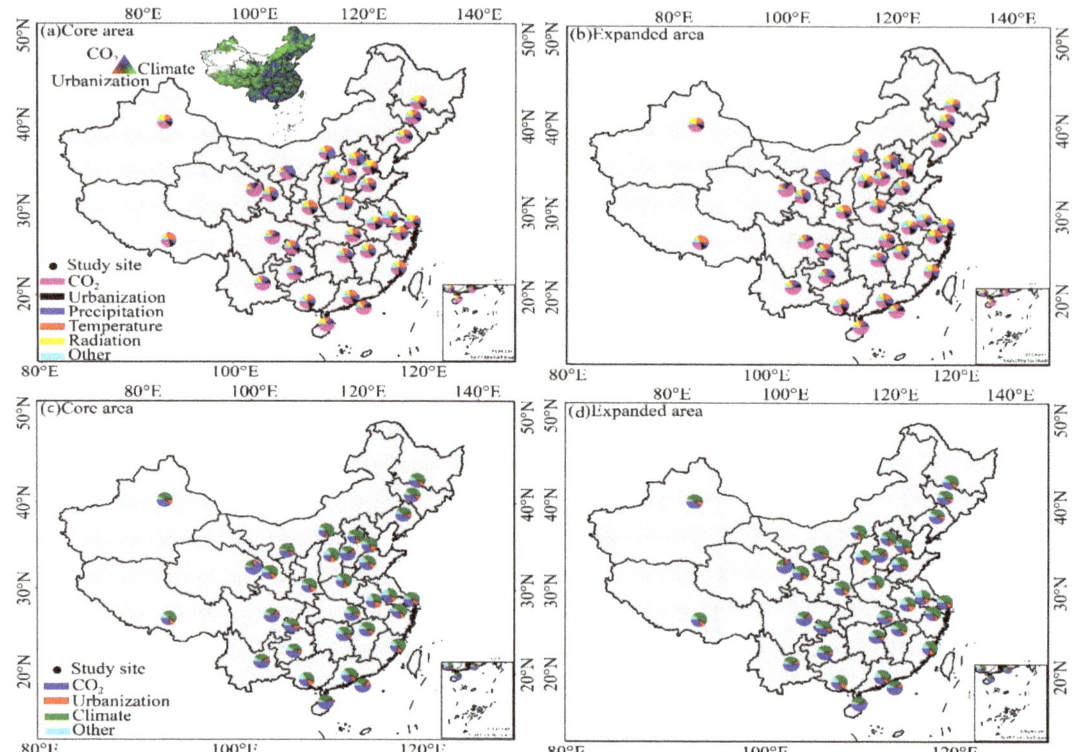

Figure 8. Driving factor contributions in 32 major cities in China: (**a**) pie chart of five driving factor contributions in the core region, (**b**) pie chart of five driving factor contributions in the expansion region, (**c**) pie chart of three major driving factor contributions in the core region, and (**d**) pie chart of three major driving factor contributions in the expansion region.

The sum of precipitation, temperature, and radiation contributions was greater than that of CO_2, which became the main driving factor of urban vegetation. We found that the contribution of climate factors to vegetation growth was nearly 50%, which applied to the core and expansion areas of 32 cities simultaneously. The combination of climate factors and CO_2 explained more than 70% of vegetation growth, indicating that the contribution of climate factors and carbon dioxide cannot be ignored in the process of vegetation growth. In comparison, the contribution of urbanization was relatively small and inly explained approximately 10% of vegetation growth overall.

The contributions of driving factors had similar regular trends but also exhibited differences. The main factors affecting vegetation growth in the 32 major Chinese cities were revealed using spatial distribution maps. Figure 9 shows that the FVC in 32 cities in China was still mainly affected by CO_2 and climate factors. According to the leading factors in the core area, 32 cities were divided into three categories: cities with climate as the dominant factor, cities with CO_2 as the dominant factor, and cities with multiple factors acting together. The cities with climate as the main factor included Hohhot, Beijing, Zhengzhou, Shanghai, Lhasa, and Fuzhou; the cities under significant CO_2 control included Shijiazhuang, Xining, Changsha, Chengdu, Guiyang, Kunming, and Shenzhen; the cities influenced by multiple driving factors included Urumqi, Harbin, Tianjin, and Taiyuan.

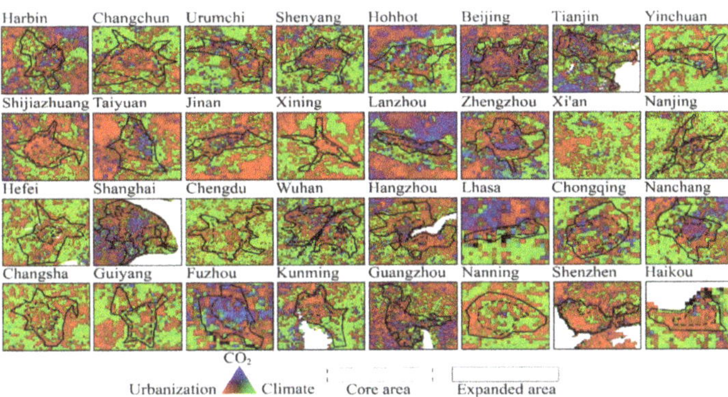

Figure 9. Spatial distribution of driving factor contributions in 32 major urban cities in China.

In addition, the study also found that there were differences in the dominant factors underlying FVC change between the urban core area and the urban expansion area. For example, the core area of Xi'an was dominated by climate factors while the expansion area was dominated by CO_2. The core area of Nanjing was affected by many factors, while the extended area was dominated by CO_2. The core area of Nanchang was dominated by temperature, while the expansion area was dominated by urbanization. In some cities, the driving factors of FVC change in core and extended areas were consistent. For example, the change in FVC in the Changchun core area and the extended area was mainly affected by CO_2. The dominant factor of FVC change in the core and extended areas of Lhasa was climate.

3.3.2. Contributions of Major Drivers

Table 3 showed the importance of driving factors to FVC changes in 32 cities. The driving factors had different effects on the FVC in different urban core areas. In Xining, Chengdu, Haikou, Shijiazhuang, and Kunming, CO_2 contributed greatly and explained 66.3%, 55.2%, 53.8%, 51.6%, and 49.6% of FVC changes, respectively. Cities such as Hefei, Tianjin, Taiyuan, Nanning, and Xi'an were mainly impacted by urbanization, with a minimum value of 13.2% and a maximum value of 16.5%. The top five cities affected by precipitation were Yinchuan, Hohhot, Beijing, Lanzhou, and Guiyang, with contributions ranging from 21.2% to 37.6%. Temperature had the greatest impact on vegetation in Lhasa, Zhengzhou, Xi'an, Hangzhou, Fuzhou, and Kunming and explained approximately 21.8–32.7% of the vegetation coverage growth. Radiation played an important role in Tianjin, Shanghai, Urumqi, Harbin, Taiyuan, and other cities, with a maximum contribution of 22.4% and minimum contribution of 18.7%.

The relative contribution of driving factors to FVC change in the extended areas differed from that in the core areas. CO_2 contributed the most in Shijiazhuang, Nanning, Chengdu, Xining, and Kunming and accounted for 52.3–67.2% of vegetation growth. Urbanization contributed the most in Nanchang, Tianjin, Yinchuan, Nanjing, and Shenyang and accounted for 14.5–19.3% of the FVC changes. Precipitation contributed the most in Lanzhou, Yinchuan, Hohhot, Beijing, and Kunming and accounted for 17.7–32.6% of the FVC changes. Temperature had the greatest impact on vegetation in Zhengzhou, Lhasa, Hohhot, Guizhou, and Hangzhou and explained 18.3–42.3% of the vegetation coverage variation. Radiation played an important role in Tianjin, Urumqi, Harbin, Taiyuan, Shenyang, and other areas and accounted for 15.9–26.4%.

Table 3. Importance of driving factors to FVC changes in 32 cities. Red shading denotes greater importance, and blue shading represents lower importance (Unit: %). An asterisk indicates that the GLM regression coefficient is greater than zero.

Drivers / City	CO_2 C	CO_2 E	Urbanization C	Urbanization E	Precipitation C	Precipitation E	Temperature C	Temperature E	Radiation C	Radiation E	Other C	Other E
Harbin	32.5	35.2	10.4 *	13.8 *	10.3	7.8 *	16.2 *	14.0 *	20.1	20.5	10.4	8.7
Changchun	38.2 *	49.1 *	8.4 *	8.9	15.8 *	12.0 *	14.9 *	12.4 *	12.2	10.0	10.6	7.7
Urumchi	37.9 *	35.0 *	11.7	5.9	7.1 *	16.0 *	16.0 *	16.6 *	20.6	20.8	6.7	5.6
Shenyang	41.9 *	29.6 *	10.0 *	14.5	9.6 *	14.9 *	13.5 *	16.7 *	14.8	15.9	10.2	8.5
Hohhot	22.2	28.1	8.7 *	7.5 *	30.3 *	28.7 *	19.3 *	23.4	9.3 *	7.5 *	10.1	4.7
Beijing	26.4 *	29.6 *	9.4	14.2	27.5 *	23.9 *	15.5	13.6 *	12.0	10.3	9.1	8.5
Tianjin	28.2	28.5	15.9 *	15.2 *	6.1 *	10.3 *	16.3 *	10.5 *	22.4	26.4	11.0	8.9
Yinchuan	36.4 *	35.6 *	8.5	15.1	37.6 *	29.7 *	9.4 *	10.9	4.2	5.4 *	3.9	3.2
Shijiazhuang	51.6	67.2	7.7 *	9.6	6.6 *	4.6 *	10.3 *	6.7 *	11.1	6.5	12.7	5.4
Taiyuan	21.0	32.9	14.7 *	12.0 *	9.1	6.7 *	19.0	16.4	18.7	20.3	17.5	11.7
Jinan	33.8	41.8 *	10.9	13.9 *	12.1	10.2 *	18.8	17.7 *	9.1	6.7	15.3	9.7
Xining	66.3 *	53.8 *	8.4 *	14.4 *	7.4 *	15.3 *	8.5 *	6.9	4.4	4.5 *	5.0	5.1
Lanzhou	28.9	31.2 *	12.1 *	10.4 *	23.2 *	32.6 *	13.6	9.5 *	11.5 *	7.5	10.7	8.7
Zhengzhou	25.0	33.1 *	9.4 *	7.8	12.6 *	5.1 *	28.0	42.3	10.6 *	8.2 *	14.3	3.5
Xi'an	30.0	51.1 *	13.2 *	8.2	7.3 *	7.0 *	26.7	14.1 *	13.8	14.1 *	9.0	5.6
Nanjing	31.2	42.6	11.4 *	14.8 *	13.2 *	10.9	10.0 *	9.6 *	12.6 *	11.0	21.6	11.2
Hefei	27.5	49.2	16.5 *	10.4	11.2	10.3	11.5 *	13.7 *	14.9	9.6	18.3	6.8
Shanghai	32.5	44.0	11.6 *	8.9 *	12.6 *	10. 5 *	12.0 *	17.1	21.3	11.8	10.0	7.8
Chengdu	55.2	55.9 *	8.3 *	6.5	7.5 *	11.3 *	11.9	17.0 *	6.7	3.4	10.3	5.9
Wuhan	40.7	40.3	9.1 *	12.1	8.2 *	8.5	15.9 *	17.3 *	12.8 *	11.8 *	13.3	9.9
Hangzhou	28.9	42.8	9.0 *	6.4 *	12.5	16.3 *	24.5	18.3	16.8	9.4	8.3	6.8
Lhasa	27.9	27.4	11.8 *	8.1 *	8.0 *	10.6 *	32.7	35.6 *	7.7 *	8.3	11.9	10.0
Chongqing	38.8	45.9 *	11.2	10.6	12. 3 *	8.8 *	18.0 *	16.4	9.0	9.3 *	10.7	9.0
Nanchang	33.7 *	36.2	9.7	19.3 *	14.3	10.4	15.5	11.2	15.0	15.4	11.7	7.5
Changsha	39.3	50.7	7.6 *	9.2 *	17.2 *	8.4 *	14.8	17.1 *	6.4 *	5.8	14.7	8.7
Guiyang	43.7	45.6 *	8.3 *	11.4	21.2 *	14.5 *	11.0	10.0 *	4.3 *	8.3 *	11.4	10.2
Fuzhou	24.0 *	48.1	11.4	6.6	13.9	16.0	23.3	8.6	14.2	13.4	13.2	7.3
Kunming	49.6	52.3 *	8.8 *	7.4	15.4 *	17.7 *	8.1	7.5	9.4 *	10.2	8.6	5.0
Guizhou	31.9	34.3	10.8	11.6	11.8	11.0	21.8	19.3	9.3 *	11.9 *	14.5	11.9
Nanning	30.1	63.3 *	14.1 *	5.2	11.2	10.0 *	17.5	7.3	11.1	7.2 *	16.0	7.0
Shenzhen	50.1 *	39.4 *	11.0 *	10.8 *	6.9 *	8.8 *	14.9 *	17.0 *	7.6	11.8	9.5	12.1
Haikou	53.8	51.4 *	4.7 *	9.7 *	9.2 *	10.4 *	11.5	7.7 *	15.0	14.4	5.8	6.4

The impacts of the same driving factors on vegetation in the same urban core area and expansion area were different. Among them, the difference in CO_2 in Nanning, Fuzhou, Hefei, Xi'an, and Shijiazhuang was the largest. Among the top five cities, the largest difference was in Nanning, where it accounted for 33.3%, and the smallest difference was in Shijiazhuang, where it accounted for 15.6%. The contribution of urbanization in Nanchang, Nanning, Yinchuan, Hefei, Xining, and other cities varied greatly and ranged from 6.1% to 9.6%. The precipitation gaps in Lanzhou, Urumqi, Changsha, Yinchuan, Xining, and other cities were relatively large, with the maximum difference reaching 9.4% and the minimum difference reaching 7.9%. In Fuzhou, Zhengzhou, Xi'an, Nanning, Hangzhou, and other cities, there was a large difference in the contribution of temperature, with a difference between 6.2% and 14.7%. Radiation differed in Shanghai, Hangzhou, Hefei, Shenzhen, and other cities, and the difference in contribution was 4.2–9.5%.

In the core area, the dominant factor of FVC in 28 cities was CO_2, among which the decrease of FVC in 20 cities was inhibited by CO_2, the increase of FVC in 6 cities was promoted by CO_2, and the growth trend of FVC in 2 cities was opposite to the effect of CO_2, namely Nanchang and Fuzhou. The FVC in Nanchang showed a downward trend, while CO_2 as the leading factor played a role in promoting it. In this case, it was considered to be caused by the inhibition of other factors except CO_2. The FVC of Fuzhou core area showed a decreasing trend, while CO_2 as the leading factor played a promoting role. In this case, it was considered to be caused by the inhibition of other factors except CO_2. The rest of the four cities' leading factor was not CO_2, including the Hohhot, Beijing, Zhengzhou, and Lhasa. Among them, the FVC of Hohhot showed a trend of decline, and it mainly promoted by precipitation. The FVC of Beijing showed an increasing trend, and precipitation as the leading factor played a promoting role. The FVC of Zhengzhou showed a decreasing trend, and temperature, as the dominant factor, played an inhibiting role. The FVC in Lhasa core area showed a decreasing trend, and temperature was the dominant factor, which restricted the vegetation growth.

In the expansion area, the dominant factor of FVC in 29 cities was CO_2, among which the decrease of FVC in 12 cities was inhibited by CO_2, the increase of FVC in 10 cities was promoted by CO_2, and the growth trend of FVC in 7 cities was contrary to the effect of CO_2, including Shenyang, Taiyuan, Xining, Xi'an, Chengdu, and Chongqing. Among them, the FVC of Shenyang showed a decreasing trend, while CO_2 played a promoting role as the leading factor. In this case, it was considered to be caused by the inhibition of urbanization, radiation, and other factors. The FVC of Xi'an showed a decreasing trend, while CO_2 played a promoting role as the leading factor. In this case, urbanization, temperature, radiation, and other factors were considered to cause the inhibition. The rest of the three cities' leading factor was not CO_2, including Hohhot, Beijing, and Zhengzhou. Among them, the FVC of Hohhot showed an increasing trend, although CO_2 was still an inhibition, with precipitation as the dominant factor, which played an important role in promoting the FVC. The FVC of Beijing showed an increasing trend, and precipitation as the leading factor played a promoting role. The FVC of Zhengzhou showed a decreasing trend, and the temperature as the dominant factor played a restraining role.

4. Discussion

From 2001 to 2018, we found that the vegetation in China generally showed an increasing trend from 2001 to 2018, although the vegetation in cities prevalently decreased. Eastern China showed increases in FVC and decreases in city clusters, especially in the Yangtze River Delta urban agglomeration [60]. This finding was in accordance with previous research results [18,20,61]. In addition, we found that the vegetation coverage in 21 of 32 cities showed a decreasing trend based on an analysis of changes in the FVC in China's major cities. This result is also reflected in relevant studies [35]. The FVC of 32 urban expansion areas was higher than that of core areas. It is worth noting that due to the management of urban green space in recent years, the vegetation in the core areas and expansion areas of some cities recovered in the later stage of development.

Urbanization explained 10.6% of the variation in vegetation dynamics, which indicated that the indirect effects generated by urban expansion should not be ignored. We further confirmed that urbanization could exert both positive and negative impacts on vegetation. We found that the vegetation coverage in 21 of 32 cities showed a decreasing trend by analyzing the changes in FVC in China's major cities. This result is also reflected in relevant studies [35]. However, an increasing trend in vegetation was detected in recent years in several cities. The difference might be explained by the changes during the urban development period [62]. Cities in the early stage of development might increase impervious surfaces to support city development at the cost of reducing vegetation coverage [60]. With the increased demands for better living conditions, more developed cities might increase urban vegetation [63].

Our results suggested that climate was the main driving factor of vegetation growth and explained 40.6% of the vegetation variation. Precipitation, temperature, and radiation explained 13.2%, 15.7%, and 11.7% of vegetation growth, respectively. Although the impacts of precipitation, temperature, and radiation on vegetation were similar, the responses of vegetation to precipitation, temperature, and radiation exhibited strong spatial heterogeneity. We found that the contribution of precipitation increased from wetter regions to drier regions (from 9.2–10.4%% in Haikou to 28.7–30.3% in Hohhot) (Table 3). Our results were similar to those of previous studies in that vegetation in arid and semiarid regions was dominated by precipitation [64]. However, the impact of vegetation was not significantly different between the core areas and expanded areas. Possible explanations included urban green space management, strengthened irrigation, and other measures, which decreased the differences in precipitation between the core area and the expansion area. We also found that the contribution degree generally showed an increasing trend with increasing latitude and coastline distance [12]. However, a significant difference was not found between the core area and the expanded area. A possible reason was that large impervious surfaces in the urban core and expansion areas increased albedo and temperature; therefore, temperature was no longer the limiting factor [36]. In our research, radiation mainly affected the vegetation in coastal areas and high altitude areas, which may be because radiation was stronger at high altitudes. The intensity of radiation was also strengthened due to the influence of reflectivity in coastal areas.

In addition, the contribution of CO_2 to vegetation variation cannot be ignored. CO_2 was the dominant factor affecting urban vegetation, and the average driving contribution of CO_2 to urban vegetation was 39.2%. This result was also reported in previous studies [22,65]. By observing the contribution of urban driving factors, we found that CO_2 dominated vegetation growth in 29 cities. The other three cities, namely Hohhot, Zhengzhou, and Lhasa, were dominated by precipitation, temperature, and radiation. The reason may be that under favorable hydrothermal conditions, CO_2 became the main growth factor of vegetation by affecting the carboxylation reaction of vegetation. However, Hohhot is located in arid and semiarid areas, and the influence of precipitation was more important. The temperature in Lhasa was low, which was the main factor limiting vegetation growth. In Zhengzhou, temperature was the dominant factor, followed by CO_2, which may be related to increases in temperature under increased albedo. We also found that in the core area, the dominant factor of FVC in 28 cities was CO_2, among which the decrease of FVC in 20 cities was inhibited by CO_2, the increase of FVC in 6 cities was promoted by CO_2, and the growth trend of FVC in 2 cities was opposite to the effect of CO_2. In the expansion area, the dominant factor of FVC in 29 cities was CO_2, among which the decrease of FVC in 12 cities was inhibited by CO_2, the increase of FVC in 10 cities was promoted by CO_2, and the growth trend of FVC in 7 cities was contrary to the effect of CO_2. These findings indicated that CO_2 was essential for the growth of vegetation, and its role varies from city to city.

We focused on China's 32 major cities and quantified the relative contributions of climate, urbanization, and CO_2 to FVC change. However, some uncertainties remained in this study. First, the interaction between human activities and climate may not have been fully considered in our research. For example, human activities have changed the types of underlying surfaces in cities, thereby increasing reflectivity and temperature and causing changes in vegetation growth. To date, the complex interaction mechanism between human activities and climate change still needs further discussion [47,66]. Second, although the effects of climate, CO_2, and urbanization on vegetation growth were considered, some factors, such as nitrogen deposition, topography, tree age, and other driving factors, were not considered but also affect vegetation growth [21,22,67,68]. Third, it is very complex to extract urban boundaries from lighting data, as can be seen from the review of urban mapping technology system based on NTL data in Zhou et al. 's research [69]. At present, a number of scholars have developed a variety of urban extent mapping methods, including the cluster-based method [70], method based on the NTL gradient [71], automatic delineation framework and morphology combined method [72], random forest classifier method [31], and stepwise-partitioning framework method [73], which have improved the deficiency of using a single threshold in the past [74–76] and helped to carry out research on a global scale. In our study, we learned the related ideas of multi threshold extraction, and extracted the boundaries of 32 cities in China by using the characteristics of data gradient and divergence. However, this result may be affected by the resolution of the NTL data [69]. Thus, in the future, finer lighting data can be used to achieve more accurate urban boundary extraction. Finally, due to the lack of CO_2 data covering China in a long time series, this study only used the data of the WLG site as the data of China for calculation. Figure 10 showed the status of CO_2 observations at the WLG site from 2001 to 2018. The data at the WLG site shows a continuous increase of 2.2 ppm per year. CO_2 has spatio-temporal differences among cities, and the difference in concentrations between Eastern and Western China ranges from 2 to 4 ppm, meaning the difference is less than 1% [77,78]. It should be pointed out that the impact of CO_2 on vegetation growth obtained in this study is only a preliminary impact result based on the statistical level of GLM regression equation. In fact, the specific positive promotion or negative inhibition effects should be further considered and evaluated by introducing the actual vegetation dynamic growth process model.

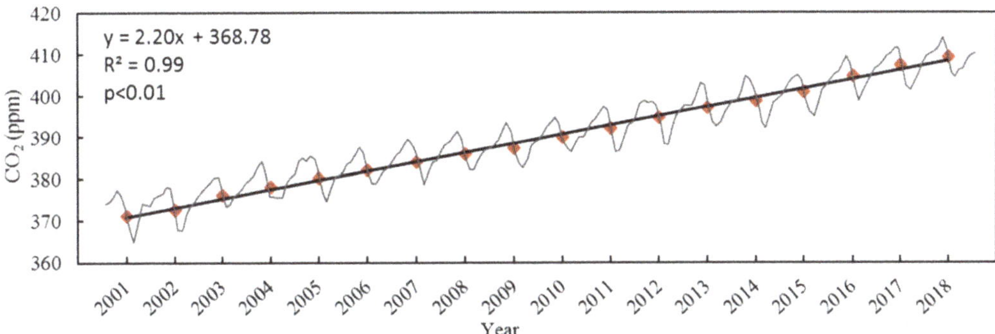

Figure 10. Curve of CO_2 observation value at Waliguan station from 2001 to 2018. The gray curve represents the monthly mean value of CO_2, the red dot represents the annual mean value of CO_2, and the black line represents the change trend of annual mean CO_2.

5. Conclusions

This study quantified the relative importance of precipitation, temperature, radiation, urbanization, and CO_2 on vegetation dynamics over China's 32 major cities from 2001 to 2018. First, we found that the vegetation in China generally showed an increasing trend from 2001 to 2018. Nevertheless, the vegetation prevalence in cities decreased, with the vegetation coverage in 21 of the 32 cities showing a decreasing trend, among which the FVC in the core area decreased in 23 cities. The FVC in the expansion area decreased in 21 cities. Second, the changes in NTL data and land cover data indicated that urban areas continued to expand from 2001 to 2018. Night light data showed that the expansion areas of 32 cities have increased to varying degrees. A comparison of the area statistics of expansion and core areas showed that the area increased by more than six times, with an average increase of approximately 168%. Land cover data showed that various land cover types changed to artificial surface types over 18 years, thereby increasing the area of urban impervious surfaces. Third, China has experienced rapid urbanization; however, the vegetation in China's 32 major cities was still mainly dominated by climate factors and CO_2 rather than urbanization. The relative contributions of climate, CO_2, and urbanization to FVC variations in China's 32 major cities were 40.6%, 39.2%, and 10.6%, respectively.

This study evaluated the vegetation change trend and then quantified the contributions of driving factors (such as precipitation, temperature, radiation, CO_2, and urbanization) to vegetation growth. Further research should consider the contribution of additional influencing factors on vegetation growth, such as topography. This study performed a change and attribution analysis of vegetation coverage in China over a long time series, and it enriches the research on vegetation and driving factors and has reference value for explaining the complex biophysical mechanism between vegetation and the environment. Moreover, it represents crucial theoretical research and provides important scientific information for environmental protection.

Author Contributions: Conceptualization, B.M. and X.Z.; Formal analysis, B.M., X.Z. and L.S.; Funding acquisition, X.Z.; Methodology, X.Z.; Writing—original draft, B.M.; Writing—review and editing, B.M., X.Z., J.Z., N.L., L.S., Q.W., N.S., M.S., Y.G. and S.Z. All authors have read and agreed to the published version of the manuscript.

Funding: This study was funded by the National Natural Science Foundation of China (42090012): the National Key Research and Development Program of China (No. 2016YFB0501404 and No. 2016YFA0600103), and the Natural Science Foundation of Anhui province of China (2008085QD167).

Institutional Review Board Statement: Not applicable.

Informed Consent Statement: Not applicable.

Data Availability Statement: Fractional vegetation cover dataset is available at http://www.geodata.cn/, accessed on 10 December 2021. Nighttime light dataset is available at https://figshare.com/articles/dataset/, accessed on 10 December 2021. Climate-related dataset is available at http://data.tpdc.ac.cn/, accessed on 10 December 2021. Globe30 dataset is available at http://www.globallandcover.com/, accessed on 10 December 2021. CO_2 dataset is available at http://www.ncdc.ac.cn/ and https://www.esrl.noaa.gov/, accessed on 10 December 2021.

Acknowledgments: We appreciate the "National Earth System Science Data Center, National Science& Technology Infrastructure of China" for providing fractional vegetation cover data. We appreciate the "National Tibetan Plateau Data Center" for providing the climate-related data. We also appreciate the "National Cryosphere Desert Data Center" and "NOAA Earth System Research Laboratories (ESRL) "for providing the CO_2 data. We thank the anonymous reviewers and editor for their valuable comments on the manuscript.

Conflicts of Interest: The authors declare no conflict of interest.

References

1. Duncan, J.; Boruff, B.; Saunders, A.; Sun, Q.; Hurley, J.; Amati, M. Turning down the heat: An enhanced understanding of the relationship between urban vegetation and surface temperature at the city scale. *Sci. Total Environ.* **2019**, *656*, 118–128. [CrossRef]
2. Rumora, L.; Majić, I.; Miler, M.; Medak, D. Spatial video remote sensing for urban vegetation mapping using vegetation indices. *Urban Ecosyst.* **2021**, *24*, 21–33. [CrossRef]
3. Hashim, H.; Abd Latif, Z.; Adnan, N.A.; Hashim, I.C.; Zahari, N.F. Vegetation extraction with pixel based classification approach in urban park area. *Plan. Malays.* **2021**, *19*, 1. [CrossRef]
4. Zhang, Y.; Shao, Z. Assessing of urban vegetation biomass in combination with LiDAR and high-resolution remote sensing images. *Int. J. Remote Sens.* **2021**, *42*, 964–985. [CrossRef]
5. Paschalis, A.; Chakraborty, T.; Fatichi, S.; Meili, N.; Manoli, G. Urban forests as main regulator of the evaporative cooling effect in cities. *AGU Adv.* **2021**, *2*, e2020AV000303. [CrossRef]
6. Ayub, M.A.; Farooqi, Z.U.R.; Umar, W.; Nadeem, M.; Ahmad, Z.; Fatima, H.; Iftikhar, I.; Anjum, M.Z. Role of urban vegetation: Urban forestry in micro-climate pollution management. In *Examining International Land Use Policies, Changes, and Conflicts*; IGI Global: Hershey, PA, USA, 2021; pp. 231–251.
7. Perini, K.; Pérez, G. Ventilative cooling and urban vegetation. In *Innovations in Ventilative Cooling*; Springer: Berlin/Heidelberg, Germany, 2021; pp. 213–234.
8. Liao, J.; Tan, X.; Li, J. Evaluating the vertical cooling performances of urban vegetation scenarios in a residential environment. *J. Build. Eng.* **2021**, *39*, 102313. [CrossRef]
9. Patel, S. The potential for urban vegetation to mitigate ambient air pollution threats to public health. *Topophilia* **2020**, 53–62. [CrossRef]
10. Meili, N.; Acero, J.A.; Peleg, N.; Manoli, G.; Burlando, P.; Fatichi, S. Vegetation cover and plant-trait effects on outdoor thermal comfort in a tropical city. *Build. Environ.* **2021**, *195*, 107733. [CrossRef]
11. Dissanayaka, C.; Weerasinghe, U.; Wijesundara, K. Urban vegetation and morphology parameters affecting microclimate and outdoor thermal comfort in warm humid cities—A review of research in the past decade. In Proceedings of the International Conference on Climate Change, Colombo, Sri Lanka, 18–19 February 2021; pp. 1–17.
12. Nemani, R.R.; Keeling, C.D.; Hashimoto, H.; Jolly, W.M.; Piper, S.C.; Tucker, C.J.; Myneni, R.B.; Running, S.W. Climate-driven increases in global terrestrial net primary production from 1982 to 1999. *Science* **2003**, *300*, 1560–1563. [CrossRef]
13. Guo, B.; Zhang, J.; Meng, X.; Xu, T.; Song, Y. Long-term spatio-temporal precipitation variations in China with precipitation surface interpolated by ANUSPLIN. *Sci. Rep.* **2020**, *10*, 1–17. [CrossRef]
14. Fensholt, R.; Langanke, T.; Rasmussen, K.; Reenberg, A.; Prince, S.D.; Tucker, C.; Scholes, R.J.; Le, Q.B.; Bondeau, A.; Eastman, R. Greenness in semi-arid areas across the globe 1981–2007—an Earth Observing Satellite based analysis of trends and drivers. *Remote Sens. Environ.* **2012**, *121*, 144–158. [CrossRef]
15. Jeong, S.; Ho, C.; Kim, K.; Jeong, J. Reduction of spring warming over East Asia associated with vegetation feedback. *Geophys. Res. Lett.* **2009**, *36*, 1–5. [CrossRef]

16. Lucht, W.; Prentice, I.C.; Myneni, R.B.; Sitch, S. Climatic control of the high-latitude vegetation greening trend and pinatubo effect. *Science* **2002**, *296*, 1687–1689. [CrossRef] [PubMed]

17. Ho, C.H.; Lee, E.J.; Lee, I.; Jeong, S.J. Earlier spring in Seoul, Korea. *Int. J. Climatol. A J. R. Meteorol. Soc.* **2010**, *26*, 2117–2127. [CrossRef]

18. Hua, W.; Chen, H.; Zhou, L.; Xie, Z.; Qin, M.; Li, X.; Ma, H.; Huang, Q.; Sun, S. Observational quantification of climatic and human influences on vegetation greening in China. *Remote Sens.* **2017**, *9*, 425. [CrossRef]

19. Trumbore, S. Carbon respired by terrestrial ecosystems–recent progress and challenges. *Glob. Chang. Biol.* **2006**, *12*, 141–153. [CrossRef]

20. Chi, C.; Park, T.; Wang, X.; Piao, S.; Xu, B.; Chaturvedi, R.K.; Fuchs, R.; Brovkin, V.; Ciais, P.; Fensholt, R. China and India lead in greening of the world through land-use management. *Nat. Sustain.* **2019**, *2*, 122–129.

21. Zhu, Z.; Piao, S.; Myneni, R.B.; Huang, M.; Zeng, Z.; Canadell, J.G.; Ciais, P.; Sitch, S.; Friedlingstein, P.; Arneth, A. Greening of the Earth and its drivers. *Nat. Clim. Chang.* **2016**, *6*, 791–795. [CrossRef]

22. Piao, S.; Yin, G.; Tan, J.; Cheng, L.; Huang, M.; Li, Y.; Liu, R.; Mao, J.; Myneni, R.B.; Peng, S. Detection and attribution of vegetation greening trend in China over the last 30 years. *Glob. Chang. Biol.* **2015**, *21*, 1601–1609. [CrossRef]

23. Wang, S.; Zhang, Y.; Ju, W.; Chen, J.M.; Ciais, P.; Cescatti, A.; Sardans, J.; Janssens, I.A.; Wu, M. Recent global decline of CO_2 fertilization effects on vegetation photosynthesis. *Science* **2020**, *370*, 1295–1300. [CrossRef]

24. Singh, C.M.; Singh, P.; Tiwari, C.; Purwar, S.; Kumar, M.; Pratap, A.; Singh, S.; Chugh, V.; Mishra, A.K. Improving Drought Tolerance in Mungbean (*Vigna radiata* L. Wilczek): Morpho-Physiological, Biochemical and Molecular Perspectives. *Agronomy* **2021**, *11*, 1534. [CrossRef]

25. Flach, M.; Brenning, A.; Gans, F.; Reichstein, M.; Sippel, S.; Mahecha, M.D. Vegetation modulates the impact of climate extremes on gross primary production. *Biogeosci. Discuss.* **2020**, *18*, 39–53. [CrossRef]

26. Yuan, M.; Wang, L.; Lin, A.; Liu, Z.; Li, Q.; Qu, S. Vegetation green up under the influence of daily minimum temperature and urbanization in the Yellow River Basin, China. *Ecol. Indic.* **2020**, *108*, 105760. [CrossRef]

27. Liang, Z.; Wang, Y.; Sun, F.; Jiang, H.; Huang, J.; Shen, J.; Wei, F.; Li, S. Exploring the combined effect of urbanization and climate variability on urban vegetation: A multi-perspective study based on more than 3000 cities in China. *Remote Sens.* **2020**, *12*, 1328. [CrossRef]

28. Zhang, Q.; Seto, K.C. Mapping urbanization dynamics at regional and global scales using multi-temporal DMSP/OLS nighttime light data. *Remote Sens. Environ.* **2011**, *115*, 2320–2329. [CrossRef]

29. Fu, P.; Weng, Q. A time series analysis of urbanization induced land use and land cover change and its impact on land surface temperature with Landsat imagery. *Remote Sens. Environ.* **2016**, *175*, 205–214. [CrossRef]

30. Xu, D.; Yang, F.; Yu, L.; Zhou, Y.; Li, H.; Ma, J.; Huang, J.; Wei, J.; Xu, Y.; Zhang, C. Quantization of the coupling mechanism between eco-environmental quality and urbanization from multisource remote sensing data. *J. Clean. Prod.* **2021**, *321*, 128948. [CrossRef]

31. Cao, W.; Zhou, Y.; Li, R.; Li, X.; Zhang, H. Monitoring long-term annual urban expansion (1986–2017) in the largest archipelago of China. *Sci. Total Environ.* **2021**, *776*, 146015. [CrossRef]

32. Ma, T.; Zhou, C.; Pei, T.; Haynie, S.; Fan, J. Quantitative estimation of urbanization dynamics using time series of DMSP/OLS nighttime light data: A comparative case study from China's cities. *Remote Sens. Environ.* **2012**, *124*, 99–107. [CrossRef]

33. United Nations, United Nations Department of Economic and Social Affairs, Popululation Division. *World Urbanization Prospects: The 2014 Revision*; United Nations: New York, NY, USA, 2015; pp. 1–18.

34. Wu, Y.; Tang, G.; Gu, H.; Liu, Y.; Yang, M.; Sun, L. The variation of vegetation greenness and underlying mechanisms in Guangdong province of China during 2001–2013 based on MODIS data. *Sci. Total Environ.* **2018**, *653*, 536–546. [CrossRef]

35. Zhou, D.; Zhao, S.; Liu, S.; Zhang, L. Spatiotemporal trends of terrestrial vegetation activity along the urban development intensity gradient in China's 32 major cities. *Sci. Total Environ.* **2014**, *488*, 136–145. [CrossRef] [PubMed]

36. Mishra, N.B.; Mainali, K.P. Greening and browning of the Himalaya: Spatial patterns and the role of climatic change and human drivers. *Sci. Total Environ.* **2017**, *587*, 326–339. [CrossRef] [PubMed]

37. Arnfield, A.J. Two decades of urban climate research: A review of turbulence, exchanges of energy and water, and the urban heat island. *Int. J. Climatol. A J. R. Meteorol. Soc.* **2003**, *23*, 1–26. [CrossRef]

38. Li, D.; Wu, S.; Liang, Z.; Li, S. The impacts of urbanization and climate change on urban vegetation dynamics in China. *Urban For. Urban Green.* **2020**, *54*, 126764. [CrossRef]

39. Jin, K.; Wang, F.; Li, P. Responses of vegetation cover to environmental change in large cities of China. *Sustainability* **2018**, *10*, 270. [CrossRef]

40. Huang, B.; Li, Z.; Dong, C.; Zhu, Z.; Zeng, H. Effects of urbanization on vegetation conditions in coastal zone of China. *Prog. Phys. Geogr. Earth Environ.* **2020**, *45*, 564–579. [CrossRef]

41. Wang, J.; Wang, K.; Zhang, M.; Zhang, C. Impacts of climate change and human activities on vegetation cover in hilly southern China. *Ecol. Eng. J. Ecotechnol.* **2015**, *81*, 451–461. [CrossRef]

42. Zhou, Q.; Zhao, X.; Wu, D.; Tang, R.; Peng, Y. Impact of Urbanization and Climate on Vegetation Coverage in the Beijing–Tianjin–Hebei Region of China. *Remote Sens.* **2019**, *11*, 2452. [CrossRef]

43. Gottfried, M.; Pauli, H.; Futschik, A.; Akhalkatsi, M.; Barancok, P.; Alonso, J.B.; Coldea, G.; Dick, J.; Erschbamer, B.; Calzado, M.F. Continent-wide response of mountain vegetation to climate change. *Nat. Clim. Chang.* **2012**, *2*, 111–115. [CrossRef]

44. Pearson, R.G.; Phillips, S.J.; Loranty, M.M.; Beck, P.; Damoulas, T.; Knight, S.J.; Goetz, S.J. Shifts in Arctic vegetation and associated feedbacks under climate change. *Nat. Clim. Chang.* **2013**, *3*, 673–677. [CrossRef]

45. Peng, S.; Piao, S.; Ciais, P.; Myneni, R.B.; Chen, A.; Chevallier, F.D.R.; Dolman, A.J.; IJanssens, V.A.; Peñuela, J.; Zhang, G. Asymmetric effects of daytime and night-time warming on Northern Hemisphere vegetation. *Nature* **2013**, *501*, 88. [CrossRef] [PubMed]

46. Tang, H.; Yu, K.; Hagolle, O.; Jiang, K.; Geng, X.; Zhao, Y. A cloud detection method based on a time series of MODIS surface reflectance images. *Int. J. Digit. Earth* **2013**, *6*, 157–171. [CrossRef]

47. Jia, K.; Liang, S.; Liu, S.; Li, Y.; Xiao, Z.; Yao, Y.; Jiang, B.; Zhao, X.; Wang, X.; Xu, S. Global land surface fractional vegetation cover estimation using general regression neural networks from MODIS surface reflectance. *IEEE Trans. Geosci. Remote Sens.* **2015**, *53*, 4787–4796. [CrossRef]

48. Yang, L.; Jia, K.; Liang, S.; Liu, J.; Wang, X. Comparison of four machine learning methods for generating the GLASS fractional vegetation cover product from MODIS data. *Remote Sens.* **2016**, *8*, 682. [CrossRef]

49. Jia, K.; Liang, S.; Wei, X.; Yao, Y.; Yang, L.; Zhang, X.; Liu, D. Validation of Global LAnd Surface Satellite (GLASS) fractional vegetation cover product from MODIS data in an agricultural region. *Remote Sens. Lett.* **2018**, *9*, 847–856. [CrossRef]

50. Li, X.; Zhou, Y.; Zhao, M.; Zhao, X. A harmonized global nighttime light dataset 1992–2018. *Sci. Data* **2020**, *7*, 1–9. [CrossRef]

51. Yang, K.; He, J.; Tang, W.; Qin, J.; Cheng, C.C. On downward shortwave and longwave radiations over high altitude regions: Observation and modeling in the Tibetan Plateau. *Agric. For. Meteorol.* **2010**, *150*, 38–46. [CrossRef]

52. He, J.; Yang, K.; Tang, W.; Lu, H.; Qin, J.; Chen, Y.; Li, X. The first high-resolution meteorological forcing dataset for land process studies over China. *Sci. Data* **2020**, *7*, 1–11. [CrossRef]

53. Jun, C.; Ban, Y.; Li, S. Open access to Earth land-cover map. *Nature* **2014**, *514*, 434. [CrossRef]

54. Du, X.; Zhao, X.; Zhou, T.; Jiang, B.; Xu, P.; Wu, D.; Tang, B. Effects of climate factors and human activities on the ecosystem water use efficiency throughout Northern China. *Remote Sens.* **2019**, *11*, 2766. [CrossRef]

55. Tao, S.; Fang, J.; Zhao, X.; Zhao, S.; Shen, H.; Hu, H.; Tang, Z.; Wang, Z.; Guo, Q. Rapid loss of lakes on the Mongolian Plateau. *Proc. Natl. Acad. Sci. USA* **2015**, *112*, 2281–2286. [CrossRef] [PubMed]

56. Calcagno, V.; De Mazancourt, C. glmulti: An R package for easy automated model selection with (generalized) linear models. *J. Stat. Softw.* **2010**, *34*, 1–29. [CrossRef]

57. Lopatin, J.; Dolos, K.; Hernández, H.; Galleguillos, M.; Fassnacht, F. Comparing generalized linear models and random forest to model vascular plant species richness using LiDAR data in a natural forest in central Chile. *Remote Sens. Environ.* **2016**, *173*, 200–210. [CrossRef]

58. Ravindra, K.; Rattan, P.; Mor, S.; Aggarwal, A.N. Generalized additive models: Building evidence of air pollution, climate change and human health. *Environ. Int.* **2019**, *132*, 104987. [CrossRef] [PubMed]

59. Virtanen, R.; Luoto, M.; Rämä, T.; Mikkola, K.; Hjort, J.; Grytnes, J.-A.; Birks, H.J.B. Recent vegetation changes at the high-latitude tree line ecotone are controlled by geomorphological disturbance, productivity and diversity. *Glob. Ecol. Biogeogr.* **2010**, *19*, 810–821. [CrossRef]

60. Yuan, J.; Xu, Y.; Xiang, J.; Wu, L.; Wang, D. Spatiotemporal variation of vegetation coverage and its associated influence factor analysis in the Yangtze River Delta, eastern China. *Environ. Sci. Pollut. Res.* **2019**, *26*, 32866–32879. [CrossRef]

61. Piao, S.; Wang, X.; Park, T.; Chen, C.; Myneni, R.B. Characteristics, drivers and feedbacks of global greening. *Nat. Rev. Earth Environ.* **2019**, *1*, 1–14. [CrossRef]

62. Fu, W.; Lü, Y.; Harris, P.; Comber, A.; Wu, L. Peri-urbanization may vary with vegetation restoration: A large scale regional analysis. *Urban For. Urban Green.* **2018**, *29*, 77–87. [CrossRef]

63. Liu, Q.; Yang, Y.; Tian, H.; Zhang, B.; Lei, G.U. Assessment of human impacts on vegetation in built-up areas in China based on AVHRR, MODIS and DMSP_OLS nighttime light data, 1992–2010. *Chin. Geogr. Sci.* **2014**, *24*, 231–244. [CrossRef]

64. Zhao, X.; Tan, K.; Zhao, S.; Fang, J. Changing climate affects vegetation growth in the arid region of the northwestern China. *J. Arid Environ.* **2011**, *75*, 946–952. [CrossRef]

65. Norby, R.J.; Delucia, E.H.; Gielen, B.; Alfapietra, C.C.; Giardina, C.P.; King, J.S.; Ledford, J.; Mccarthy, H.R.; Moore, D.J.; Ceulemans, R. Forest response to elevated CO_2 is conserved across a broad range of productivity. *Proc. Natl. Acad. Sci. USA* **2005**, *102*, 18052–18056. [CrossRef] [PubMed]

66. Pourghasemi, H.R.; Rossi, M. Landslide susceptibility modeling in a landslide prone area in Mazandarn Province, north of Iran: A comparison between GLM, GAM, MARS, and M-AHP methods. *Theor. Appl. Climatol.* **2017**, *130*, 1–25. [CrossRef]

67. Mao, J.; Ribes, A.; Yan, B.; Shi, X.; Thornton, P.E.; Séférian, R.; Ciais, P.; Myneni, R.B.; Douville, H.; Piao, S. Human-induced greening of the northern extratropical land surface. *Nat. Clim. Chang.* **2016**, *6*, 959–963. [CrossRef]

68. Johnson, S.E.; Abrams, M.D. Age class, longevity and growth rate relationships: Protracted growth increases in old trees in the eastern United States. *Tree Physiol.* **2009**, *29*, 1317–1328. [CrossRef]

69. Li, X.; Zhou, Y. Urban mapping using DMSP/OLS stable night-time light: A review. *Int. J. Remote Sens.* **2016**, *38*, 6030–6046. [CrossRef]

70. Zhou, Y.; Smith, S.J.; Elvidge, C.D.; Zhao, K.; Thomson, A.; Imhoff, M. A cluster-based method to map urban area from DMSP/OLS nightlights. *Remote Sens. Environ.* **2014**, *147*, 173–185. [CrossRef]

71. Zhao, M.; Zhou, Y.; Li, X.; Cheng, W.; Zhou, C.; Ma, T.; Li, M.; Huang, K. Mapping urban dynamics (1992–2018) in Southeast Asia using consistent nighttime light data from DMSP and VIIRS. *Remote Sens. Environ.* **2020**, *248*, 111980. [CrossRef]

72. Li, X.; Gong, P.; Zhou, Y.; Wang, J.; Bai, Y.; Chen, B.; Hu, T.; Xiao, Y. Mapping global urban boundaries from the global artificial impervious area (GAIA) data. *Environ. Res. Lett.* **2020**, *15*, 094044. [CrossRef]
73. Zhao, M.; Cheng, C.; Zhou, Y.; Li, X.; Shen, S.; Song, C. A global dataset of annual urban extents (1992–2020) from harmonized nighttime lights. *Earth Syst. Sci. Data Discuss.* **2021**, *7*, 1–25. [CrossRef]
74. Cao, X.; Chen, J.; Imure, H.; Higashi, O. A SVM-based method to extract urban areas from DMSP-OLS and SPOT VGT data. *Remote Sens. Environ.* **2009**, *113*, 2205–2209. [CrossRef]
75. Frolking, S.; Milliman, T.; Seto, K.C.; Friedl, M.A. A globa lfingerprint of macro-scale changes in urban structure from 1999 to 2009. *Environ. Res. Lett.* **2013**, *8*, 024004. [CrossRef]
76. Liu, Z.; He, C.; Zhang, Q.; Huang, Q.; Yang, Y. Extracting the dynamics of urban expansion in China using DMSP-OLS nighttime light data from 1992 to 2008. *Landsc. Urban Plan.* **2012**, *106*, 62–72. [CrossRef]
77. Liu, F.; Tang, L.; Liao, K.; Ruan, L.; Liu, P. Spatial distribution and regional difference of carbon emissions efficiency of industrial energy in China. *Sci. Rep.* **2021**, *11*, 1–14. [CrossRef] [PubMed]
78. Wang, S.; Fang, C.; Wang, Y.; Huang, Y.; Ma, H. Quantifying the relationship between urban development intensity and carbon dioxide emissions using a panel data analysis. *Ecol. Indic.* **2015**, *49*, 121–131. [CrossRef]

Article

The Surface Urban Heat Island and Key Mitigation Factors in Arid Climate Cities, Case of Marrakesh, Morocco

Abdelali Gourfi [1,2,3,*], **Aude Nuscia Taïbi** [1], **Salima Salhi** [1,2], **Mustapha El Hannani** [1] and **Said Boujrouf** [2]

[1] ESO, UMR CNRS 6590, 5 bis Bd. Lavoisier, Université d'Angers, 49045 Angers, France
[2] Laboratoire des Etudes sur les Ressources, Mobilité et Attractivité (LERMA), Faculté des Lettres et des Sciences Humaines, Université Cadi Ayyad, Marrakech 40030, Morocco
[3] Laboratoire de Géoressources, Géoenvironnement et Génie Civil (L3G), Faculté des Sciences et Techniques, Université Cadi Ayyad, Marrakech 40000, Morocco
* Correspondence: abdelali.gourfi@edu.uca.ac.ma

Abstract: The use of vegetation is one of the effective methods to combat the increasing Urban Heat Island (UHI). However, vegetation is steadily decreasing due to urban pressure and increased water stress. This study used air temperature measurements, humidity and an innovative advanced earth system analysis to investigate, at daytime, the relationship between green surfaces, built-up areas and the surface urban heat island (SUHI) in Marrakesh, Morocco, which is one of the busiest cities in Africa and serves as a major economic centre and tourist destination. While it is accepted that UHI variation is generally mitigated by the spatial distribution of green spaces and built-up areas, this study shows that bare areas also play a key role in this relationship. The results show a maximum mean land surface temperature difference of 3.98 °C across the different city neighbourhoods, and bare ground had the highest correlation with temperature (r = 0.86). The correlation between the vegetation index and SUHI is decreasing over time, mainly because of the significant changes in the region's urban planning policy and urban growth. The study represents a relevant overview of the factors impacting SUHI, and it brings a new perspective to what is known so far in the literature, especially in arid climate areas, which have the specificity of large bare areas playing a major role in SUHI mitigation. This research highlights this complex relationship for future sustainable development, especially with the challenges of global warming becoming increasingly critical.

Keywords: urban heat island; land surface temperature; vegetation; built-up; bare soils; Marrakesh

Citation: Gourfi, A.; Taïbi, A.N.; Salhi, S.; Hannani, M.E.; Boujrouf, S. The Surface Urban Heat Island and Key Mitigation Factors in Arid Climate Cities, Case of Marrakesh, Morocco. *Remote Sens.* **2022**, *14*, 3935. https://doi.org/10.3390/rs14163935

Academic Editors: Yuyu Zhou, Elhadi Adam, John Odindi and Elfatih Abdel-Rahman

Received: 26 June 2022
Accepted: 8 August 2022
Published: 13 August 2022

Publisher's Note: MDPI stays neutral with regard to jurisdictional claims in published maps and institutional affiliations.

1. Introduction

Today, cities are facing many new challenges and issues. They are increasingly urbanised [1], dense and sprawling in order to accommodate demographic growth and the concentration of activities. This has led to a degradation of the urban microclimate due to an increase in urban temperatures [2,3]. This phenomenon, known as the urban heat island (UHI) effect, is intensified by the effects of global warming [4]. Moreover, several studies [5–7] argue that extreme heat will pose an increased threat to public health, such as the severe increase in heat-related mortality in urban areas, explained in part by the urban heat island effect.

Studies on UHI are of central interest to all scientific disciplines, as it affects human activities, health [8] and ecosystems [9]. UHI has several causes, but it is strongly linked to a reduction in latent heat at the expense of sensible variation in heat and the removal of vegetation cover, particularly in urban areas where there is low vegetation cover [10,11].

Vegetation lowers ground surface temperatures by providing shade [12,13]. Trees and vegetation decrease ambient air temperature through the process of evapotranspiration, where they release water vapour into the atmosphere [14,15]. In fact, areas with very high temperatures are characterised either by an absence of vegetation or by predominantly artificial impermeable surfaces, such as roads, buildings, pavement, etc. Thus, the properties of materials used in the construction of urban structures, such as solar reflectance, heat capacity and thermal emissivity, play a major role in the formation of UHIs. Additionally, the waste heat generated by factories, air conditioners and motor vehicles, which are ubiquitous in the city, is an important factor [16]. These negative effects of UHI can be significantly controlled through sustainable development combined with mitigation or adaptation measures. The 'strategic' planting of trees and vegetation in urban areas is one of the most effective methods to reduce the effects of UHI, as vegetation cover increases evaporative cooling while providing shade, reducing solar radiation from the heat-exposed soil [17].

In their study, [18] found that water bodies and vegetation had the lowest land surface temperatures, whereas bare ground had the highest ones. However, bare soils are commonly considered as all urbanised surfaces, whereas in reality it is necessary to distinguish bare areas from built up areas because they do not behave the same way when it comes to heat islands. There are very few studies combining UHI formation with vegetation and built-up areas, particularly in cities of the Global South and at the neighbourhood level. Therefore, in this study, we will present the relationship and impact of vegetation and built-up areas on the presence or absence of UHI, providing insight into this phenomenon.

Two kinds of UHI exist: the canopy layer heat island (CLHI) and the surface urban heat island (SUHI). While CLHI refers to the warming of the urban air, SUHI describes changes in surface temperature [19,20] and, influenced mainly by the albedo, land use and building typology and materials [21]. The most common method used in the literature for SUHI analysis is the retrieval of land surface temperature (LST) from satellite images [22–29].

Solar radiation warms the earth's surface, while longwave infrared radiation mostly heats the atmosphere from the ground up. Given that the land surface's energy balance is a complicated process that depends on several factors, the link between LST and air temperature (Tair) may change over time and space (e.g., surface roughness, cloud cover, soil moisture and wind speed). In a recent study, when [30] compared LST and Tair from 2007 to 2013, they found that LST and Tair at daytime are strongly correlated; this correlation is generally stronger than that of the night time. Satellite LST images can be used to derive air temperature [31]. On this basis, it can be concluded that the detailed study of LST can provide a lot of information on the state of UHI and its spatial variation.

In contrast to in situ measurements, which provide sparsely distributed data, satellite imagery products allow monitoring of the urban heat island with global spatial coverage, ensuring better analysis of intra-urban spatial variability of UHI, which is closely related to the distribution of buildings, surface materials and green spaces. Satellite remote sensing is an excellent tool to study the earth's surface properties and the effects of UHI [32,33]. In this study, we use an exploratory analysis to investigate the SUHI during the daytime and its relationship with different factors in order to understand the mitigating factors for the city of Marrakesh, Morocco in light of the challenges related to climate, environment, demography and water needs.

2. Study Area

Marrakesh is a semi-arid continental city located in central Morocco (Figure 1). The area has undergone radical change since it was founded. It was desert land before the occupation, a garden city before colonisation, and now, it has become a tourist city. Marrakesh has also experienced a rapid urban demographic transition in the last decades as a result of its economic development. Currently, the city is part of the Marrakesh–Safi region, which covers 230 km^2, with 5 districts (Marrakesh–Medina, Guéliz, Menara, Ennakhil and Sidi Youssef Ben Ali/SYBA) and 25 neighbourhoods. It is home to 928,850 inhabitants with a density of 350 inhabitants per km^2 [34]. The Medina and Mechewer districts constitute the centuries-old centre of the city, next to the colonial neighbourhoods of Hivernage and Guéliz, as well as the military district in colonial times. These French neighbourhoods, examples of the French colonial urban model with public parks and gardens, squares and public places, and trees aligned along the streets [35], were created with a hygienist objective, synonymous with modernity, civilisation and social progress, and an attractive reassuring framework and pleasant environment for French people [36]. The other districts were built after independence. SYBA, an informal working-class neighbourhood from the 1960s, was built around old foundations and later regularised. In the south, Hay Jadid underwent a similar process after the 1990s. Daoudiate is a planned middle-class neighbourhood from the 1970s, as are Issil and Amerchich from the 1980s. Mhamid, Massira, Iziki, Azli, Hay Hassani and Riad Salam are high-rise working-class neighbourhoods planned in the 1990s and 2000s and which spill over into predominantly rural neighbourhoods such as Sidi Ghanem, El Inara, Maata Allah, Azkjjour, Bouaakaz and Al Izdihar. The districts of Ennakhil North and South, which developed in the palm grove mainly after 2000, are home to luxury villas and hotels and are relatively unurbanized; however, the palm grove has completely lost its traditional features, retaining only one layer of low-density palm trees, leaving large areas of bare soils.

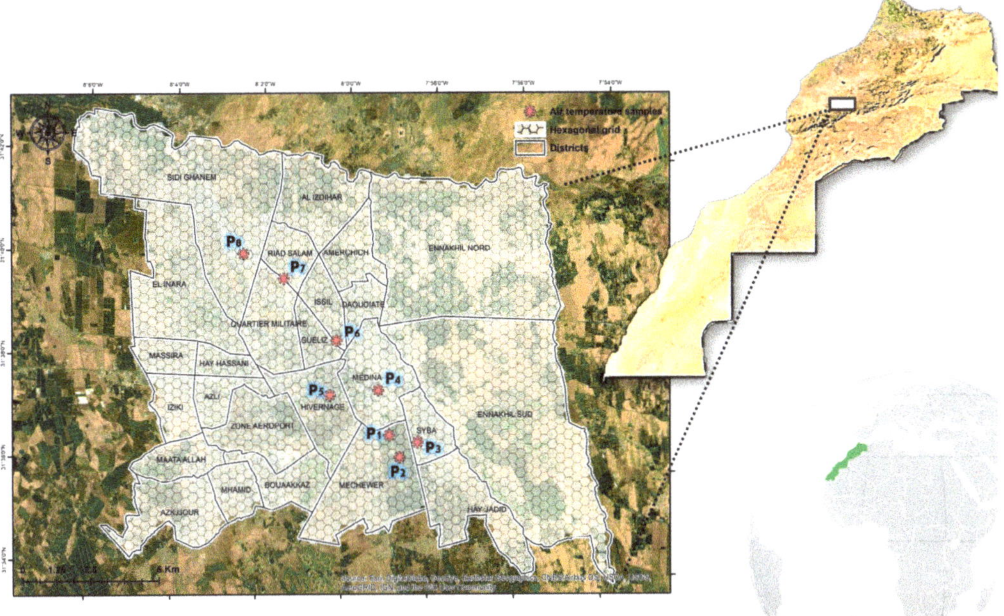

Figure 1. Location of the study area, air temperature samples, districts and hexagonal grid used.

Marrakesh is characterised by a semi-arid continental climate. Its average annual temperature is around 20 °C with peaks of around 40 °C. Rainfall periods are between November and April and reach a climatological mean annual total that fluctuates between 150 and 350 mm/year. Outside the rainy season, the atmosphere remains dry with a high evaporative demand.

Today, the urban growth in Marrakesh has led to an increase in demand for housing, basic infrastructure, parks, tourist attractions and real estate, often at the expense of vegetated areas.

The vegetation in the city corresponds largely to the heritage of the colonial period and is today in poor condition, with the exception of a few emblematic public and private parks and gardens, widely recognised in the countries of the North as former colonial metropolises.

3. Material and Methods

In order to understand the relationship between and spatial variation in vegetation, the heat island and built-up areas, we studied the area at the borough, neighbourhood and hexagonal scales. The borough scale was used because population distribution data are available at this unique level, and it was compared with the neighbourhood scale, which is more accurate. We also used a hexagonal grid with polygons of 10 ha, in order to extract more detailed spatial information about the extent of the vegetation, heat island and built-up areas in 2020.

The methodological approach can be divided into three steps:

The first step was to measure air temperature and humidity at the near-surface using a XIAOMI Mijia thermometer (LYWSD03MMC). The temperature measurement range was from 0 °C to 60 °C (resolution ± 0.1 °C), and the humidity range was from 0% to 99% RH (resolution ± 1% RH). The sampling points were taking during midday (between 12:00 am and 14:00 am Greenwich Mean Time) and were carefully selected on the basis of one sample per location (Table 1) to cover the different land uses of the city. The sample point P7 was taken at the level of bare ground in Riad Salam district. The other specimen point, P5, was taken at the level of dense vegetation at Hivernage district, and the sample points P1 and P2 were taken at the level of sparse vegetation at Mechewer district. Additionally, the specimen points P3 and P4 were taken at the level of an alley in the SYBA and Medina districts, respectively. The alleyways in the Medina and SYBA districts of Marrakesh are characterized by narrow streets between dense, neighbouring buildings (ground floor plus one or two floors) and with an open sky view, often sheltered from the sun. The sample point P7 is taken at the level of an alley in Issil. The alleys of the Issil district are characterized by fairly wide alleys; the construction is of the ground floor type plus two storeys, with an open sky view, often exposed to solar radiation.

Table 1. Air temperature and humidity sampling points.

Sample	District	X	Y	Air Temperature (°C)	Humidity (%)	LST (°C)
P1	Mechewer	31.60954	−7.9825	31.5	48	31.26
P2	Mechewer	31.6026	−7.97826	33.4	46	30.51
P3	SYBA	31.60755	−7.97137	33	45	30.96
P4	Medina	31.62393	−7.98717	30.5	49	30.69
P5	Hivernage	31.62195	−8.00576	30.2	50	27.29
P6	Issil	31.64318	−8.0075	34.4	44	31.39
P7	Riad Salam	31.65922	−8.02481	37.4	38	35.47
P8	Sidi Ghanem	31.66696	−8.04049	36.9	40	34.93

The samples were taken during the same daytime period in order to obtain a clear result on the spatial variation in air temperatures with different land uses.

The thermometer used was covered by a box to protect the temperature sensor from being influenced by direct or reflected sunlight. The wood box was painted white to better reflect the sun's rays. The sides with openings allowed outside air to flow around the thermometer. The box was raised to a height of 50 cm to measure the temperature of the air above the ground.

The second step consisted of identifying vegetation cover and built-up and bare soil areas (Figure 2). In this step, the Sentinel 2A was used for more accurate results in estimating land use land cover than those generated from the Landsat time series [37]. In this step, three different stages were carried out: (1) Atmospheric correction and radiometric calibration were performed for each Sentinel-2A spectral band; in this stage, radiance, reflectance and brightness temperatures were optimized and dark subtraction was used to remove the effects of atmospheric scattering from an image. ENVI software was used because of the very powerful scripts used for this purpose, through the "FLAASH Atmospheric correction" and "Radiometric calibration" functions. (2) On the corrected and preprocessed dataset, principal component analysis was performed in order to improve the image presentation by using a data compression technique to separate the noise components, lower the dimensionality of datasets and provide uncorrelated output bands [38]. The ArcGIS software was used to perform the Principal Component Analysis on a set of raster bands and generate a single multiband raster as output. (3) The supervised classification using the Maximum Likelihood Algorithm was used for vegetation sensing, with the samples selected on the basis of the false colour multispectral composite image where the vegetation appears in red colour. For built-up sensing, the samples were chosen using Google Earth satellite imagery [39]. Finally, bare ground refers to stony mineral areas that have not been built on and not vegetated; basically, the entire area with the exception of vegetation and buildings.

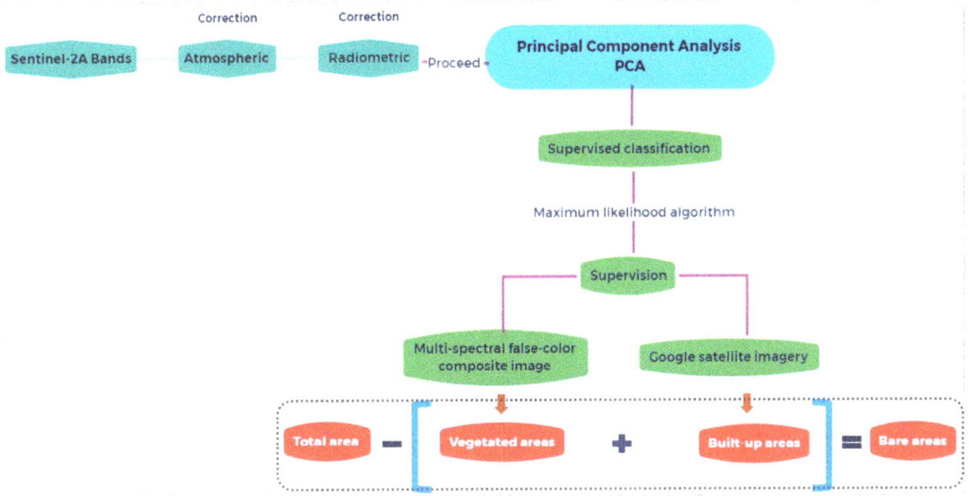

Figure 2. Methodology for the determination of the vegetation cover, built-up areas and bare areas.

The third step concentrated on the calculation of the spatial distribution of LST (land surface temperature) and NDVI (Normalized Difference Vegetation Index).

In this study, we will calculate these two parameters and their linear regression for the years studied (1985–1990–1995–2000–2005–2010–2015–2020) using Landsat satellite images.

Images from the Landsat satellite (5, 7, 8) and Sentinel-2A satellite were used in this study (Table 2). They consist of optical images from 2020 for Sentinel-2A, mainly used to map vegetation areas and built-up areas, and Landsat, used mainly to map NDVI and LST indexes. All the images were acquired from the USGS Global Visualization Viewer website (GloVis) [40]. All the Landsat data were acquired in the morning (between 10:00 am and 12:00 am Greenwich Mean Time).

Table 2. Satellite products used for the study.

Satellite	Sensor	No. of MS Bands (Nominal Resolution)	Period
Landsat 5	TM	6 (30 m)	27 July 1985 9 July 1990 8 August 1995 19 August 2005 16 July 2010
Landsat 7	ETM+	6 (30 m)	13 August 2000
Landsat 8	OLI/TIRS	8 (30 m)	14 July 2015 12 August 2020
Sentinel-2A		13 (10–20 m)	14 August 2020

The images were taken during the dry months (June–July) in order to identify the permanent vegetation cover, which is irrigated during this dry season in Marrakesh. Only images with little or no cloud cover were selected to improve the accuracy of the classifications.

The NDVI values range from −1 to +1. A higher NDVI value indicates healthy and dense vegetation; a lower NDVI value indicates sparse vegetation [41]. It is expressed as follows:

$$NDVI = (NIR - RED)/(NIR + RED) \tag{1}$$

where RED and NIR represent spectral reflectance measurements acquired in the red (visible) and near infrared regions, respectively.

The LST index is the key factor for calculating the highest and lowest temperatures of a specific place [42]. It is calculated by using Land Surface Emissivity, top of atmosphere brightness temperature and wavelength of emitted radiance.

The LST ($^\circ$C) is calculated using two methodologies depending on the sensor used: TM, ETM+ or OLI.

For the TM and ETM+ sensor, LST is calculated using the following equation:

$$LST = K2/(\ln(K1/L\lambda + 1)) - 273.15 \tag{2}$$

where K1 is the calibration constant 1 in Watts/(m^2 × sr × µm) and K2 is the calibration constant 1 in Kelvin. K1 and K2 are calculated based on the band 6 depending on the sensor used (Table 3):

Table 3. Values of K1 and K2 for Landsat 5 (TM), Landsat 7 (ETM+) and Landsat 8 (OLI).

Sensor	K1	K2
Landsat 5 TM	Band 6 607.76	Band 6 1260.56
Landsat 7 ETM+	Band 6 666.09	Band 6 1282.71
Landsat 8 OLI	Band 10 774.89	Band 10 1321.08

Lλ (Watts/(m^2 × sr × µm)) is the TOA spectral radiance, calculated as follows:

$$L\lambda = \frac{LMAX_\lambda - LMIN_\lambda}{QCALMAX - QCALMIN} \times (QCAL - QCALMIN) + LMIN_\lambda \tag{3}$$

where $QCAL$ = Band 6; $LMAX_\lambda$: Radiance Maximum Band in Watts/(m^2 × sr × µm); $LMIN_\lambda$ = Radiance Minimum Band in Watts/(m^2 × sr × µm); $QCALMIN$ = Minimum quantized calibrated pixel value of band 6 and $QCALMAX$ = Maximum quantized calibrated pixel value of band 6.

Recurring data for the determination of K1, K2, $QCALMIN$, $QCALMAX$, $LMAX_\lambda$ and $LMIN_\lambda$ could be found in the MTL file in the Landsat image folder.

For the OLI sensor, LST ($^\circ$C) is calculated using the following equation:

$$LST = BT/(1+ (W \times BT/14380) \times \ln(E)) \tag{4}$$

where W = Wavelength of Emitted Radiance of band 10 (11.5 um); BT = Top of Atmosphere Brightness Temperature in $^\circ$C; W = Wavelength of Emitted Radiance and E = Land Surface Emissivity.

The top of atmosphere Brightness Temperature (BT) is calculated as follows:

$$BT = K2/(\ln(K1/L\lambda + 1) - 273.15 \tag{5}$$

where K1 = calibration constant 1 in Watts/(m^2 × sr × µm); K2 = calibration constant 1 in Kelvin and Lλ = TOA Spectral Radiance (Watts/(m^2 × sr × µm)).

K1 and K2 are calculated based on the band 10 of the sensor Landsat 8 (Table 3). TOA Spectral Radiance (Lλ) can be calculated as follows:

$$L\lambda = ML \times Qcal + AL - Oi \tag{6}$$

where ML = radiance multiplicative band (3.342 10–4); Qcal = quantized and calibrated standard product pixel values (Band 10); AL = radiance Add Band (AL = 0.1) and Oi=correction value for band 10 (Oi = 0.29).

E, the Land Surface Emissivity, can be calculated as follows:

$$E = 0.986 + PV \times 0.004 \tag{7}$$

where PV = proportion of vegetation, which can be calculated as follows:

$$PV = [(NDVI - NDVImin)/(NDVImax + NDVImin)]^2 \tag{8}$$

where PV = proportion of vegetation; NDVI = DN values from NDVI image (Calculated from Equation (1)); NDVImin = minimum DN values from NDVI image and NDVImax = maximum DN values from NDVI image.

Recurring data for determination of ML, AL, K1 and K2 could be found in the MTL file in the Landsat image folder.

4. Results

4.1. Initial Results

The Urban Community of Marrakesh has a surface area of 213.31 km^2, where vegetation currently represents 43.32 km^2 (20% of the total surface area), the built-up surface area 39.99 km^2 (19% of the total surface area) and the remaining 130 km^2 (61%) represents bare areas (Figure 3).

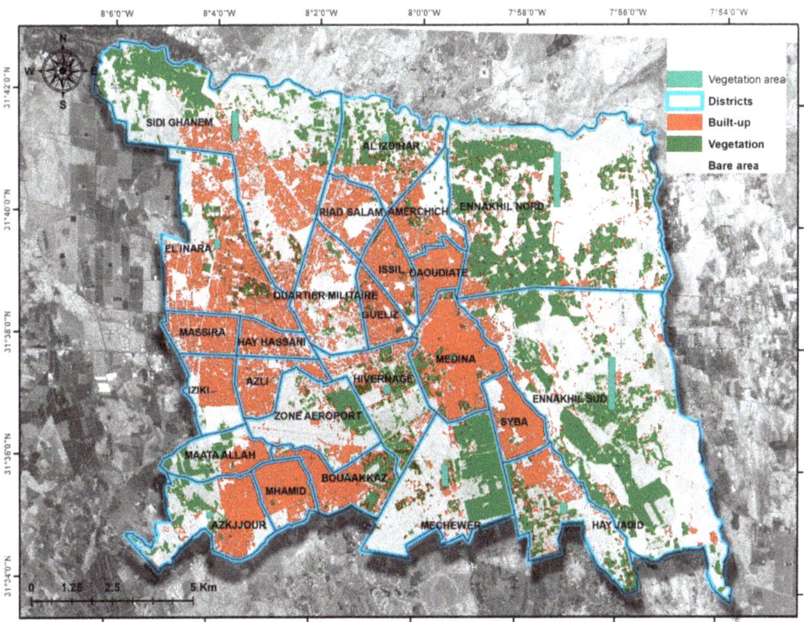

Figure 3. Variation in vegetation in relation to built-up and bare surfaces.

As regards the total vegetation area, the Ennakhil North district has the largest vegetated area, with 995.67 ha. This district also has the second largest surface area (3217.13 ha) and large surfaces of bare ground (2074.11ha, ~65%). The Mhamid district has the lowest vegetated area with 2.04 ha, and almost the entire area is occupied by built-up area (209.75 ha, 68% of total area) and bare ground (95.20 ha, 31% of total area).

As regards the total built-up area, the Medina district is in first place, with an area of 460.09 ha, while the least built-up district is the Mechewer district (47.92 ha).

As regards vegetation density, the analysis of the variation in vegetation on a hexagon scale showed that the most densely vegetated district is Mechewer (33.54%), with a mean of 3.17 ha per 10 ha (hexagon), and it is also the district with the highest standard deviation (3.44 ha). The district with the lowest vegetation density is the Mhamid district (0.66%), with 0.077 ha per 10 ha (hexagon).

Built-up surface density also varies greatly. The Daoudiate district is the densest one in terms of construction, with a percentage of approximately 72%, equivalent to 6.93 ha per 10 ha. The least dense district is Ennakhil Sud (4%), equivalent to 6.38 ha per 10 ha.

4.2. Spatial Variation of LST and NDVI

Figure 4 displays the spatial variation in land surface temperature in the study area. A minimum surface temperature of 25.28 °C was recorded at the water basins used for irrigation, which are located in the Menara and Agdal gardens, and the maximum was 41.45 °C recorded in the large area of bare ground located in the northeastern part of Marrakesh. The range was 16.17 °C, the mean was 33.90 °C and the standard deviation was 2.55 °C.

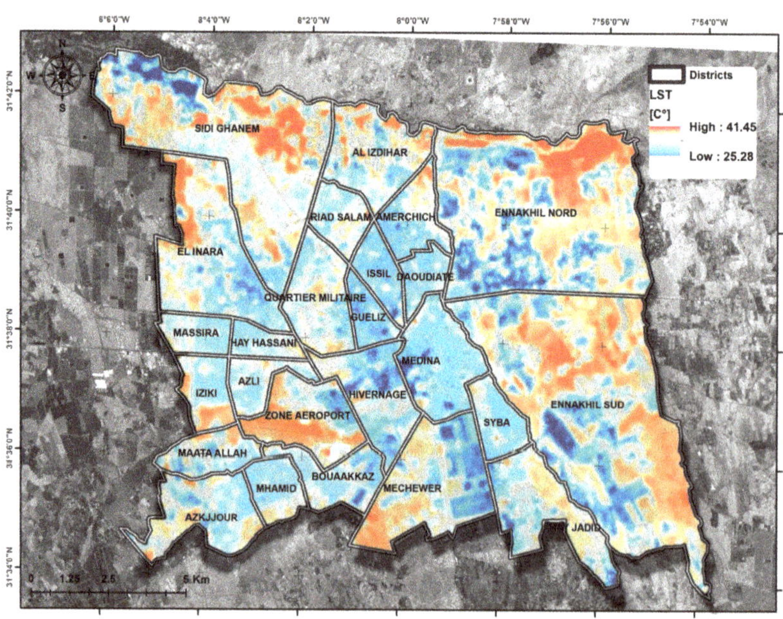

Figure 4. Spatial variation in land surface temperature.

We can see that the highest temperatures were recorded on bare soils, and the temperatures were medium to low in built-up areas and low in vegetated areas.

At the level of the districts (Figure 5), the highest mean LST was observed at Zone Airport with a mean of 35.30 °C (std = 2.33 °C), the lowest mean at Guéliz with a mean of

31.31 °C (std = 1.02 °C) and the highest LST standard deviation was observed at Ennakhil North with std = 2.52 °C.

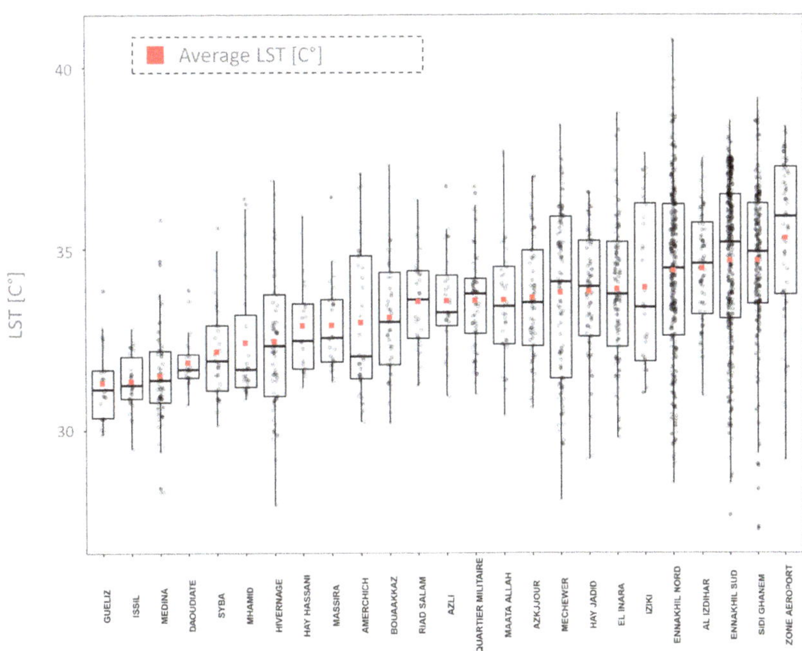

Figure 5. Average temperature by district.

Concerning the NDVI, essentially related to the density, type and health of the vegetation, the Ennakhil North district had the highest average (NDVI = 0.15), and the Mhamid district recorded the lowest average (NDVI = 0.07).

4.3. Relationship between LST, NDVI, Bare Areas, Vegetated Areas and Built-Up Areas

Figure 6 shows the heat map of the correlation matrix of the different factors studied, and the correlation is made by studying LST, NDVI, bare area, vegetated area and built-up area for each hexagon. According to the heat map, NDVI had a close relationship with vegetation, with a very high positive correlation index of r = 0.94, which proves that the NDVI calculation processes are correct. The second strongest correlation was between LST and bare area, with a positive correlation coefficient of r = 0.86, which implies that the more soil ground there is, the higher the ground temperature will be. The correlation between LST and NDVI and vegetation areas showed negative mean values of r = −0.48 and r = −0.53, respectively, and, notably, of r = −0.44 with built-up areas.

4.4. Relationship between LST and NDVI from 1985 to 2020

In order to have more detailed information on the relationship between NDVI and LST, we tracked these two indices over time (Figure 7). In 1985, a strong negative correlation of r = −0.71 was observed, and by 2020, this correlation had decreased significantly, to r = −0.48. The correlation between NDVI and LST is getting weaker over the years, which could be explained by the change in land use of some hexagons from bare or vegetated areas to built-up areas. This change in land use has resulted in hexagons with medium to low LST and low NDVI, which is mainly due to the fact that built-up areas have a different NDVI than bare and vegetated areas.

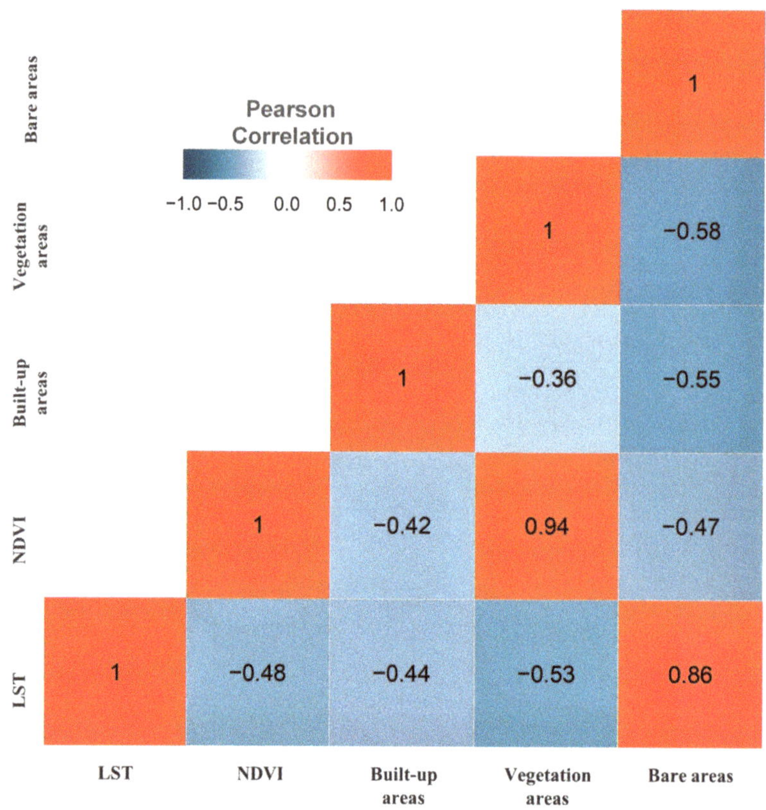

Figure 6. Relationship between LST, NDVI, bare areas, vegetated areas and built-up areas.

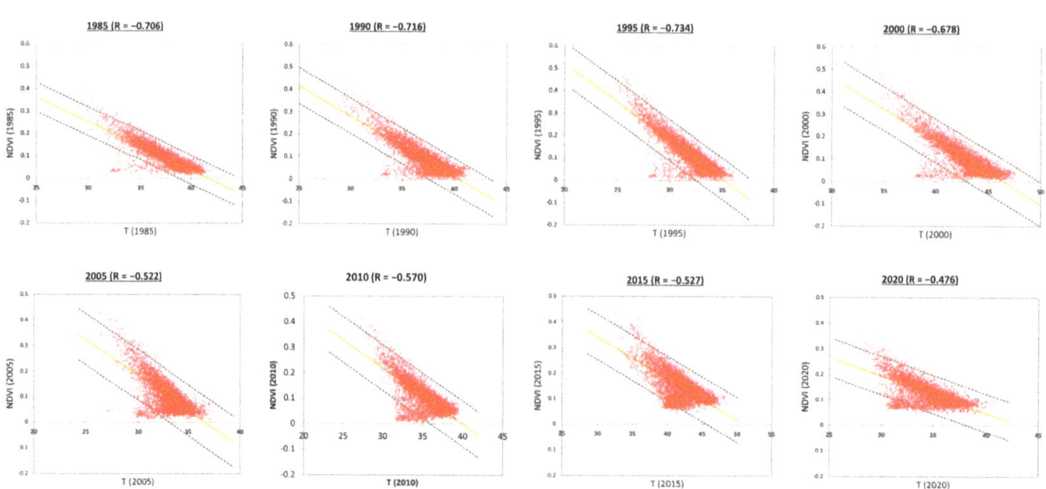

Figure 7. Relationship between LST and NDVI from 1985 to 2020.

On the other hand, the detailed analysis of the relationship between LST and NDVI (Figures 8 and 9) for the year 2020 shows that the hexagons experiencing significant change and with a low mean LST index and a low NDVI index are hexagons with very high urbanisation (>60%), such as SYBA, Mhamid and Medina.

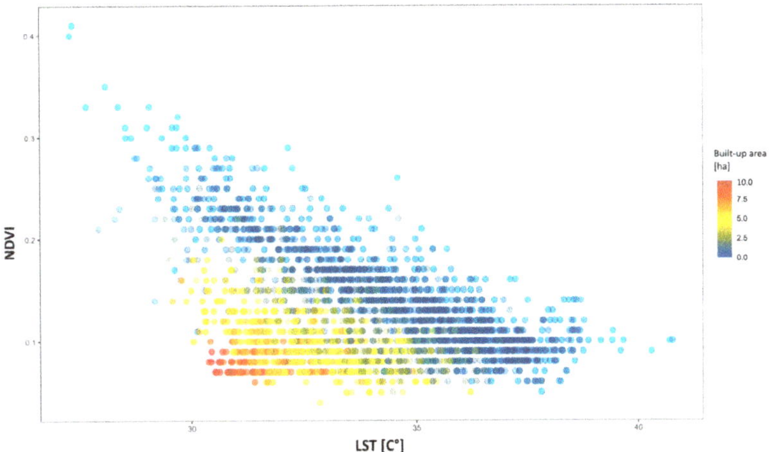

Figure 8. Analysis of the relationship between LST and NDVI per built-up area (2020).

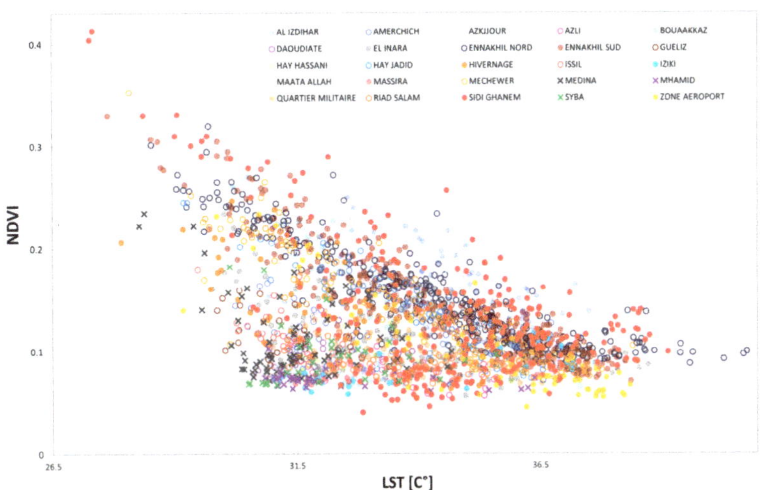

Figure 9. Analysis of the relationship between LST and NDVI by neighbourhood (2020).

4.5. Relationship between Observed Air Temperature and LST

The results of measurement of air temperatures (Table 1) show that the point P5, characterizing a typical green space in the Guéliz district, recorded the lowest temperature (30.20 °C). The highest value is recorded at the point P7, characterizing a bare area from Riad Salam district (37.4 °C). It should be noted that the point characterizing Agdal (P2), with sparse vegetation—mainly olive trees—has recorded relatively medium air temperatures (33.4 °C), while it was expected to have a lower value. On the other hand, the sample taken in the Medina district (P4), with old buildings, recorded the second lowest temperature (30.5 °C), while the sample taken in the industrial buildings in the Sidi Ghanem district (P8) recorded the second highest temperature (36.9 °C).

The comparison between air temperature-observed data with the simulated data of the land surface temperature LST (Figure 10) shows a very high coefficient of determination (R^2 = 0.8), which explains the close relationship and the connection between them. These two factors are also connected to the air humidity: the higher the temperature the lower the humidity, and vice versa.

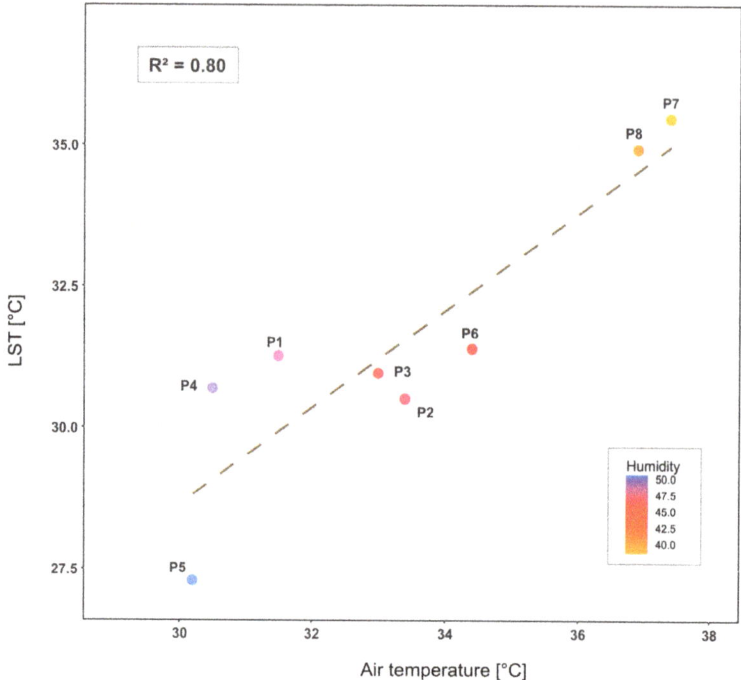

Figure 10. Comparison between observed air temperature and LST.

5. Discussion

Understanding the factors that minimise the adverse effects of UHI is important in the global context of climate change and the increasing frequency of heat waves.

5.1. Relationship LST, Vegetation, Urban and Bares Areas

The vegetation is recognised as being able to attenuate the urban warming effect, since plants use part of the energy absorbed for vital processes and then cool the environment and the surrounding air by producing shade and releasing water vapour [32].

Therefore, the NDVI is widely used as a proxy for monitoring vegetation health and abundance [43–45]. In Marrakesh, the high negative relationship found between NDVI and LST suggests that green areas in the city can help mitigate the SUHI effect. Areas with high NDVI values and relatively healthy and dense vegetation such as grass or woodlands generally have the lowest temperatures. Medium NDVI values represent unhealthy and degraded green areas, and the lowest NDVI values represent bare ground without vegetated areas.

The most practical example is the royal territory in the north of the Mechewer district, composed mainly of the large centuries-old Agdal Garden and the private royal golf course (Figure 11). In the private golf course area, composed mainly of grassy areas and some cypress trees, the vegetation index is high (NDVI > 0.35) and the LST is low, with temperatures below 30 °C. The Agdal Garden, composed mainly of olive trees, relatively spaced in the eastern part, has NDVI values of 0.1–0.2 in the eastern part and between 0.2 and 0.35 in the western part of the garden, with temperatures between 32 and 30 °C.

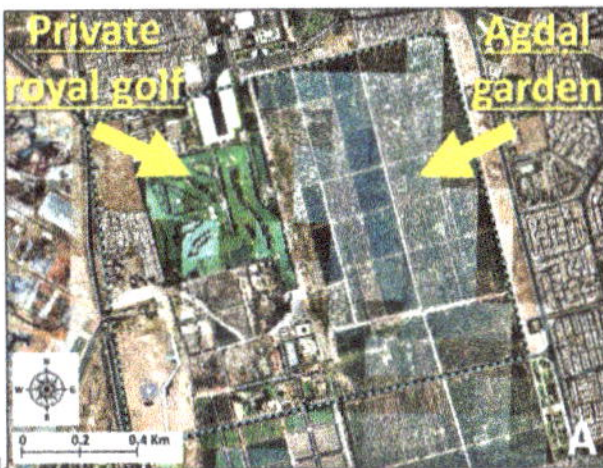

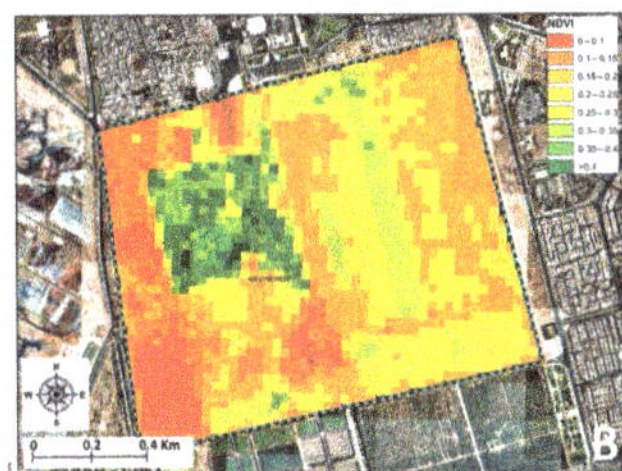

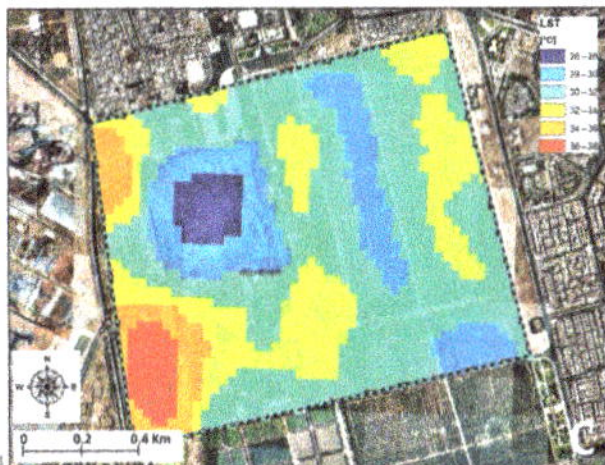

Figure 11. The satellite image (**A**), NDVI (**B**) and LST (**C**) of the Agdal Garden and the royal private golf course.

On this basis, many studies have attempted to understand the relationship and spatial distribution of LST, built-up areas and vegetation in various metropolitan areas worldwide [46–50].

In many investigations, UHI variation has been compared to built-up areas and vegetation [28,46,51–54]. However, although bare soils sometimes occupy large areas of land, notably, in arid climate cities, bare areas have rarely been considered.

Many other studies have investigated these relationships using indexes such as NDVI (Normalized Difference Vegetation Index) and NDBI (The Normalized Difference Built-up Index) on LST [28,55–57]. On the other hand, [58] and [59] indicated that, due to the complexity of the spectral response patterns of vegetation and buildings, NDBI is unable to separate urban areas from certain barren lands, and also that NDVI can result in high values for some types of built-up areas. Therefore, this study presents an innovative method to overcome these drawbacks and study the effects of vegetation and built-up areas as well as bare areas on LST and their interactions. The results show that there is a large spatial and structural variation in neighbourhoods in terms of total area, bare soil areas, vegetation areas and built-up areas, which directly influences LST, thus explaining the wide variation in LST at the neighbourhood level, which is consistent with the findings of [60,61].

The neighbourhood of the Zone Airport (Figure 12) recorded the highest mean LST (35.30° C) and is considered the hottest neighbourhood in the city, despite the presence of a vast and stable green space, the Menara Garden, originating in the 12th century. This can be explained by the presence of large areas of bare ground, the second highest average (7.34/10 ha) in this airport district of Marrakesh, much of which is covered by asphalt pavements with an albedo of about 0.05, which means that only 5% of the light is reflected. The remaining 95% is absorbed, which is why asphalt pavements are considered to be a major factor contributing to the increase in SUHI [62].

In contrast, the neighbourhood of Guéliz (Figure 13), with the lowest mean LST (31.32 °C), is characterised by medium to low occupancy of bare ground and medium to high occupancy of vegetation (23.49 ha) and built-up areas (102.70 ha). This combination has made this district the coolest in the city. Guéliz, historically called "the European neighbourhood", built during the French colonial period to the detriment of the palm grove, was part of the urban policy of Hubert Lyautey (the first French Resident–General of Morocco in the 1910s), which consisted of creating new modern cities for the large French community that settled in Morocco during the colonial period [63]. The area was planned based on the urban French colonial model, featuring real estate complexes and luxury villas with small grass gardens for each building accompanied by alignment trees in the streets. The goal was to provide the city with the necessary hygiene and quality of life [35]. This translates into a relatively small standard deviation in vegetation variation in that neighbourhood (Std = 1.06 per 10 ha), which is due to the uniform distribution of vegetation in the Guéliz district. As far as the type of construction is concerned, the district has undergone changes over time in terms of zoning, with certain single unit zones converted into R +5 building zones (buildings with no more than five floors above the ground floor).

Figure 12. Neighbourhood of the Zone Airport (upper right, the Menara Garden with a water basin).

Figure 13. Panoramic view of the centre of Guéliz.

In their study on the effect of park proximity on the urban heat island effect, [64] found that smaller scale green spaces can reduce SUHI with a magnitude close to that of larger parks. Therefore, small-scale green spaces such as those in Guéliz, which consume less water and take up little urban space, significantly mitigate SUHI in a cost-effective and sustainable way, explaining why Guéliz is the coolest neighbourhood in Marrakesh.

Concerning the impact of bare ground areas on surface temperature, very little is discussed in the literature [65–67], while the correlation analysis between the five parameters of LST, NDVI, bare area, vegetated area and built-up area showed that LST is mostly related to bare areas, with a high correlation of r = 0.86. Bare ground areas presented the highest mean surface temperatures, and the LST map also shows that the bare ground surface area plays a key role in the LST and built-up areas' relationship; the larger the bare areas, the higher the temperatures will be, thus cancelling out the temperature-lowering effect of small vegetated areas in many cases. The Ennakhil neighbourhood, for example, which should induce low temperatures, on the contrary posts high average temperatures. These results have been recently reported in a few studies [18,68].

Furthermore, it is generally accepted that LSTs have an inverse correlation with NDVI, meaning that vegetation spaces help mitigate the LST, and the denser and larger the space, the more the SUHI effect decreases [28,69–71]. However, the correlation of LST with vegetation and NDVI is not strong enough (r < 0.6), with a negative correlation coefficient of r = −0.53 and r = −0.48, although visually (Figure 4), it can be seen that the green areas are those characterised by the lowest temperatures.

Moreover, the analysis of the spatial and temporal variation in the LST-NDVI correlation from 1985 to 2020 shows that, over time, this correlation is reduced, mainly due to the expansion of urbanisation that has a different trend from the LST-NDVI. In other words, the more buildings the district contains, the lower the correlation.

5.2. Impact of Construction Type on LST

The nature of the construction is another factor influencing heat. Thus, the Quartier Industriel (Industrial district) in the Sidi Ghanem district (Figure 14), a commercial centre and artisanal enclave on the outskirts of the city, filled with local designers' workshops and trendy boutiques, posts high temperatures (34.68 °C) despite the density of built-up areas. Indeed, the industrial buildings are composed mainly of load-bearing masonry (bricks, blocks, stones), main frame structures supporting the main floors and roof, and secondary structural elements including balconies, canopies and metal walkways. This type of construction has a low albedo, absorbs higher amounts of solar radiation and converts it to thermal energy, favouring the storage of solar energy [53] and, consequently, increasing the heat islands despite the density of the buildings. Therefore, several studies have recommended increasing the albedo of the area to reduce the accumulated heat by introducing reflective materials with high albedo, especially light-coloured coatings for building roofs [72–74].

One of the most interesting elements that has emerged from this study is the low LST of densely built and poorly vegetated neighbourhoods such as the Medina (31.51° C) and SYBA (32.16 °C). Most researchers theorise that the gradual replacement of natural surfaces by built surfaces, through urbanisation, constitutes the main cause of UHIs [46,75,76]. However, in Marrakesh, the downtown neighbourhoods post some of the lowest temperatures in the city.

The Medina of Marrakesh (Figure 15), classified as a UNESCO World Heritage Site, was designed centuries ago, mostly in the 12th century, with an urban morphology consisting of an assemblage of house courtyards (patios) linked by a hierarchical street network. Residential houses have massive facades with few openings, yet the courtyards connect the houses with the outside environment. This compact organic rather than geometric urban fabric thoroughly adapts to the immediate hot environment by providing shade in all the narrow and winding streets.

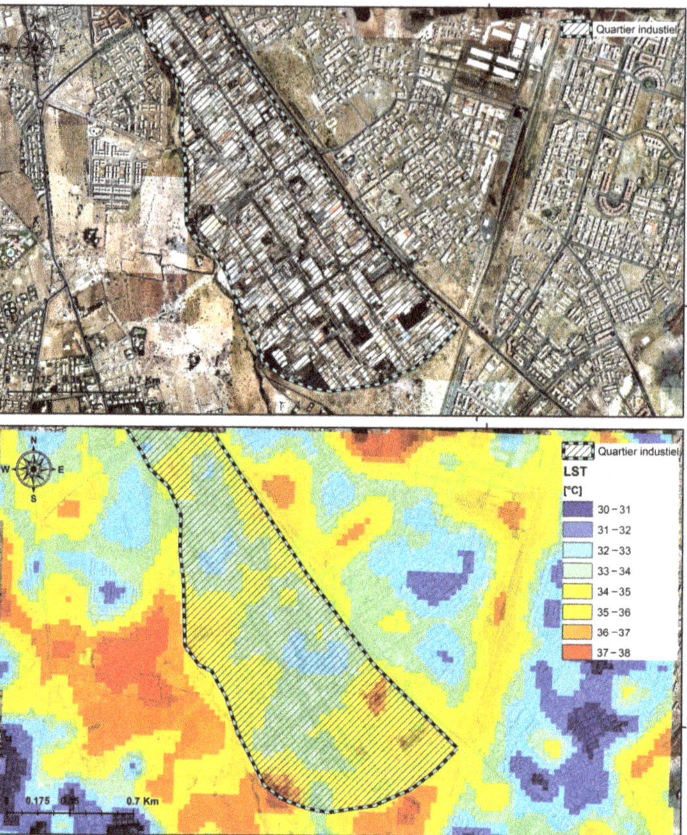

Figure 14. The Quartier Industrial in the Sidi Ghanem district.

Figure 15. The SYBA and Medina district.

The Sidi Youssef Ben Ali (SYBA) (Figure 15) neighbourhood, located in the south of the Medina between the Issil River and the ramparts of the Agdal Garden, developed out of a self-construction phase starting in the 1950s by neo-urban people who could not find places to live in the saturated Medina in a context of strong rural exodus. The urbanisation of this sector was clandestine at the beginning, with multi-family dwellings built on small plots of land (60 to 100 m^2), single-storey at the beginning and rebuilt with two or three storeys in recent decades. This spontaneous extension of the typo-morphology of the Medina in the northern part of the SYBA district features houses with patios squeezed together like traditional fabrics, with high densities and a lack of infrastructure. The area has a compact and dense organisation which has developed in the absence of appropriate urban planning. This mainly concerns the old district, which had experienced a high density as a result of large immigration flows, and since then, this district has shown a definite need for urban recovery, equipment and infrastructure. Due to its geographical location in the southeast of the city of Marrakesh, close to the Medina and in the heart of three major natural components, namely The Agdal gardens and their ramparts, Oued Issil and the vast Hassan II Oasis farmlands to the south, the SYBA district has undergone a programmed urban evolution to limit its spontaneous expansion and diversify its housing through projects carried out by ERAC. A number of public and socio-economic facilities have also been built. Sidi Youssef Ben Ali is a part of the city that some describe as a countryside within an urban perimeter, as it has, in the past, reflected all the images of marginalisation and social exclusion. The latter is closely linked to poverty, unemployment and destitution, which are reflected in individual (public services and facilities) and/or collective deprivation and in the lack of adequate and dignified housing and fairly paid jobs. Despite all these elements of social and urban degradation and downgrade, this neighbourhood demonstrates real qualities in terms of reduction in LST. This is due to the fact that the dense nature of the buildings in these two districts prevents the heat from reaching the street canyon, provides shade and also keeps the sky view open. Another reason could be the construction materials with which the buildings are built, which are often natural and traditional materials such as natural stones or raw earth.

5.3. Synthesis on the Variation in LST in Marrakesh

Marrakesh is a city characterised by an arid climate, where history and social inequality have contributed to differences in the city's districts; these differences mainly concern urban management, vegetation and the urban development plan. These different parameters have had a direct influence on surface heat islands and their presence in districts (Figure 16).

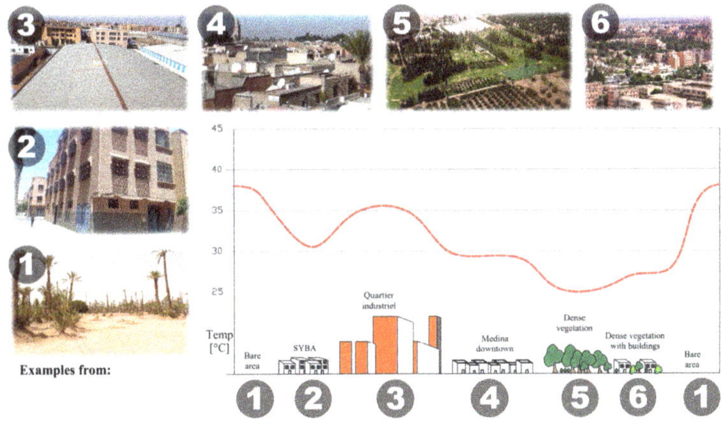

Figure 16. Summary of the spatial variation in LST in the city during the daytime.

In cities with an arid climate, during the daytime, the presence of LST is mainly related to the presence of bare land [65–67]. For example, in Marrakesh, the Airport Zone and Sidi Ghanem districts are considered as the hottest districts of Marrakesh, at 35.30 °C and 34.68 °C, respectively. In these two districts, in addition to large areas of bare lands, there are asphalt pavements (in the case of the Airport Zone) and metal construction (in the case of Sidi Ghanem) which retain even more heat due to their low albedo [72–74].

On the other hand, vegetation and built-up areas contribute to the attenuation of the LST [60,61], as in the case of the districts of Guéliz and Médine. In the case of Guéliz, which has a relatively different organisation due to its colonial history, it is characterised by R + 5 buildings, preventing heat from reaching the street gutter, providing shade and also keeping the view to the sky open. The Guéliz district is also characterised by well-distributed vegetation, reflected in luxury villas and high-class hotels with a large number of grassy gardens and small courtyards [63]. This combination has been shown to minimize temperatures and mitigate the LST [64]. It is important to mention that the presence of vegetation does not always mean low temperatures; the typology of the vegetation, its health and density play an essential role in the mitigation (for example, the Agdal Garden and the Royal Private Golf Course).

In the case of the Medina and SYBA districts, also characterised by low average temperatures of 31.51 °C and 32.16 °C, respectively, the very dense urban architecture as well as the very narrow alleys prevent the sun's heat from reaching the roof of the street and providing more shade [77]. The nature of the buildings as well as their construction materials, mainly natural materials such as straw and clay bricks, act as thermal insulation and prevent light heat from reaching the surface, thus mitigating UHI [78].

6. Conclusions

Most of the studies dealing with the urban heat island effect take into consideration the parameters of vegetation and built-up areas, since it is accepted that vegetation helps reduce UHI while built-up areas have the opposite effect. In this study, we introduced another factor, bare ground, which is rarely taken into account in studies and which is even more important in the semi-arid and arid climate where Marrakesh is located. This study shows that bare areas play a key role in the variation in SUHI, increasing surface temperatures at daytime. Vegetation, depending on its typology and distribution, also comes into play in reducing air and ground temperatures. In fact, well distributed smaller scale green spaces can reduce SUHI with a magnitude close to that of larger parks (case of the Guéliz district).

On the other hand, the traditional dense urban structures characteristic of Arab Muslim cities such as Marrakesh, with dense buildings and narrow winding streets, help reduce temperatures. This is the case of the centuries-old Medina district, but also of the newer SYBA neighbourhood, whose structure imitates that of the old city. In contrast, the 'industrial' buildings generate a higher SUHI effect due to their albedo. The analysis by neighbourhood shows an imbalance at the city scale with strong inequalities. The neighbourhoods of dense vegetation and buildings are generally the coolest.

These results will be of great interest to city managers and planners in arid zones faced with the future challenge of global warming. Although revegetation is of great interest in reducing SUHI, it is also confronted with the scarcity of water in Marrakesh. In this case, the choice of urban structures that mitigate the SUHI effect is crucial.

Author Contributions: A.G. conceived the paper's concept and performed data collection, data pre-processing and processing, and the field mission. A.N.T. and S.S.: writing—original draft preparation; M.E.H. and S.B.: writing—review and editing. All authors have read and agreed to the published version of the manuscript.

Funding: This work, carried out at ESO-Angers, UMR 6590 CNRS (France), is funded by French Partenariat Hubert Curien "Maghreb" (38MAG20 project 4363PH) and the ATLAS short-term postdoctoral mobility program in the Humanities and Social Sciences, proposed by the Arab Council for Social Sciences (ACSS) and the Fondation Maison des Sciences de l'Homme (FMSH).

Data Availability Statement: The data presented in this study are available on request from the corresponding author.

Acknowledgments: The authors wish to acknowledge the constructive comments of the anonymous reviewers.

Conflicts of Interest: The authors declare no conflict of interest.

References

1. Balk, D.; Leyk, S.; Montgomery, M.R.; Engin, H. Global Harmonization of Urbanization Measures: Proceed with Care. *Remote Sens.* **2021**, *13*, 4973. [CrossRef]
2. Congedo, L.; Macchi, S. The demographic dimension of climate change vulnerability: Exploring the relation between population growth and urban sprawl in Dar es Salaam. *Curr. Opin. Environ. Sustain.* **2015**, *13*, 1–10. [CrossRef]
3. Mohsen, H.; Raslan, R.; El-bastawissi, I.Y. the Impact of Changes in Beirut Urban Patterns on the Microclimate: A Review of Urban Policy and Building. *Arch. Planing J.* **2020**, *25*. Available online: https://digitalcommons.bau.edu.lb/apj/vol25/iss1/2/ (accessed on 29 May 2022).
4. Keppas, S.C.; Papadogiannaki, S.; Parliari, D.; Kontos, S.; Poupkou, A.; Tzoumaka, P.; Kelessis, A.; Zanis, P.; Casasanta, G.; De'Donato, F.; et al. Future Climate Change Impact on Urban Heat Island in Two Mediterranean Cities Based on High-Resolution Regional Climate Simulations. *Atmosphere* **2021**, *12*, 884. [CrossRef]
5. Périard, J.D.; Travers, G.J.S.; Racinais, S.; Sawka, M.N. Cardiovascular adaptations supporting human exercise-heat accli-mation. *Auton. Neurosci. Basic Clin.* **2016**, *196*, 52–62. [CrossRef] [PubMed]
6. Pascal, M.; Wagner, V.; Alari, A.; Corso, M.; Le Tertre, A. Extreme heat and acute air pollution episodes: A need for joint public health warnings? *Atmos. Environ.* **2021**, *249*, 118249. [CrossRef]
7. Jung, J.; Uejio, C.K.; Adeyeye, T.E.; Kintziger, K.W.; Duclos, C.; Reid, K.; Jordan, M.; Spector, J.T.; Insaf, T.Z. Using social security number to identify sub-populations vulnerable to the health impacts from extreme heat in Florida, U.S. *Environ. Res.* **2021**, *202*, 111738. [CrossRef] [PubMed]
8. Santamouris, M. Recent progress on urban overheating and heat island research. Integrated assessment of the energy, environmental, vulnerability and health impact. Synergies with the global climate change. *Energy Build.* **2020**, *207*, 109482. [CrossRef]
9. Wang, J.; Xiang, Z.; Wang, W.; Chang, W.; Wang, Y. Impacts of strengthened warming by urban heat island on carbon sequestration of urban ecosystems in a subtropical city of China. *Urban Ecosyst.* **2021**, *24*, 1165–1177. [CrossRef]
10. Kong, J.; Zhao, Y.; Carmeliet, J.; Lei, C. Urban Heat Island and Its Interaction with Heatwaves: A Review of Studies on Mesoscale. *Sustainability* **2021**, *13*, 10923. [CrossRef]
11. Gourfi, A.; Daoudi, L.; Rhoujjati, A.; Benkaddour, A.; Fagel, N. Use of bathymetry and clay mineralogy of reservoir sediment to reconstruct the recent changes in sediment yields from a mountain catchment in the Western High Atlas region, Morocco. *CATENA* **2020**, *191*, 104560. [CrossRef]
12. Edmondson, J.L.; Stott, I.; Davies, Z.G.; Gaston, K.J.; Leake, J.R. Soil surface temperatures reveal moderation of the urban heat island effect by trees and shrubs. *Sci. Rep.* **2016**, *6*, 33708. [CrossRef] [PubMed]
13. Palmer, B.J.; Fulbright, T.E.; Grahmann, E.D.; Hernández, F.; Hehman, M.W.; Wester, D.B. Vegetation Structural Attributes Providing Thermal Refugia for Northern Bobwhites. *J. Wildl. Manag.* **2021**, *85*, 543–555. [CrossRef]
14. Nedbal, V.; Láska, K.; Brom, J. Mitigation of Arctic Tundra Surface Warming by Plant Evapotranspiration: Complete Energy Balance Component Estimation Using LANDSAT Satellite Data. *Remote Sens.* **2020**, *12*, 3395. [CrossRef]
15. Gourfi, A.; Daoudi, L.; de Vente, J. A new simple approach to assess sediment yield at a large scale with high landscape diversity: An example of Morocco. *J. Afr. Earth Sci.* **2020**, *168*, 103871. [CrossRef]
16. Iddo, A.G.; Paramasivam, V.; Selvaraj, S.K. Design and Techno-economic analysis of power generating unit from waste heat (Preheater and grate cooler) of cement factory in Ethiopia. *Mater. Today: Proc.* **2021**, *46*, 7825–7838. [CrossRef]
17. Tan, P.Y.; Wong, N.H.; Tan, C.L.; Jusuf, S.K.; Chang, M.F.; Chiam, Z.Q. A method to partition the relative effects of evaporative cooling and shading on air temperature within vegetation canopy. *J. Urban Ecol.* **2018**, *4*, juy012. [CrossRef]
18. Sayão, V.M.; dos Santos, N.V.; Mendes, W.D.S.; Marques, K.P.; Safanelli, J.L.; Poppiel, R.R.; Demattê, J.A. Land use/land cover changes and bare soil surface temperature monitoring in southeast Brazil. *Geoderma Reg.* **2020**, *22*, e00313. [CrossRef]
19. Nichol, J.E.; Fung, W.Y.; Lam, K.-S.; Wong, M.S. Urban heat island diagnosis using ASTER satellite images and 'in situ' air temperature. *Atmos. Res.* **2009**, *94*, 276–284. [CrossRef]
20. Saaroni, H.; Ben-Dor, E.; Bitan, A.; Potchter, O. Spatial distribution and microscale characteristics of the urban heat island in Tel-Aviv, Israel. *Landsc. Urban Plan.* **2000**, *48*, 1–18. [CrossRef]
21. Cheval, S.; Dumitrescu, A.; Irașoc, A.; Paraschiv, M.-G.; Perry, M.; Ghent, D. MODIS-based climatology of the Surface Urban Heat Island at country scale (Romania). *Urban Clim.* **2021**, *41*, 101056. [CrossRef]

22. Hartz, D.; Prashad, L.; Hedquist, B.; Golden, J.; Brazel, A. Linking satellite images and hand-held infrared thermography to observed neighborhood climate conditions. *Remote Sens. Environ.* **2006**, *104*, 190–200. [CrossRef]

23. Lu, D.; Weng, Q. Spectral mixture analysis of ASTER images for examining the relationship between urban thermal features and biophysical descriptors in Indianapolis, Indiana, USA. *Remote Sens. Environ.* **2006**, *104*, 157–167. [CrossRef]

24. Meng, F.; Liu, M. Remote-sensing image-based analysis of the patterns of urban heat islands in rapidly urbanizing Jinan, China. *Int. J. Remote Sens.* **2013**, *34*, 8838–8853. [CrossRef]

25. Pu, R.; Gong, P.; Michishita, R.; Sasagawa, T. Assessment of multi-resolution and multi-sensor data for urban surface temperature retrieval. *Remote Sens. Environ.* **2006**, *104*, 211–225. [CrossRef]

26. Zakšek, K.; Oštir, K. Downscaling land surface temperature for urban heat island diurnal cycle analysis. *Remote Sens. Environ.* **2012**, *117*, 114–124. [CrossRef]

27. Chen, X.-L.; Zhao, H.-M.; Li, P.-X.; Yin, Z.-Y. Remote sensing image-based analysis of the relationship between urban heat island and land use/cover changes. *Remote Sens. Environ.* **2006**, *104*, 133–146. [CrossRef]

28. Jamei, Y.; Rajagopalan, P.; Sun, Q. Spatial structure of surface urban heat island and its relationship with vegetation and built-up areas in Melbourne, Australia. *Sci. Total Environ.* **2018**, *659*, 1335–1351. [CrossRef] [PubMed]

29. Chen, L.; Wang, X.; Cai, X.; Yang, C.; Lu, X. Seasonal Variations of Daytime Land Surface Temperature and Their Underlying Drivers over Wuhan, China. *Remote Sens.* **2021**, *13*, 323. [CrossRef]

30. Marzban, F.; Sodoudi, S.; Preusker, R. The influence of land-cover type on the relationship between NDVI–LST and LST-T_{air}. *Int. J. Remote Sens.* **2017**, *39*, 1377–1398. [CrossRef]

31. Wloczyk, C.; Borg, E.; Richter, R.; Miegel, K. Estimation of instantaneous air temperature above vegetation and soil surfaces from Landsat 7 ETM+ data in northern Germany. *Int. J. Remote Sens.* **2011**, *32*, 9119–9136. [CrossRef]

32. Anniballe, R.; Bonafoni, S.; Pichierri, M. Spatial and temporal trends of the surface and air heat island over Milan using MODIS data. *Remote Sens. Environ.* **2014**, *150*, 163–171. [CrossRef]

33. Lazaar, A.; El Hammouti, K.; Naiji, Z.; Pradhan, B.; Gourfi, A.; Andich, K.; Monir, A. The manifestation of VIS-NIRS spectroscopy data to predict and map soil texture in the Triffa plain (Morocco). *Kuwait J. Sci.* **2020**, *48*, 8012. [CrossRef]

34. HCP Recensement Général de la Population et de l'Habitat de 2014, le Haut Commissariat au Plan du Maroc. 2014. Available online: https://www.hcp.ma/Recensement-General-de-la-Population-et-de-l-Habitat-RGPH-2014_a2945.html (accessed on 29 May 2022).

35. Taïbi, A.N.; Hannani, M. *El Le Végétal dans l'Espace Public des Villes Coloniales de Marrakech, Alger, Antananarivo et To-liara. Enjeux Socio-Environnementaux d'un «Patrimoine» vert dans un Contexte Urbain en Mutation*; CEAUP—Centro de Estudos Africanos da Universidade do Porto: Porto, Portugal, 2020; pp. 27–38.

36. Gillot, G. La ville Nouvelle Coloniale au Maroc: Moderne, Salubre, Verte, Vaste.; La Maison.; les Villes Nouvelles Écrit par François Leimdorfer. 2014. Available online: https://halshs.archives-ouvertes.fr/halshs-01272511/document (accessed on 5 May 2022).

37. Nasiri, V.; Deljouei, A.; Moradi, F.; Sadeghi, S.M.M.; Borz, S.A. Land Use and Land Cover Mapping Using Sentinel-2, Landsat-8 Satellite Images, and Google Earth Engine: A Comparison of Two Composition Methods. *Remote Sens.* **2022**, *14*, 1977. [CrossRef]

38. Richards, J.A. *Remote Sensing Digital Image Analysis: An introduction*; Springer: Cham, Switzerland, 2013; ISBN 9783642300622.

39. Vigneshwaran, S.; Kumar, S.V. Extraction of built-up area using high resolution sentinel-2A and google satellite imagery. *ISPRS—Int. Arch. Photogramm. Remote Sens. Spat. Inf. Sci.* **2018**, *42*, 165–169. [CrossRef]

40. USGS. Landsat Data Users Handbook. Available online: https://www.usgs.gov/ (accessed on 5 January 2022).

41. Weier, J.; Herring, D. *Measuring Vegetation (NDVI & EVI)—Normalized Difference Vegetation Index (NDVI)*; Nasa Earth Observatory: Springdale, UT, USA, 2000.

42. PMuhammad, Y.; Sheng, H.; Sami, U.; Rahman, I.; Atif, Z.; Asif, M. Estimation of Land Surface Temperature using LAND-SAT-8 Data-A Case Study of District Malakand, Khyber Pakhtunkhwa, Pakistan. *J. Lib. Arts Humanit.* **2020**, *1*, 140–148.

43. Gallo, K.P.; Owen, T.W. Satellite-Based Adjustments for the Urban Heat Island Temperature Bias. *J. Appl. Meteorol.* **1999**, *38*, 806–813. [CrossRef]

44. Gallo, K.P.; McNab, A.L.; Karl, T.R.; Brown, J.F.; Hood, J.J.; Tarpley, J.D. The Use of NOAA AVHRR Data for Assessment of the Urban Heat Island Effect. *J. Appl. Meteorol.* **1993**, *32*, 899–908. [CrossRef]

45. Petralli, M.; Prokopp, A.; Morabito, M.; Bartolini, G.; Torrigiani, T.; Orlandini, S. Ruolo Delle Aree Verdi Nella Mitiga-zione Dell'Isola Di Calore Urbana: Uno Studio Nella Città Di Firenze. *Riv. Ital. di Agrometeorol.* **2006**, *1*. Available online: https://www.mendeley.com/catalogue/1901752a-f70d-3806-acec-3fe702778ac0/?utm_source=desktop&utm_medium=1.19.8 &utm_campaign=open_catalog&userDocumentId=%7B932e2722-18fd-3001-9592-f1826dffbc22%7D (accessed on 29 May 2022).

46. dos Santos, A.R.; de Oliveira, F.S.; da Silva, A.G.; Gleriani, J.M.; Gonçalves, W.; Moreira, G.L.; Silva, F.G.; Branco, E.R.F.; Moura, M.M.; da Silva, R.G.; et al. Spatial and temporal distribution of urban heat islands. *Sci. Total Environ.* **2017**, *605–606*, 946–956. [CrossRef] [PubMed]

47. Fujibe, F. Urban warming in Japanese cities and its relation to climate change monitoring. *Int. J. Clim.* **2011**, *31*, 162–173. [CrossRef]

48. Sodoudi, S.; Shahmohamadi, P.; Vollack, K.; Cubasch, U.; Che-Ani, A.I. Mitigating the urban heat island effect in megacity tehran. *Adv. Meteorol.* **2014**, *2014*, 547974. [CrossRef]

49. Vargo, J.; Stone, B.; Habeeb, D.; Liu, P.; Russell, A. The social and spatial distribution of temperature-related health impacts from urban heat island reduction policies. *Environ. Sci. Policy* **2016**, *66*, 366–374. [CrossRef]

50. Zhou, D.; Zhang, L.; Hao, L.; Sun, G.; Liu, Y.; Zhu, C. Spatiotemporal trends of urban heat island effect along the urban development intensity gradient in China. *Sci. Total Environ.* **2016**, *544*, 617–626. [CrossRef] [PubMed]

51. Al-Saadi, L.M.; Jaber, S.H.; Al-Jiboori, M.H. Variation of urban vegetation cover and its impact on minimum and maximum heat islands. *Urban Clim.* **2020**, *34*, 100707. [CrossRef]

52. Farhadi, H.; Faizi, M.; Sanaieian, H. Mitigating the urban heat island in a residential area in Tehran: Investigating the role of vegetation, materials, and orientation of buildings. *Sustain. Cities Soc.* **2019**, *46*, 101448. [CrossRef]

53. Senanayake, I.; Welivitiya, W.; Nadeeka, P. Remote sensing based analysis of urban heat islands with vegetation cover in Colombo city, Sri Lanka using Landsat-7 ETM+ data. *Urban Clim.* **2013**, *5*, 19–35. [CrossRef]

54. Stache, E.; Schilperoort, B.; Ottelé, M.; Jonkers, H. Comparative analysis in thermal behaviour of common urban building materials and vegetation and consequences for urban heat island effect. *Build. Environ.* **2021**, *213*, 108489. [CrossRef]

55. Guha, S.; Govil, H.; Dey, A.; Gill, N. Analytical study of land surface temperature with NDVI and NDBI using Landsat 8 OLI and TIRS data in Florence and Naples city, Italy. *Eur. J. Remote Sens.* **2018**, *51*, 667–678. [CrossRef]

56. Sharifi, E.; Sivam, A.; Karuppannan, S.; Boland, J. Landsat Surface Temperature Data Analysis for Urban Heat Resilience: Case Study of Adelaide. In *Lecture Notes in Geoinformation and Cartography*; Springer: New York, NY, USA, 2017; pp. 433–447. [CrossRef]

57. Nakata-Osaki, C.M.; Souza, L.C.L.; Rodrigues, D.S. THIS—Tool for Heat Island Simulation: A GIS extension model to calculate urban heat island intensity based on urban geometry. *Comput. Environ. Urban Syst.* **2018**, *67*, 157–168. [CrossRef]

58. Zha, Y.; Gao, J.; Ni, S. Use of normalized difference built-up index in automatically mapping urban areas from TM imagery. *Int. J. Remote Sens.* **2003**, *24*, 583–594. [CrossRef]

59. He, C.; Shi, P.; Xie, D.; Zhao, Y. Improving the normalized difference built-up index to map urban built-up areas using a semiautomatic segmentation approach. *Remote Sens. Lett.* **2010**, *1*, 213–221. [CrossRef]

60. Chakraborty, T.; Lee, X. A simplified urban-extent algorithm to characterize surface urban heat islands on a global scale and examine vegetation control on their spatiotemporal variability. *Int. J. Appl. Earth Obs. Geoinf. ITC J.* **2018**, *74*, 269–280. [CrossRef]

61. Kim, G. Assessing Urban Forest Structure, Ecosystem Services, and Economic Benefits on Vacant Land. *Sustainability* **2016**, *8*, 679. [CrossRef]

62. EPA Urban Heat Island Basics. Reducing Urban Heat Islands: Compendium of Strategies. In *Heat Island Effect*; US EPA: Washington, DC, USA, 2008.

63. El Hannani, M.; Taïbi, A.N.; Brabra, N.; Giffon, S. Les enjeux du végétal dans une ville du « Sud » The Importance of Plant Life in a City of the "South"—The Case of Marra-kesh and the End of "Garden-City" Model. *Proj. Paysage* **2017**, *16*, 5862. [CrossRef]

64. Algretawee, H.; Rayburg, S.; Neave, M. Estimating the effect of park proximity to the central of Melbourne city on Urban Heat Island (UHI) relative to Land Surface Temperature (LST). *Ecol. Eng.* **2019**, *138*, 374–390. [CrossRef]

65. Bourscheidt, V. Análise da influência do uso do solo nas variações de temperatura utilizando imagens MODIS e LAND-SAT 8. In Proceedings of the XVII Simpósio Bras. Sensoriamento Remoto—SBSR, João Pessoa, Brazil, 25–29 April 2015.

66. Abir, F.A.; Saha, R. Assessment of land surface temperature and land cover variability during winter: A spatio-temporal analysis of Pabna municipality in Bangladesh. *Environ. Chall.* **2021**, *4*, 100167. [CrossRef]

67. Dissanayake, D.; Morimoto, T.; Ranagalage, M.; Murayama, Y. Land-Use/Land-Cover Changes and Their Impact on Surface Urban Heat Islands: Case Study of Kandy City, Sri Lanka. *Climate* **2019**, *7*, 99. [CrossRef]

68. Knight, J.H.; Minasny, B.; McBratney, A.; Koen, T.B.; Murphy, B.W. Soil temperature increase in eastern Australia for the past 50 years. *Geoderma* **2018**, *313*, 241–249. [CrossRef]

69. Deng, Y.; Wang, S.; Bai, X.; Tian, Y.; Wu, L.; Xiao, J.; Chen, F.; Qian, Q. Relationship among land surface temperature and LUCC, NDVI in typical karst area. *Sci. Rep.* **2018**, *8*, 64. [CrossRef]

70. Ibrahim, F.; Rasul, G. Urban Land Use Land Cover Changes and Their Effect on Land Surface Temperature: Case Study Using Dohuk City in the Kurdistan Region of Iraq. *Climate* **2017**, *5*, 13. [CrossRef]

71. Raynolds, M.K.; Comiso, J.C.; Walker, D.A.; Verbyla, D. Relationship between satellite-derived land surface temperatures, arctic vegetation types, and NDVI. *Remote Sens. Environ.* **2008**, *112*, 1884–1894. [CrossRef]

72. Synnefa, A. Cool-colored coatings fight the urban heat-island effect. *SPIE Newsroom* **2007**. [CrossRef]

73. Akbari, H.; Pomerantz, M.; Taha, H. Cool surfaces and shade trees to reduce energy use and improve air quality in urban areas. *Sol. Energy* **2001**, *70*, 295–310. [CrossRef]

74. Urban, B.; Roth, K. *Guidelines for Selecting Cool Roofs*; U.S. Department of Energy: Washington, DC, USA, 2010.

75. Debbage, N.; Shepherd, J.M. The urban heat island effect and city contiguity. *Comput. Environ. Urban Syst.* **2015**, *54*, 181–194. [CrossRef]

76. Taslim, S.; Parapari, D.M.; Shafaghat, A. Urban Design Guidelines to Mitigate Urban Heat Island (UHI) Effects In Hot-Dry Cities. *J. Teknol.* **2015**, *74*. [CrossRef]

77. Sari, D.P. A Review of How Building Mitigates the Urban Heat Island in Indonesia and Tropical Cities. *Earth* **2021**, *2*, 653–666. [CrossRef]

78. Österreicher, D.; Sattler, S. Maintaining Comfortable Summertime Indoor Temperatures by Means of Passive Design Measures to Mitigate the Urban Heat Island Effect—A Sensitivity Analysis for Residential Buildings in the City of Vienna. *Urban Sci.* **2018**, *2*, 66. [CrossRef]

Article

"Cool" Roofs as a Heat-Mitigation Measure in Urban Heat Islands: A Comparative Analysis Using Sentinel 2 and Landsat Data

Terence Mushore [1,2], John Odindi [1,*] and Onisimo Mutanga [1]

[1] Discipline of Geography, School of Agricultural, Earth and Environmental Sciences, University of KwaZulu-Natal, Scottsville, Pietermaritzburg 3209, South Africa

[2] Department of Space Science and Applied Physics, Faculty of Science, University of Zimbabwe, 630 Churchill Avenue, Mt Pleasant, Harare 00263, Zimbabwe

* Correspondence: odindi@ukzn.ac.za

Abstract: Urban growth, characterized by expansion of impervious at the cost of the natural landscape, causes warming and heat-related distress. Specifically, an increase in the number of buildings within an urban landscape causes intensification of heat islands, necessitating promotion of cool roofs to mitigate Urban Heat Islands (UHI) and associated impacts. In this study, we used the freely available Sentinel 2 and Landsat 8 data to determine the study area's Land Use Land Covers (LULCs), roof colours and Land Surface Temperature (LST) at a 10-m spatial resolution. Support Vector Machines (SVM) classification algorithm was adopted to derive the study area's roof colours and proximal LULCs, and the Transformed Divergence Separability Index (TDSI) based on Jeffries Mathussitta distance analysis was used to determine the variability in LULCs and roof colours. To effectively relate the Landsat 8 thermal characteristics to the LULCs and roof colours, the Gram–Schmidt technique was used to pan-sharpen the 30-m Landsat 8 image data to 10 m. Results show that Sentinel 2 mapped LULCs with over 75% accuracy. Pan-sharpening the 30-m-resolution thermal data to 10 m improved the spatial resolution and quality of the Land Surface map and the correlation between LST and Normalized Difference Vegetation Index (NDVI) used as proxy for LULC. Green-colour roofs were the warmest, followed by red roofs, while blue roofs were the coolest. Generally, black roofs in the study area were cool. The study recommends the need to incorporate other roofing properties, such as shape, and further split the colours into different shades. Furthermore, the study recommends the use of very high spatial resolution data to determine roof colour and their respective properties; these include data derived from sensors mounted on aerial platforms such as drones and aircraft. The study concludes that with appropriate analytical techniques, freely available image data can be integrated to determine the implication of roof colouring on urban thermal characteristics, useful for mitigating the effects of Urban Heat Islands and climate change.

Keywords: cool roofs; urban heat islands; land surface temperatures; roof colour; mitigation; urban growth

Citation: Mushore, T.; Odindi, J.; Mutanga, O. "Cool" Roofs as a Heat-Mitigation Measure in Urban Heat Islands: A Comparative Analysis Using Sentinel 2 and Landsat Data. *Remote Sens.* **2022**, *14*, 4247. https://doi.org/10.3390/rs14174247

Academic Editor: Anthony Brazel

Received: 5 July 2022
Accepted: 25 August 2022
Published: 28 August 2022

Publisher's Note: MDPI stays neutral with regard to jurisdictional claims in published maps and institutional affiliations.

1. Introduction

Urbanization, and the associated urban land use and land cover (LULC) spatial structure transformations influence the urban thermal characteristics [1–3]. This process is typified by transformation from natural to impervious surfaces such as buildings and other urban fabrics that alter surface and near-surface temperatures [4,5]. The increase in temperatures attributable to urban growth are associated with a range of challenges that include adverse effects on human health, increased water and energy demand and air pollution [6–8]. As such, urbanization and consequent thermal elevation has been known to exacerbate in- and out-door ambient thermal discomfort that diminish the quality of urban life [9,10]. Hence, it is increasingly becoming desirable to adopt climate-smart approaches that could enhance sustainable urban living.

Remotely sensed data offer an opportunity to determine urban spatio-temporal variations and their respective thermal characteristics [11,12]. Additionally, remotely sensed data allows for analysis at a range of time-scales that include sub-seasonal patterns. Over the years, technological advancement has facilitated the acquisition of both optical and thermal data on the same sensor platforms (e.g., Landsat, ASTER and MODIS), valuable for urban landscape transformation and thermal analysis [13–16]. Hence, these moderate resolution sensors have been widely used to determine the influence of LULCs on the thermal environment e.g., [17–22]. However, such moderate resolution datasets suffer from the mixed pixel problem, especially in urban areas characterized by landscape heterogeneity, which compromises their value for detailed surface analysis such as the detection of individual houses and their thermal properties [23].

Fortunately, recent sensor developments and advancements in computational power offer an opportunity for improved land surface analysis. For instance, whereas the Landsat series has over the years improved in spectral and radiometric properties, new generation sensors such as Sentinel 2 offer data with improved spatial resolution [24–26]. The sensors' 10-m spatial resolution for instance, ref. [27] allows for analysis of complex environments such as urban areas with reduced mixed pixel effect and high mapping accuracy. Whereas the Sentinel 2s platform lacks a thermal sensor, its integration with high quality data such as Landsat has potential to improve our knowledge of the relationship between urban LULCs and surface temperatures. Recently, Mushore et al. [28] showed that pan-sharpening of Landsat thermal data improves its Land Surface Temperature (LST) mapping accuracy, while Kaplan and Avdan [26] used Sentinel 2's pan-sharpened 10-m to improve 20-m resolution bands. However, whereas the Sentinel's 10-m spatial resolution optical data can be used to derive detailed urban surface features, Landsat thermal data need to be at a similar spatial resolution for optimal analysis and mapping accuracy.

Several studies have demonstrated that built-up areas absorb and store large amounts of heat when compared to other LULC types, e.g., [22,27–30]. The thermal effect is enhanced by increased building densities that result in large surface areas for heat absorption. Furthermore, dense high-rise buildings increase heat storage capacity as walls present even larger surface areas for heat absorption. Buildings also concentrate heat in an area by retarding its removal by winds [31]. To date, a significant number of studies have dwelt on the effect of buildings on temperature. For instance, the effect of building density and height have been widely demonstrated in both the developed and the developing world, e.g., [28,32–35]. Besides density and height, building materials and other properties such as roof characteristics influence a built environment's thermal properties. For example, Mackey et al. [36] demonstrated that cool roofs surpassed green roofs, street trees and green spaces in cooling effects in Chicago. However, the adoption of remotely sensed data to understand the influence of roofing properties on temperature remains limited. Emphasis has been largely placed on understanding the influence at a broad scale and general LULC classes on the thermal environment. Focus on localized phenomena that include the effect of individual houses and their characteristics such as roof properties using freely available remotely sensed data has remained a grey area.

Studies on the effect of roofs on buildings thermal characteristics have mainly focused on rooftops with vegetation (i.e., 'green roofs') and commonly use data derived from installed meteorological instruments and analytical models [37–40]. Other studies have investigated roof characteristics such as roof angle; for instance, Tian et al. [41] compared the thermal characteristics of curved and flat roofs. Studies on roof colour have established that white roofs have more cooling effect than grey, red and black roofs [42–44], while coating coloured roofs with highly reflective materials can increase thermal performance and energy efficiency of buildings [45]. For instance, Libbra et al. [45] found that the use of cool roofs can reduce air conditioning energy consumption by 70%. For the same roof type, variations such as colour and age may also influence their interactions with heat [43,46,47].

However, due to limitations of the new generation sensors' spatial resolution, literature on the influence of roof colour on thermal performance of buildings remains scarce.

Zhao et al. [48] examined daytime and nighttime effects of roof footprints and configurations using high resolution airborne LIDAR and Quickbird satellite data (2.4-m resolution) and MODIS/ASTER simulated airborne 7-m-resolution surface temperature data. They observed that rooftop spectral attributes, slope, aspect and surrounding trees affected roof surface temperature. Although they accurately delineated roof configurations, they did not segment the roofs by colour. Furthermore, while sensors on aerial platforms such as drones and airplanes can provide data for detailed analysis of effects of roofs on thermal characteristics, such data remain expensive and not viable for studies over large spatial extents. Hence, there is a need to test the value of freely available moderate resolution optical and thermal datasets to enhance our understanding on the influence of building roof colour on thermal characteristics, especially in growing cities of developing countries. Such efforts are necessary to determine the potential adoption of roof type and colour to mitigate heat islands.

According to Alchapar and Correa [46], roof coating is the most influential morphological determinant of roof thermal behavior, while Libbra et al. [45] notes that roof colour controls the absorption of heat during the day and its emission at night. As such, it is necessary to consider "cool" roofs for UHI mitigation. Hence, in relation to adjacent LUCLs, this study sought to determine the value of Sentinel 2 10-m resolution and pan-sharpened Landsat image data in differentiating the influence of roof colour on surface thermal values.

2. Methodology

2.1. Description of the Study Area

The study was carried out in a low-density residential area close to the Central Business District (CBD) of the capital city of Zimbabwe, Harare (Figure 1). Since the study sought to determine the influence of roof colour on urban thermal characteristics, it was restricted to a small spatial extent to limit excessive heterogeneity that typifies urban landscapes. Also, a large area could have introduced additional variables (e.g., elevation and slope) that influence thermal characteristics. The area is in a low-to-medium-density residential type, however, some of the houses, especially towards the CBD, have been turned into offices. Low-to-medium-density residential areas in Zimbabwe are characterized by spacious housing units, high land value and higher vegetation density when compared to high-density residential areas, which are predominantly occupied by the low-income strata. Since house units in the low-to-medium-density residential areas are generally large, they are potentially discriminable using 10-m or higher spatial resolution image data. Hence, based on the 10-m spatial resolution image data, the area was chosen to minimize the mixed pixel problem that characterizes the high-density residential areas. Furthermore, the area is dominated by houses with tiles, thus eliminating the effect of other roof types such as concrete, zinc, aluminium or thatch on the area's thermal characteristics. This enabled the study to determine the variability in temperature based on roof tiling of different colours.

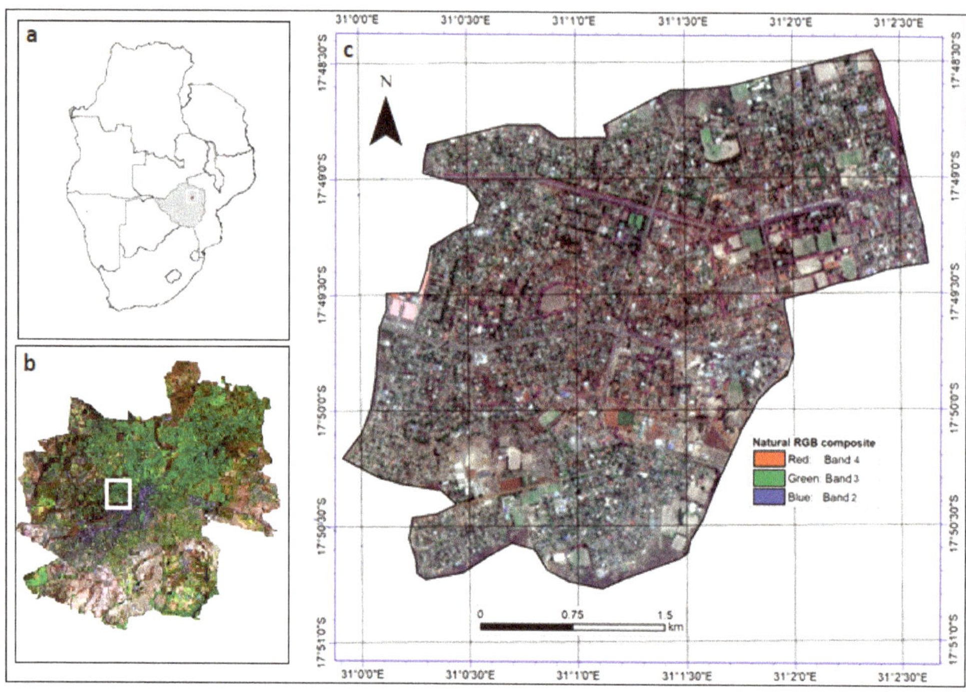

Figure 1. Location of the study area in Southern Africa, Zimbabwe and Harare (**a**), Harare and the study area (**b**) and 10-m resolution natural colour composite showing variations of LULC in the study area (**c**).

2.2. Field and Remotely Sensed Datasets

A field survey was conducted to identify the LULCs and roof colours in the area. The survey revealed that the major LULC classes are grasslands, buildings with heterogeneous tiled roof colouring, bare soil and roads. The building class was further split into roof colours in line with the main objective of the study, and a stratified random sampling approach used to collect the ground control points (GCPs). Non-tiled roofs were categorized into "Other LULCs". For each identified category, coordinates of representative covers were collected using a handheld Global Positioning System. To maximize the spectral and thermal variability, the hot dry season (mid-September to mid-November) was chosen for the collection of the well-distributed LULCs' GCPs as it presents a period of maximum solar energy with no rainfall cooling effect. The LULC types were verified using a GoogleEarth image, which was also used to verify the roof colours and to shift the GCPs to the roof center for classification and validation purposes. The data were split: 70% to be used for classification and 30% for validation.

Landsat and Sentinel 2 data were downloaded from the United States Geological Survey's earth explorer portal at no cost. To minimize variation between field data and image scenes, cloud-free imagery was collected on dates close to field data collection. Two Landsat images (scene capture dates: 16 September 2021 and 3 November 2021) and Sentinel images (scene capture dates: 18 September and 2 November 2021) were used in this study. The dates were chosen as they correspond to the period of maximum radiation in the hot season and the proximity of the two sensor dates. The wind was calm and cloudless, presenting similar weather conditions when data from the same sensor were acquired. Given that the period is dry, vegetation conditions were assumed to be uniform and largely maintained by irrigation/watering throughout the periods. Landsat data acquired on the 16th of September were matched with Sentinel data of the 18th of September, a short enough period to assume that LULCs did not change. Similarly, Landsat data acquired on the 3rd of November were matched with Sentinel data of the 2nd of November. This gave the best compromise to enable relating and blending multi-sensor data with different spatial and temporal resolutions. The two Landsat images were used to compute the average LST to minimize randomness associated with a single-date image, while the two Sentinel 2 image datasets were used for LULC classification. Multi-spectral optical 10-m resolution Sentinel 2 and 3-m resolution Landsat 8 data for each acquisition date were merged into a multi-layer files using the 'Layer stacking' tool in ENVI software. This was done separately for Landsat 8 and Sentinel 2 data. In order to eliminate the effect of aerosols on reflectance values, atmospheric correction was done using the Fast Line-of-sight Atmospheric Analysis of Spectral Hypercubes (FLAASH) module in the ENVI software. Due to the proximity to the CBD and location in an urban area, the urban aerosol mode was used in FLAASH, which produced multi-layer reflectance files. Multilayer 10-m resolution Sentinel 2 reflectance data were required to provide multi-spectral information to enhance separation of features in supervised image classification. On the other hand, spectral reflectivity bands in the near-infrared and red range from Landsat 8 were needed for the computation of normalized difference vegetation index (NDVI), useful for emissivity correction in LST retrieval.

2.3. Separability Analysis

Sentinel 2 10-m resolution bands and the 70% of the field-collected GPS points for the LULCs and roof colour categories were overlaid in an ENVI version 4.7 environment. Surface separability was done using the Transformed Divergence Separability Index (TDSI) based on Jeffries Mathussitta distance analysis. For each paired classes, TDSI ranges between 0 and 2, with values greater than 1 indicating that two classes are distinguishable and values close to 2 implying very high separability. Values below 1 and close to 0 suggests that the classes should be merged. The TDS analysis was necessary to test whether different roof colours and LULCs could be separated before classification.

2.4. Land Use/Cover Mapping and Retrievals of Roof Colours

The LULCs were derived from Sentinel 2's 10-m bands based on the 70% GCPs using the Support Vector Machines (SVM) algorithm in ENVI version 4.7 software. Default settings of 0.083 and 100 were used for Gamma in kennel function and penalty parameter, respectively. The SVM uses two classes of training samples within a multidimensional feature space to fit an optimum dividing hyperplane. It aims to maximize the variability between the most proximal training samples (support vectors) and the hyperplane [49,50]. To achieve our objective, we chose a Gaussian radial-basis kennel function as it is ideal for working in an infinite-dimensional space and has a single parameter [49–51]. We classified the images into eight classes, namely, Roads and Bare, Trees, Grassland, Red roof, Blue roofs, Green colour roofs, Black roofs and Grey roofs. To display the roof colours from other LULCS, the Roads and Bare, Trees and Grassland classes were amalgamated into "Other LULC". Thereafter, a confusion matrix was generated. A confusion matrix compares the assigned class labels on the classified map with the location's actual LULC class observed in the field (ground truth). The confusion matrix was used to derive the most widely used accuracy indicators, namely, Overall Accuracy (OA) and Kappa (k) [52].

2.5. Land Surface Temperature Retrieval from Landsat 8 Data

Band 10 of Landsat 8 was used to retrieve LST from thermal infrared data using Planck's radiation law-based equation for single-channel Landsat thermal data [53]. Initially, thermal infrared digital numbers were converted to surface-leaving radiance using Equation (1);

$$L_I = M_I\, Q_{CAL} + A_L \tag{1}$$

where, L_l is spectral radiance at Top of the Atmosphere measured in Watts/m^2/srad/μm, M_l is Band-specific multiplicative rescaling factor, Q_{CAL} represents pixel values (Digital Numbers) and A_L is the Band-specific additive rescaling factor. M_l, A_L and Q_{CAL} are obtained from the metadata downloaded together with the Landsat 8 data. As described by U.S. Geological Survey [54], the coefficients for converting digital numbers to thermal radiances were obtained from the metadata file accompanying Landsat 8 data download.

Mumtaz et al. [55] provides an in-depth description of steps for land surface temperature retrieval. The procedures include conversion of thermal radiances to blackbody/brightness temperature followed by emissivity correction to obtain surface temperatures. As such, derived radiances were used in Equation (2) to determine brightness/blackbody temperature.

$$T_B = \frac{K_2}{ln\left(\frac{K_1}{L_{II}} + 1\right)} \tag{2}$$

where, T_B is the brightness temperature (in degrees Kelvins), K_2 and K_1 area conversion constants for the thermal band (in this case Band 10), also obtained from the metadata file. Since brightness temperature over surfaces is calculated by assuming emissivity to be equal to 1, further analysis must consider actual emissivity which varies with LULC type. This was achieved through emissivity correction, which converted brightness temperatures to actual surface temperatures using Equation (3) [53,55].

$$T_S = \left(\frac{T_B}{1 + \left(\frac{\lambda \times T_B}{\alpha}\right) ln\ \varepsilon}\right) - 273.16 \tag{3}$$

where T_S is the LST in Degree Celsius, λ is the central wavelength of emitted radiance (10.9 μm for band 10 of Landsat 8), ε is the emissivity and α is a constant (1.438×10^{-2} mK). Due to its simplicity, Equation (4) was used to estimate emissivity from Normalized Difference Vegetation Index (NDVI) using [55–57];

$$\varepsilon = a + b\ ln(NDVI) \tag{4}$$

where $a = 1.0094$ and $b = 0.047$. Developed in Botswana, which is close to the study area, the equation was chosen due to ease of computation, parsimony and proven applicability in a tropical environment [55]. The *NDVI* was retrieved using reflectance in the Near Infrared (Band 5) and Red (Band 4) of Landsat 8 in Equation (5) [53,58];

$$NDVI = \frac{(NIR - RED)}{(NIR + RED)} \tag{5}$$

where *NIR* and *RED* are reflectance in the near-infrared and red ranges [59] derived from Band 5 and Band 4 of Landsat 8, respectively. The steps above obtained LST at a resampled resolution of 30 m, requiring further enhancement for analysis of roofs thermal properties at a local scale.

2.6. Gram-Schmidt Pan-Sharpening Based Method for LST Image Data Pan-Sharpening

Improvement of LST data from 30-m to 10-m spatial resolution was achieved using the Gram–Schmidt pan-sharpening technique. The Gram–Schmidt method uses weighted addition of multi-spectral bands to produce a replicated pan-sharpened low-resolution image. Gram–Schmidt orthogonalization is then used to make all bands of the multi-spectral low-resolution data orthogonal and scalar products are computed and turned into covariances [60]. For each band of the low-resolution multispectral data, angles between the band and the simulated low-resolution panchromatic are computed. Gain and bias of the high-resolution panchromatic band is used to simulate each low-resolution panchromatic band. The process is reversed using the same transformation coefficients, and high resolution pan-sharpened bands are produced [60,61]. Using Gram–Schmidt transformation, the colours of the composite RGB pan-sharpened bands are near similar to the respective original images, thus there is minimal distortion of spatial patterns. The method was chosen because all transform coefficients are computed in the low MS resolution, hence are more robust to spatial misalignment of the bands than most other pan-sharpening methods [60]. In this study, the Sentinel 2's 10-m resolution Band 2 was used to improve the Landsat data. The purpose was mainly aimed at producing thermal data for retrieval of LST at 10-m resolution to match with the products from supervised image classification.

2.7. Intensity Analysis for In-Depth Characterization of Local Climate Zones Changes

The LST spatial configurations before and after pan-sharpening were compared to assess the effect of improving spatial resolution on image quality. The root mean-square error was also used to check the difference after resampling the LSTs to 30-m resolution to assess the effect on values per pixel. A 30-m resolution Landsat scene was used to derive NDVI and its correlation with 30-m LST (after resampling using a bicubic convolution) was obtained using the "Zonal Statistics as a Table" tool in ArcGIS version 10.2, ESRI, Redlands, California, USA. Similarly, NDVI was calculated using 10-m resolution near infrared and red Sentinel 2 and correlated with pan-sharpened 10-m resolution LST. The LST correlations with NDVI before and after pan-sharpening were then compared.

2.8. Linking LULC Types and Roof Colours with LST

Qualitatively, the spatial structure of LULC and roof colours was compared with that of LST using visual inspections of maps produced from the combination of Sentinel 2 10-m resolution and Landsat 8 thermal data. For quantitative assessment, field-collected points corresponding to each LULC and roof colour category were used to extract LST values using the "Extract values to points" spatial overlay function in ArcGIS version 10.2. The field-collected points were used instead of overlaying the LST with the retrieved LULC map to eliminate the effect of classification accuracy on extracted temperatures for the different categories. Box plots were used to depict the variations of LST between and within LULC and roof colour categories in the study area. The mean LSTs for the different LULC and roof colour categories were also used to compare their thermal performances. This was done to assess the effect of improving resolution on the relationship between LULC and LST using NDVI as a proxy for LULC spatial patterns.

Figure 2 summarizes the procedures from data collection to linking of roof colours to LST spatial structures in the study area.

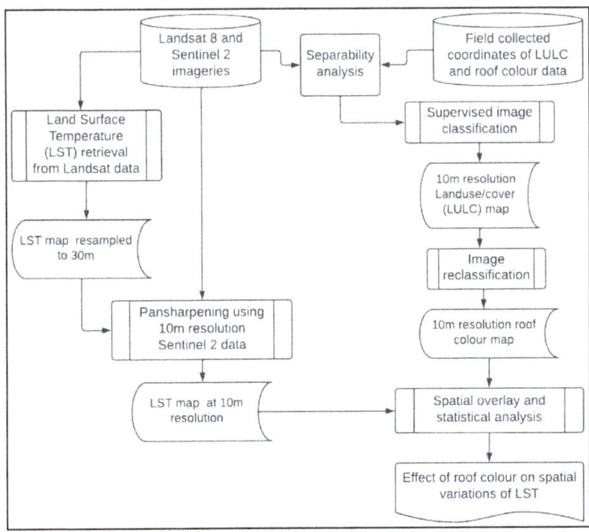

Figure 2. Summary of steps followed in the study.

3. Results

3.1. Separability of LULC and Roofs by Colour

Table 1 indicates that the TDSI values ranged between 1.74 and 2.00, implying that the LULC and roof colour categories were distinguishable using spectral signatures from 10-m resolution Sentinel 2 bands. Tarred roads and Trees were the most discriminable classes while the Green and Grey roofs were least discriminable, as indicated by TDSI values of 2.000 and 1.708, respectively. However, although Green and Grey roofs were the least separable, the TDSI value was significantly above the separability threshold of 1, hence guaranteed that the two classes were distinguishable. Among the roof colour categories, blue and red roofs were the most separable with a TDSI value of 1.995. The trees LULC category was the most separable from other cover types, with TDSI values ranging between 1.997 and 2.000. Overall, TDSI values greater than 1.7 indicate that the LULC and roof colour classes in the study area were easily distinguishable.

Table 1. Discriminability of LULC types in the study area using 10-m resolution Sentinel 2 data.

Compared LULC and Roof Classes	TDSI
Green-colour roofs and Grey roofs	1.708
Black roofs and Grey roofs	1.769
Black roofs and Green-colour roofs	1.826
Grey roofs and Red roofs	1.835
Black roofs and Tarred roads	1.874
Black roofs and Red roofs	1.920
Green-colour roofs and Red roofs	1.928
Blue roofs and Grey roofs	1.930
Red roofs and Bare areas	1.935
Blue roofs and Green-colour roofs	1.944
Grey roofs and Tarred roads	1.945
Black roofs and Blue roofs	1.955
Grey roofs and Bare areas	1.965
Black roofs and Bare areas	1.970
Grass and Bare areas	1.972
Grass and Red roofs	1.985
Red roofs and Tarred roads	1.985
Grass and Grey roofs	1.986
Green-colour roofs and Bare areas	1.990
Blue roofs and Tarred roads	1.994
Blue roofs and Red roofs	1.995
Black roofs and Trees	1.996
Grass and Green-colour roofs	1.997
Black roofs and Grass	1.998
Blue roofs and Bare areas	1.998
Grass and Tarred roads	1.998
Grey roofs and Trees	1.999
Blue roofs and Grass	1.999
Tarred roads and Bare areas	2.000
Trees and Bare areas	2.000
Blue roofs and Trees	2.000
Green-colour roofs and Trees	2.000
Grasslands and Trees	2.000
Red roofs and Trees	2.000
Tarred roads and Trees	2.000

3.2. Land Use/Cover and Roof Colour Mapping Using 10 M Resolution Sentinel 2 Data

The LULCs presented the houses surrounded by abundant vegetation, a characteristic of low-to-medium-density residential areas in Zimbabwe (Figure 3a). The study area has large grasslands, especially in the northeastern regions. The grasslands in the northeast are mainly sporting grounds. The other open grasslands within built-up areas are school grounds while the fragmented grasslands are mainly lawns around houses as well as unused land. The other abundant vegetation was in built-up areas. Figure 3b shows that due to narrow widths in relation to the 10-m resolution of the data, most roads, especially in the black roofs category were not visible.

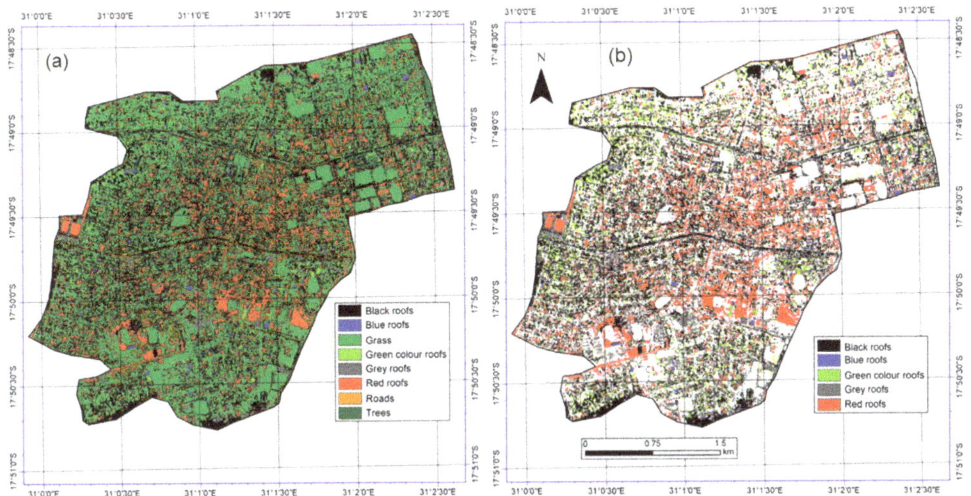

Figure 3. 10-m resolution (**a**) LULC map and (**b**) roof colour map.

3.3. Accuracy of LULC and Roof Colour Retrievals from Sentinel 2 Data

LULC and roof colour categories were mapped with Overall Accuracy (OA) of 84.5% and Kappa of 0.81. Producer Accuracies (PA) were greater than 75% except for the grey roofs and tarred roads (Table 2). User Accuracies (UA) were less than 75% for the black roofs, trees and grey roofs, while greater than 77% for the other categories. The red roofs were mapped with the highest accuracy of all the other categories (PA and OA greater than 93%).

Table 2. LULC and roof colour mapping accuracies.

LULC and Roof Colour Category	Producer Accuracy (%)	User Accuracy (%)
Black roofs	74.44	70.42
Blue roofs	95.79	98.45
Grasslands	92.38	79.66
Green-colour roofs	81.69	90.64
Grey roofs	70.38	69.02
Red roofs	93.43	94.97
Tarred roads	53.47	77.42
Trees	75.96	72.51

3.4. Comparison of 30 M Resolution with Sharpened 10 M Resolution LST Retrievals

Although the study area was small, variations in temperature were observed as some places were more than 15 °C cooler than others. Hotspots were noticed, especially on the southern half of the area where LSTs close to 49 °C were observed. The northern half was generally cooler, with the dominance of LSTs close to 41 °C. There was a general southeastward warming in the area. Comparison of Figure 4a,b shows that sharpening of LSTs to 10-m resolution by blending Landsat-derived LSTs with 10-m resolution Sentinel 2 did not compromise the spatial structure of LST and their ranges in the area. The 30-m resolution LST map was more pixelated than the 10-m resolution, indicating the latter's improved quality. When compared to the 30-m resolution, 10-m LST were retrieved with high accuracy (RMSE = 0.5 °C). Correlation between LULC and NDVI was −0.516 and −0.999 before and after pan-sharpening, respectively.

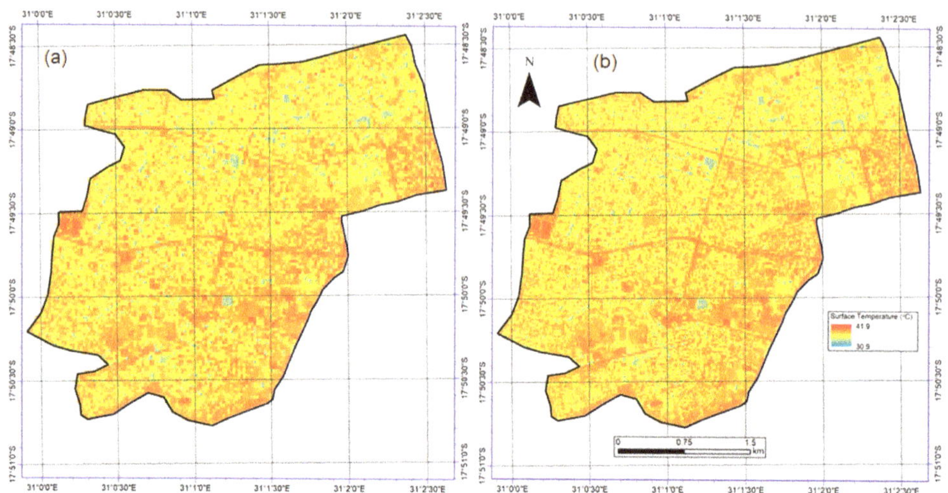

Figure 4. Spatial structure of (**a**) LST derived from Landsat thermal data at 30-m resolution (**b**) LST sharpened to 10-m resolution.

3.5. Variations of LST with LULC and Roof Colours

Although there were overlaps in temperature between different LULCs and roof colour categories, their mean LSTs were clearly distinct (Figure 5). The mean LST was lowest in the trees LULC category followed by blue roof. Highest LSTs were recorded in green-colour roofs and tarred roads areas. The grasslands LCZ showed greatest variability in LST, followed by green and red roofs. The order of roof colours from coolest to warmest based on average LST was blue (36.2 °C), black (35.8 °C), grey (36.9 °C), red (37.4 °C) and green (37.7 °C).

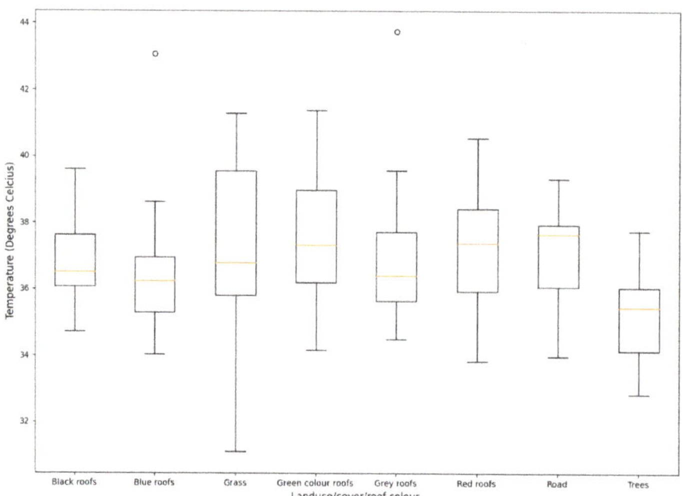

Figure 5. Observed variations of LST with LULC and roof colour classes in the study area.

4. Discussion

Separability of all classes was high, as indicated by a TDSI greater than 1.7. This was attributed to the strength of the spectral information at the 10-m resolution bands of Sentinel 2 to distinguish between different LULC and roof colours. As aforementioned, separability values close to 2 indicate that the classes are sufficiently separable using the remotely sensed image guided by GCPs [49]. The LULCs and roof colours in the study area were mapped with 75% and 0.73 accuracy and kappa, respectively. To facilitate effective separability of the heterogeneous study area, the study focused on a small spatial extent. Hence, the mapping accuracy was reasonable and comparable with other studies in urban environments such as Sithole and Odindi [62]. However, in this study, roads, (producer accuracy < 50%) were not effectively mapped. The low roads-mapping accuracy can be attributed to their narrow width, that they are largely below 10-m and along the road, and tree and tree shading, hence camouflage and/or mixed pixel with adjacent features.

Based on validation points, the roof colours were retrieved with reasonable accuracies. Producer accuracies (PA) ranged between 58 and 95%, while user accuracies (UA) were between 55 and 91%. The PA and UA values between 55 and 65% could be attributed to intra-class variabilities, which caused some similarities between different roof colours. For instance, some fading shades of black were near similar to dark shades of grey. Similarly, some shades of blue were closer to grey and black. Although not investigated, we speculate that roof ages and fading influence the similarities in roof colours. This is consistent with Alchapar and Correa [46] who noted that for a given roof colour, thermal properties can change due to age. Mapping accuracy could also be influenced by other effects such as roof shapes, reflectivity [63] and ventilation. For instance, Triano-Juárez et al. [64] observed variations in thermal properties for the same roof colour depending on reflectivity and presence of coating materials. On the other hand, Bojić et al. [65] observed differences between slanted and flat roofs. However, despite the above-named factors that could influence thermal variability based on roof colouring, our study shows that roof colours could be mapped with acceptable accuracy. We however suggest that for applications that require very high mapping accuracy (>90%), the Sentinel 2's 10-m resolution data may be insufficient. In this regard, the use of Unmanned Aerial Vehicles derived high spatial resolution data offers great potential for fine-scale mapping.

Similar LST spatial structure was observed before and after sharpening, while accuracy of retrieved 10-m resolution LST relative to the original 30-m resolution was high (RMSE of about 0.5 °C). Similar to a recent study by Mushore et al. [28], pan-sharpening also improved correlation between LST and NDVI. In this study, the LST maps effectively showed thermal variations. Spatial comparison of the LULC and LST maps showed that vegetation covers such as large grasslands and trees as well as built-up areas with abundant vegetation (which characterize most of the study area) had comparatively low temperature, an indication that even vegetation within built-up areas has heat mitigation value [62]. Zhang et al. [66] also highlighted that vegetation patches and spatial structure combine in contributing to the reduction in surface temperature of the area they occupy. This explains the surface-temperature-reduction effect of vegetation even within built-up areas. Besides latent heat transfer, the shading effect of vegetation, especially trees, lowers surface temperatures in areas they cover. As such, Zhao et al. [48] noticed the cooling effect of shadows of surrounding trees on roof-top surface temperatures during daytime.

The grey and red roofs were warmer than the black roofs, but cooler than green-colour roofs, which were the warmest (Figure 4). Contrary to expectation, black roofs were not the warmest. This could be attributed to variations in thermal characteristics in relation to, among others, roof and colour shading. For example, due to age, black roofs colouring ranged between dark black and grey. Red and green also had higher thermal values. This finding is consistent with Farhan et al. [44], who found that red roofing had higher thermal values than white roofing. Our findings show that green-colour roofs were the warmest, with average LST values close to tarred roads. On the other hand, blue roofs were the coolest, a finding consistent with Libbra et al. [45], who note that roof colour influences surface temperatures and hence could be used to mitigate heat islands. Although not investigated in this study, pigments on roof materials could have influenced their thermal behaviour. For example, it was reported that for the same roof colour, cool pigments have the potential to increase albedo by at least 20% [67]. This may have caused dark roofs to absorb less than or comparable heat to light-coloured roofs.

5. Conclusions

The 10-m resolution Sentinel 2 data mapped LULC and roofs by colour with reasonable accuracy. However, findings show that Sentinel 2's 10-m spatial resolution is still limited by the mixed pixel problem. Other roof characteristics such as age, shape and coating need to be investigated for potential improvement in mapping accuracy. Sharpening of LSTs derived from Landsat to Sentinel's 10-m resolution improved the LST spatial structure. It also increased the correlation between LST and NDVI, implying an improved relationship with LULC. Different roof colour showed variations in mean LST, which highlighted the contribution of roof colours in mitigating or intensifying the heat island effect. Due to variations in shades attributed to changes in age, black roofs were not the warmest. Blue roofs were found to be the coolest while green-colour roofs were the warmest, followed by red roofs. Grey roofs had a moderate effect, with the cooling effect increasing with lightness of the grey colour. Overall, the study showed that colour, in combination with other roof properties, determines a building unit's thermal characteristics. However, the study observed that even after pan-sharpening, Sentinel 2's 10-m spatial resolution was still coarse for urban roof mapping.

The study observed that even after pan-sharpening, Sentinel 2s 10-m spatial resolution was still coarse for urban roof mapping. This implies the need to test other higher spatial resolution datasets, for example those derived from UAVs and aircraft platforms. Future studies should also consider separating different shades of the same colour, especially in view of colour changes associated with roof aging. Additionally, the combined effects of various physical factors, which include roof coating, thickness, ventilation, and shape, should be included for in-depth analysis of the effect of roofs on the area's thermal environment. Among the factors to be included simultaneously is the presence and effect of any pigment that may affect albedo and heat absorption capacities, even for rooftops of the same colour. Given the inadequacy of freely available moderate-resolution Landsat 8 and Sentinel datasets in mapping thermal properties of rooftops, there is a need to test other higher spatial resolution datasets, for example those derived from UAVs and aircraft platforms.

Author Contributions: Conceptualization, T.M., O.M. and J.O.; methodology, T.M.; software, T.M.; validation, T.M., J.O. and O.M.; formal analysis, T.M.; investigation, T.M., O.M. and J.O.; resources, T.M., O.M. and J.O.; data curation, T.M., O.M. and J.O.; writing—original draft preparation, T.M.; writing—review and editing, T.M., O.M. and J.O.; visualization, T.M.; supervision, O.M. and J.O.; project administration, O.M. and J.O.; funding acquisition, T.M., O.M. and J.O. All authors have read and agreed to the published version of the manuscript.

Funding: The research of this article was supported by DAAD within the framework of the climapAfrica program of the Federal Ministry of Education and Research. The publisher is fully responsible for the content. The work and article processing charge was also funded by the National Research Foundation of South Africa (NRF) Research Chair in Land Use Planning and Management (Grant Number: 84157).

Data Availability Statement: Remotely sensed data used in this study can be freely downloaded from Earth Explorer website (www.earthexplorer.usgs.gov) courtesy of United States Geological Survey (USGS). Accessed on 10 January 2022.

Acknowledgments: We acknowledge the Climate Modeling Group of the climapAfrica fellowship for the support and the Discipline of Geography at the University of KwaZulu-Natal and the Department of Space Science and Applied Physics at the University of Zimbabwe for providing conducive working environment.

Conflicts of Interest: The authors declare no conflict of interest.

References

1. Meinel, G.; Winkler, M. Long-term investigation of urban sprawl on the basis of remote sensing data-Results of an international city comparison. In Proceedings of the 24th EARSeL-Symposium, Paphos, Cyprus, 13–16 September 2022; Leibniz Institute of Ecological and Regional Development (IOER): Dresden, Germany, 2005.
2. Aldwaik, S.Z.; Pontius, R.G., Jr. Landscape and Urban Planning Intensity analysis to unify measurements of size and stationarity of land changes by interval, category, and transition. *Landsc. Urban Plan.* **2012**, *106*, 103–114. [CrossRef]
3. Hamin, E.M.; Gurran, N. Urban form and climate change: Balancing adaptation and mitigation in the U.S. and Australia. *Habitat Int.* **2009**, *33*, 238–245. [CrossRef]
4. Odindi, J.O.; Bangamwabo, V.; Mutanga, O. Assessing the value of urban green spaces in mitigating multi-seasonal urban heat using MODIS land surface temperature (LST) and landsat 8 data. *Int. J. Environ. Res.* **2015**, *9*, 9–18. [CrossRef]
5. Ngie, A.; Abutaleb, K.; Ahmed, F.; Taiwo, O.J.; Darwish, A.A.; Ahmed, M. An estimation of land surface temperatures from landsat ETM+ images for Durban, South Africa. *Rwanda J.* **2017**, *1*, 1–17. [CrossRef]
6. Lee, L.; Chen, L.; Wang, X.; Zhao, J. Use of Landsat TM/ETM+ data to analyze urban heat island and its relationship with land use/cover change. In Proceedings of the 2011 International Conference on Remote Sensing, Environment and Transportation (RSETE) Engineering, Nanjing, China, 24–26 June 2011; pp. 922–927. [CrossRef]
7. Pérez-Andreu, V.; Aparicio-Fernández, C.; Martínez-Ibernón, A.; Vivancos, J.L. Impact of climate change on heating and cooling energy demand in a residential building in a Mediterranean climate. *Energy* **2018**, *165*, 63–74. [CrossRef]
8. Pantavou, K.; Theoharatos, G.; Mavrakis, A.; Santamouris, M. Evaluating thermal comfort conditions and health responses during an extremely hot summer in Athens. *Build. Environ.* **2011**, *46*, 339–344. [CrossRef]
9. Mahdavi, A.; Kumar, S. Implications of indoor climate control for comfort, energy and environment. *Energy Build.* **1996**, *24*, 167–177. [CrossRef]
10. Ormandy, D.; Ezratty, V. Thermal discomfort and health: Protecting the susceptible from excess cold and excess heat in housing. *Adv. Build. Energy Res.* **2016**, *10*, 84–98. [CrossRef]
11. Parkinson, T.; de Dear, R. Thermal pleasure in built environments: Spatial alliesthesia from air movement. *Build. Res. Inf.* **2017**, *45*, 320–335. [CrossRef]
12. Hecht, R.; Herold, H.; Meinel, G.; Buchroithner, M. Automatic Derivation of Urban Structure Types from Topographic Maps by Means of Image Analysis and Machine Learning. In Proceedings of the 26th International Cartographic Conference, Dresden, Germany, 25–30 August 2013; Volume 2013, p. 18.
13. Ketterer, C.; Matzarakis, A. Comparison of different methods for the assessment of the urban heat island in Stuttgart, Germany. *Int. J. Biometeorol.* **2015**, *59*, 1299–1309. [CrossRef]
14. Zhou, D.; Xiao, J.; Bonafoni, S.; Berger, C.; Deilami, K.; Zhou, Y.; Frolking, S.; Yao, R.; Qiao, Z.; Sobrino, J.A. Satellite Remote Sensing of Surface Urban Heat Islands: Progress, Challenges, and Perspectives. *Remote Sens.* **2019**, *11*, 48. [CrossRef]
15. Sun, Y. Retrieval and Application of Land Surface Temperature. *Geo. Utexas. Edu* **2008**, *1*, 1–27. Available online: http://www.geo.utexas.edu/courses/387H/PAPERS/Termpaper-Sun.pdf (accessed on 18 June 2022).
16. Choe, Y.J.; Yom, J.H. Improving accuracy of land surface temperature prediction model based on deep-learning. *Spat. Inf. Res.* **2020**, *28*, 377–382. [CrossRef]
17. Zhou, W.; Huang, G.; Cadenasso, M.L. Does spatial configuration matter? Understanding the effects of land cover pattern on land surface temperature in urban landscapes. *Landsc. Urban Plan.* **2011**, *102*, 54–63. [CrossRef]
18. Zhang, X.; Estoque, R.C. Capturing urban heat island formation in a subtropical city of China based on Landsat images: Implications for sustainable urban development. *Environ. Monit. Assess.* **2021**, *193*, 1–13. [CrossRef]
19. Maimaitiyiming, M. *Effects of Spatial Pattern of Greenspace on Effects of Spatial Pattern of Greenspace on Land*; Universitat Jaume I: Castellón, Spain; Westfälische Wilhelmns-Universität: Münster, Germany; Universidade Nova de Lisboa: Lisboa, Portugal, 2014.
20. Adeyeri, O.E.; Akinsanola, A.A.; Ishola, K.A. Investigating surface urban heat island characteristics over Abuja, Nigeria: Relationship between land surface temperature and multiple vegetation indices. *Remote Sens. Appl. Soc. Environ.* **2017**, *7*, 57–68. [CrossRef]
21. Yuan, F.; Bauer, M.E. Comparison of impervious surface area and normalized difference vegetation index as indicators of surface urban heat island effects in Landsat imagery. *Remote Sens. Environ.* **2007**, *106*, 375–386. [CrossRef]
22. Tyubee, B.T.; Anyadike, R.N.C. Investigating the Effect of Land Use/Land Cover on Urban Surface Temperature in Makurdi, Nigeria. In Proceedings of the ICUC9–9th International Conference on Urban Climate Jointly with 12th Symposium on the Urban Environment, Toulouse, France, 20–24 July 2015.
23. Ardiyansyah, A.; Munir, A.; Gabric, A. The Utilization of Land Surface Temperature Information as an Input for Coastal City The Utilization of Land Surface Temperature Information as an Input for Coastal City. In Proceedings of the IOP Conference Series: Earth and Environmental Science, Jakarta, Indonesia, 25–26 September 2021.
24. Waqar, M.M.; Mirza, J.F.; Mumtaz, R.; Hussain, E. Development of New Indices for Extraction of Built-Up Area & Bare Soil from Landsat Data. *Open Access Sci. Rep.* **2012**, *1*, 4. [CrossRef]
25. Heldens, W.; Del Frate, F.; Lindberg, F.; Mitraka, Z.; Latini, D.; Chrysoulakis, N.; Esch, T. Mapping urban surface characteristics for urban energy flux modelling. In Proceedings of the Mapping Urban Areas from Space Conference, Frascati, Italy, 4–5 November 2015.

26. Kaplan, G.; Avdan, U. Sentinel-2 Pan Sharpening—Comparative Analysis. *Proceedings* **2018**, *2*, 345. [CrossRef]
27. Dhau, I.; Dube, T.; Mushore, T.D. Examining the prospects of sentinel-2 multispectral data in detecting and mapping maize streak virus severity in smallholder Ofcolaco farms, South Africa. *Geocarto Int.* **2019**, *36*, 1873–1883. [CrossRef]
28. Mushore, T.D.; Mutanga, O.; Odindi, J.; Sadza, V.; Dube, T. Pansharpened landsat 8 thermal-infrared data for improved Land Surface Temperature characterization in a heterogeneous urban landscape. *Remote Sens. Appl. Soc. Environ.* **2022**, *26*, 100728. [CrossRef]
29. Shi, Y.; Lau, K.K.; Ren, C.; Ng, E. Urban Climate Evaluating the local climate zone classi fi cation in high-density heterogeneous urban environment using mobile measurement. *Urban Clim.* **2018**, *25*, 167–186. [CrossRef]
30. Wilson, J.S.; Clay, M.; Martin, E.; Stuckey, D.; Vedder-Risch, K. Evaluating environmental influences of zoning in urban ecosystems with remote sensing. *Remote Sens. Environ.* **2003**, *86*, 303–321. [CrossRef]
31. Zhou, X.; Wang, Y.C. Dynamics of Land Surface Temperature in Response to Land-Use/Cover Change. *Geogr. Res.* **2011**, *49*, 23–36. [CrossRef]
32. Klok, L.; Zwart, S.; Verhagen, H.; Mauri, E. The surface heat island of Rotterdam and its relationship with urban surface characteristics. *Resour. Conserv. Recycl.* **2012**, *64*, 23–29. [CrossRef]
33. Wang, X.; Dai, W.; Dai, J. Developing Fine-scale Urban Canopy Parameters in Guangzhou City and its Application in the WRF-Urban model. *World* **2017**, *9000*, 10000. Available online: http://www.meteo.fr/icuc9/LongAbstracts/gd5-2-3221289_a.pdf (accessed on 4 July 2022).
34. Mushore, T.D.; Dube, T.; Manjowe, M.; Gumindoga, W.; Chemura, A.; Rousta, I.; Odindi, J.; Mutanga, O. Remotely sensed retrieval of Local Climate Zones and their linkages to land surface temperature in Harare metropolitan city, Zimbabwe. *Urban Clim.* **2019**, *27*, 259–271. [CrossRef]
35. Qiu, C.; Mou, L.; Schmitt, M.; Xiang, X. Local climate zone-based urban land cover classification from multi-seasonal Sentinel-2 images with a recurrent residual network. *ISPRS J. Photogramm. Remote Sens.* **2019**, *154*, 151–162. [CrossRef]
36. Mackey, C.W.; Lee, X.; Smith, R.B. Remotely sensing the cooling effects of city scale efforts to reduce urban heat island. *Build. Environ.* **2012**, *49*, 348–358. [CrossRef]
37. Cai, M.; Ren, C.; Xu, Y. Investigating the relationship between Local Climate Zone and land surface temperature. In Proceedings of the 2017 Joint Urban Remote Sensing Event (JURSE), Dubai, United Arab Emirates, 6–8 March 2017; pp. 1–4. [CrossRef]
38. Cilek, M.U.; Cilek, A. Analyses of land surface temperature (LST) variability among local climate zones (LCZs) comparing Landsat-8 and ENVI-met model data. *Sustain. Cities Soc.* **2021**, *69*, 102877. [CrossRef]
39. Smith, K.R.; Roebber, P.J. Green roof mitigation potential for a proxy future climate scenario in Chicago, Illinois. *J. Appl. Meteorol. Climatol.* **2011**, *50*, 507–522. [CrossRef]
40. Bass, B.; Baskaran, B. Evaluating Rooftop and Vertical Gardens as an Adaptation Strategy for Urban Areas. *Natl. Res. Counc. Canada* **2001**, *NRCC-46737*, 111. Available online: http://www.nps.gov/tps/sustainability/greendocs/bass.pdf (accessed on 21 March 2022).
41. Tian, Y.; Bai, X.; Qi, B.; Sun, L. Study on heat fluxes of green roofs based on an improved heat and mass transfer model. *Energy Build.* **2017**, *152*, 175–184. [CrossRef]
42. Dominique, M.; Tiana, R.H.; Fanomezana, R.T.; Ludovic, A.A. Thermal behavior of green roof in Reunion Island: Contribution towards a net zero building. *Energy Procedia* **2014**, *57*, 1908–1921. [CrossRef]
43. Granja, A.D.; Labaki, L.C. Influence of external surface colour on the periodic heat flow through a flat solid roof with variable thermal resistance. *Int. J. Energy Res.* **2003**, *27*, 771–779. [CrossRef]
44. Farhan, S.A.; Ismail, F.I.; Kiwan, O.; Shafiq, N.; Zain-Ahmed, A.; Husna, N.; Hamid, A.I.A. Effect of roof tile colour on heat conduction transfer, roof-top surface temperature and cooling load in modern residential buildings under the tropical climate of Malaysia. *Sustainability* **2021**, *13*, 4665. [CrossRef]
45. Libbra, A.; Tarozzi, L.; Muscio, A.; Corticelli, M.A. Spectral response data for development of cool coloured tile coverings. *Opt. Laser Technol.* **2011**, *43*, 394–400. [CrossRef]
46. Alchapar, N.L.; Correa, E.N. Aging of roof coatings. Solar reflectance stability according to their morphological characteristics. *Constr. Build. Mater.* **2016**, *102*, 297–305. [CrossRef]
47. Bansal, N.K.; Garg, S.N.; Kothari, S. Effect of exterior surface colour on the thermal performance of buildings. *Build. Environ.* **1992**, *27*, 31–37. [CrossRef]
48. Zhao, Q.; Myint, S.W.; Wentz, E.A.; Fan, C. Rooftop surface temperature analysis in an Urban residential environment. *Remote Sens.* **2015**, *7*, 12135–12159. [CrossRef]
49. Hsu, C.W.; Chang, C.C.; Lin, C.J. *A Practical Guide to Support Vector Classification*; BJU International: Taipei, Taiwan, 2008; Volume 101, pp. 1396–1400.
50. Adelabu, S.; Mutanga, O.; Adam, E.; Cho, M.A. Exploiting machine learning algorithms for tree species classification in a semiarid woodland using RapidEye image. *J. Appl. Remote Sens.* **2013**, *7*, 073480. [CrossRef]
51. Melgani, F.; Bruzzone, L. Classification of hyperspectral remote sensing images with support vector machines. *IEEE Trans. Geosci. Remote Sens.* **2004**, *42*, 1778–1790. [CrossRef]
52. Sheykhmousa, M.; Mahdianpari, M.; Ghanbari, H. Support Vector Machine vs. Random Forest for Remote Sensing Image Classification: A Meta-analysis and systematic review. *IEEE J. Sel. Top. Appl. Earth Obs. Remote Sens.* **2020**, *13*, 6308–6325. [CrossRef]

53. Tran, D.X.; Pla, F.; Latorre-Carmona, P.; Myint, S.W.; Caetano, M.; Kieu, H.V. Characterizing the relationship between land use land cover change and land surface temperature. *ISPRS J. Photogramm. Remote Sens.* **2017**, *124*, 119–132. [CrossRef]
54. U.S. Geological Survey. Landsat 8 Data Users Handbook. *Nasa* **2019**, *8*, 97. Available online: https://landsat.usgs.gov/documents/Landsat8DataUsersHandbook.pdf (accessed on 4 April 2022).
55. Mumtaz, F.; Tao, Y.; De Leeuw, G.; Zhao, L.; Fan, C.; Arshad, A. Modeling Spatio—Temporal Land Transformation and Its Associated Impacts on land Surface Temperature (LST). *Remote Sens.* **2020**, *12*, 2987. [CrossRef]
56. Van De Griend, A.A.; Owe, M. On the relationship between thermal emissivity and the normalized difference vegetation index for natural surfaces. *Int. J. Remote Sens.* **1993**, *14*, 1119–1131. [CrossRef]
57. Bakar, S.B.A.; Pradhan, B.; Lay, U.S.; Abdullahi, S. Spatial assessment of land surface temperature and land use/land cover in Langkawi Island. *IOP Conf. Ser. Earth Environ. Sci.* **2016**, *37*, 012064. [CrossRef]
58. Tariq, A.; Riaz, I.; Ahmad, Z.; Yang, B.; Amin, M.; Kausar, R.; Andleeb, S.; Farooqi, M.A.; Rafiq, M. Land surface temperature relation with normalized satellite indices for the estimation of spatio-temporal trends in temperature among various land use land cover classes of an arid Potohar region using Landsat data. *Environ. Earth Sci.* **2020**, *79*, 1–15. [CrossRef]
59. Ejiagha, I.R.; Ahmed, M.R.; Hassan, Q.K.; Dewan, A.; Gupta, A.; Rangelova, E. Use of Remote Sensing in Comprehending the Influence of Urban Landscape's Composition and Configuration on Land Surface Temperature at Neighbourhood Scale. *Remote Sens.* **2020**, *12*, 2508. [CrossRef]
60. Maurer, T.; Street, N.Y. How to pan-sharpen images using the gram-schmidt pan-sharpen method—A recipe. *Remote Sens. Spat. Inf. Sci.* **2013**, *XL*, 21–24. [CrossRef]
61. Parente, C.; Santamaria, R. Synthetic sensor of Landsat 7 ETM + imagery to compare and evaluate pan-sharpening methods. *Sens. Transducers* **2014**, *177*, 294.
62. Sithole, K.; Odindi, J. Determination of Urban Thermal Characteristics on an Urban/Rural Land Cover Gradient Using Remotely Sensed Data. *South Afr. J. Geomat.* **2015**, *4*, 384. [CrossRef]
63. Collins, S.; Kuoppamäki, K.; Kotze, D.J.; Lü, X. Thermal behavior of green roofs under Nordic winter conditions. *Build. Environ.* **2017**, *122*, 206–214. [CrossRef]
64. Triano-Juárez, J.; Macias-Melo, E.V.; Hernández-Pérez, I.; Aguilar-Castro, K.M.; Xamán, J. Thermal behavior of a phase change material in a building roof with and without reflective coating in a warm humid zone. *J. Build. Eng.* **2020**, *32*, 101648. [CrossRef]
65. Bojić, M.L.; Milovanović, M.; Lovedaya, D.L. Thermal behavior of a building with a slanted roof. *Energy Build.* **1997**, *26*, 145–151. [CrossRef]
66. Zhang, X.; Zhong, T.; Feng, X.; Wang, K. Estimation of the relationship between vegetation patches and urban land surface temperature with remote sensing. *Int. J. Remote Sens.* **2009**, *30*, 2105–2118. [CrossRef]
67. Chen, A. Cool Colors, Cool Roofs. In *Berkeley Lab COVID-19 Related Research and Additional Information*; Department of Energy National Laboratory, University of California: Berkeley, CA, USA, 2004.

remote sensing

Article

Cooling Effects of Urban Vegetation: The Role of Golf Courses

Thu Thi Nguyen [1,2], **Harry Eslick** [3], **Paul Barber** [2,3], **Richard Harper** [2] and **Bernard Dell** [4,*]

[1]   Faculty of Forest Resources and Environmental Management,
      Vietnam National University of Forestry (VNUF), Xuan Mai, Chuong My, Ha Noi 10018, Vietnam
[2]   Centre for Terrestrial Ecosystem Science & Sustainability, Harry Butler Institute, Murdoch University,
      Murdoch, WA 6150, Australia
[3]   ArborCarbon Pty Ltd., Murdoch University, Murdoch, WA 6150, Australia
[4]   Agricultural and Forestry Sciences, Murdoch University, Murdoch, WA 6150, Australia
*   Correspondence: b.dell@murdoch.edu.au

Abstract: Increased heat in urban environments, from the combined effects of climate change and land use/land cover change, is one of the most severe problems confronting cities and urban residents worldwide, and requires urgent resolution. While large urban green spaces such as parks and nature reserves are widely recognized for their benefits in mitigating urban heat islands (UHIs), the benefit of urban golf courses is less established. This is the first study to combine remote sensing of golf courses with Morphological Spatial Pattern Analysis (MSPA) of vegetation cover. Using ArborCam^{TM} multispectral, high-resolution airborne imagery (0.3×0.3 m), this study develops an approach that assesses the role of golf courses in reducing urban land surface temperature (LST) relative to other urban land-uses in Perth, Australia, and identifies factors that influence cooling. The study revealed that urban golf courses had the second lowest LST (around 31 °C) after conservation land (30 °C), compared to industrial, residential, and main road land uses, which ranged from 35 to 37 °C. They thus have a strong capacity for summer urban heat mitigation. Within the golf courses, distance to water bodies and vegetation structure are important factors contributing to cooling effects. Green spaces comprising tall trees (>10 m) and large vegetation patches have strong effects in reducing LST. This suggests that increasing the proportion of large trees, and increasing vegetation connectivity within golf courses and with other local green spaces, can decrease urban LST, thus providing benefits for urban residents. Moreover, as golf courses are useful for biodiversity conservation, planning for new golf course development should embrace the retention of native vegetation and linkages to conservation corridors.

Keywords: ArborCam; high-resolution airborne imagery; morphological spatial pattern analysis; land surface temperature; golf courses; vegetation structure

Citation: Nguyen, T.T.; Eslick, H.; Barber, P.; Harper, R.; Dell, B. Cooling Effects of Urban Vegetation: The Role of Golf Courses. *Remote Sens.* **2022**, *14*, 4351. https://doi.org/10.3390/rs14174351

Academic Editors: Elhadi Adam, John Odindi, Elfatih Abdel-Rahman and Yuyu Zhou

Received: 12 July 2022
Accepted: 27 August 2022
Published: 1 September 2022

Publisher's Note: MDPI stays neutral with regard to jurisdictional claims in published maps and institutional affiliations.

1. Introduction

Urban development has transformed the land cover of cities causing profound changes in the biological and physical characteristics of the transformed surfaces [1,2]. These changes often result in environmental degradation leading to negative impacts on the quality of life for city dwellers [3]. One of the consequences of urbanization is the relatively higher temperature in urban compared to surrounding peri-urban/rural areas, producing "urban heat islands" (UHIs) [4]. This is due to differences in land use/land cover resulting from human activities. The combined effect of global warming and UHIs is called urban heat [5]. Over recent decades, extreme summer heat has become more frequent across many cities in the world, making urban heat an increasingly important topic in environmental research [6,7]. It is projected that this problem will increase in many regions of the world under the influence of climate change [8] and increased urbanization.

Extreme temperatures have serious impacts on human health, such as heat rash, sunburn, fainting, and heat exhaustion [9,10], which lower the life quality of city dwellers [11].

A large number of deaths related to heat occurred during heat waves in Chicago in 1995, and in 16 European countries in 2003 [9]. Moreover, rising temperatures in urban areas create an uncomfortable environment for residents that results in increasing demand for energy for cooling systems in homes during extreme heat events [12]. Therefore, understanding the spatial distribution of temperature and underlying drivers associated with cooling effects in urban landscapes is a key concern for urban planners.

In order to deal with urban heat issues, studies have been undertaken to identify the dynamics of warming in urban areas [13,14]. In general, an increase in urban green space results in a cooling effect, while impervious land cover leads to a warming effect [15,16]. Impervious surfaces absorb and retain solar energy, with heat slowly released heat back into the atmosphere [17]. In contrast, grass, trees, and other vegetation have a natural heating and cooling cycle that is disrupted by urban structures. Vegetation cools the surrounding area by providing shade and through evapotranspiration [17–19]. Shaded surfaces may be 10–25 °C cooler than unshaded surfaces [20]. Evapotranspiration can help reduce peak summer temperatures by 1–5 °C [21]. Therefore, the amount and quality of vegetation in a city can influence the rate of atmospheric CO_2 sequestration and the amount of heat that a city retains [22–25]. Whether urban vegetation occurs as large nature reserves or as more fragmented and less functionally healthy green spaces for purposes other than conservation (such as public parks, golf courses, cemeteries, military bases, hospitals, university campuses, or streetscapes), they are critically important in cooling cities and making them more livable [26]. Therefore, livable city planning should require a flexible approach that takes advantage of all opportunities to retain green spaces, combining efforts both in formal parks and other recreational spaces.

Golf courses are a type of recreational green space established for commercial and public purposes. They are often a controversial land-use due to their heavy use of water, chemical herbicides, and exotic ornamental vegetation, and this has led to criticism from ecologists [27]. However, other studies have emphasized the ecological values of golf courses for biodiversity conservation [28,29] and for enhancing the connectivity of vegetation networks in urban landscapes [30]. Although the rough (out of play) vegetated areas and irrigated lawns in golf courses are expected to play a role in cooling cities, the ecological value of golf courses in reducing urban heat has been largely ignored by ecologists [31].

Remote sensing-based studies have allowed researchers to assess the spatial distribution of Land Surface Temperature (LST) in urban areas and to establish correlations between vegetation and urban LST models [32–35]. However, most studies have used low and moderate-resolution satellite imageries, such as MODIS or Landsat, to calculate LST as a proxy of urban heat [36,37]. These approaches do not provide information about how vegetation characteristics such as fragmentation, vertical structure, and crown health impact on the local cooling effect. The moderate resolution (30 m) satellite imagery (the Landsat Thematic Mapper (TM)) sensor limits the capacity to detect vegetation of different height classes and their associated LST variability. In contrast, airborne high spatial resolution imagery (0.3 m) has a much greater capacity to detect more detailed vegetation characteristics such as vegetation height classes [38]. Furthermore, studies that have investigated variation in LST among urban formal parks, open spaces, and residential gardens [39–41] have not included golf courses as a separate urban land use. Not surprisingly then, urban planners often lack information for planning urban development that can help to reduce heat exposure.

Native vegetation is undervalued and is often lost during urban development unless it is protected in biodiversity reserves. Remnant vegetation in golf courses is often under pressure from players who want trees removed that are close to fairways. There is a need to quantify some of the benefits of this vegetation to the broader community in golf course management plans in the future. Hence, we explore the value of golf courses and their vegetation in reducing LST. This study examines the hypotheses that golf courses in urban landscapes play a role in reducing urban heat, and that vegetation structure (height class and spatial configuration) influences variations in LST in urban landscapes. Using high-resolution airborne ArborCam imagery, we compared the land surface temperature within golf courses with those of other urban land-use categories, and determined the influence of vegetation traits and geographic location on the function of golf courses for urban heat reduction. The study was conducted in the suburbs of Perth, Australia, where many golf courses retain some native vegetation and provide green connectivity in the urban landscape [30]. The research examines the potential of using golf courses as a green space out-side of protected area networks, and will thus inform the planning of vegetation configuration and vegetation management to optimize cooling at the local and city scales.

2. Materials and Methods

2.1. Study Area

2.1.1. General Description

Perth city is located at latitude −31.953512 and longitude 115.857048 (Figure 1). The Perth Metropolitan area covers 6418 km^2, with a population density of 317.7 people per square kilometer. The area is in Australia's southwest corner, a global biodiversity "hotspot" with outstanding natural environments having the highest concentration of rare and endangered species on the entire continent [42,43].

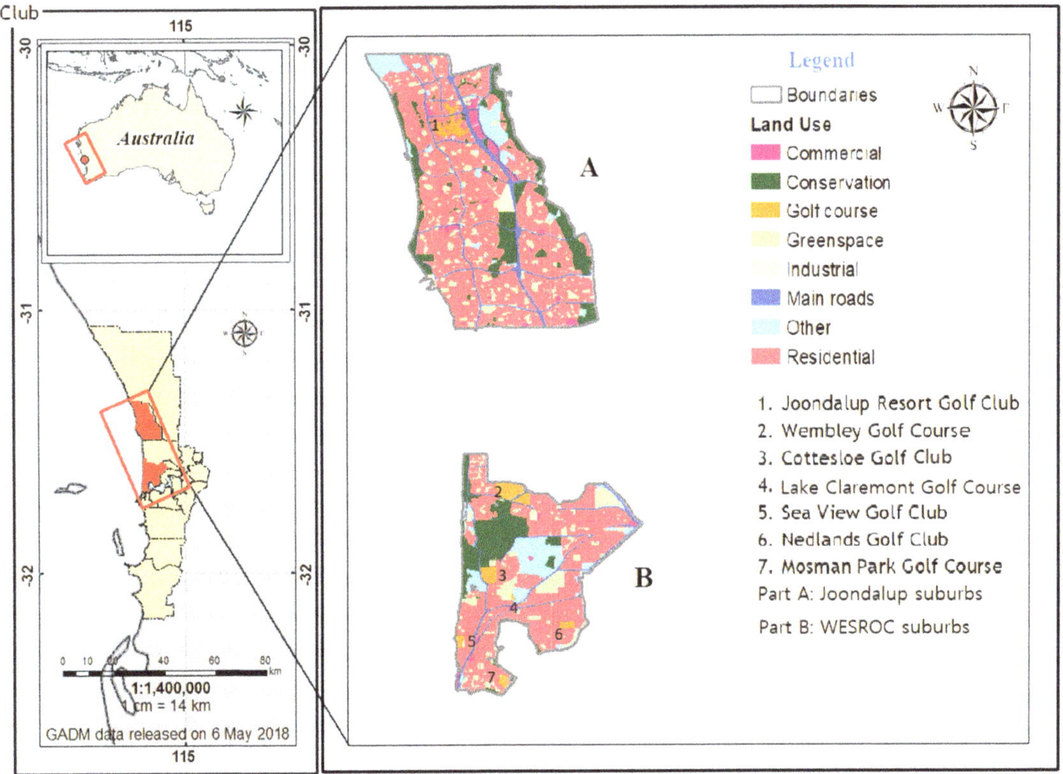

Figure 1. Map showing land use categories and the location of golf courses in the two study regions of the Perth Metropolitan Region.

Perth has a Mediterranean-type climate with hot dry summers, lasting from December to late March, and cool wet winters [22]. Extreme heat events (substantial rises in temperature, duration, and frequency of very hot days) have increased in Perth over the past two decades, and are projected to increase in coming years [44]. These events pose health risks for urban citizens especially the elderly, young, sick, and people from lower socio-economic areas [45].

Perth has experienced extensive urban expansion since the 1960s and this has caused sustainability concerns due to the large-scale conversion of natural land to impervious surfaces [46], which can contribute to an increasingly warming urban environment. It is projected that by 2030 the annually averaged warming of this region will be about 0.5 to 1.2 °C above 1986–2005 levels [47]. Therefore, the Western Australian government issued a long-term strategic guide for the development of Perth by 2050, which identified reducing urban heat as one of sixteen aspirations under the strategy for the Planning Commission, and State and Local Government by expanding the tree canopy in high urban heat risk areas [45,48].

2.1.2. Spatial Subdivision

This study focuses on the western suburbs (WESROC suburbs in the south) and the Joondalup suburbs (in the north), covering 16,205 ha (Figure 1). The WESROC suburbs are a group of old suburbs established prior to the first urban development planning of Perth (i.e., pre-1950s), and are located west of the city's central business district and north of the Swan River. These suburbs are characterized by low to moderate-density residential areas, recreation areas, nature reserves, and wetlands. Joondalup is a younger urban area that was developed as a result of northerly urban expansion following extensive urban development in the 1990s, and is characterized by dense commercial and residential areas. The suburbs of the two subdivisions, established through different times in the history of Perth's planning with different urban designing styles, are representative of residential suburbs across Perth's sprawling urban landscape. There are six golf courses in the WESROC suburbs and one in the Joondalup region (Figure 1). The golf courses vary in ownership (public, private, and semi-private), size (small < 40 ha, moderate 40–70 ha, and large > 70 ha), number of holes, and the linkage of golf courses to other vegetation (Table 1). The Lake Claremont Golf Course was converted to parkland in recent years. Images of vegetation and environment are readily available online for the respective golf courses used in this study. Figure 2 shows a typical scene. In general, the tees, fairways, and greens are reticulated and irrigated during the dry season (November to April) from underground aquifers. With declining groundwater supplies and a warming climate, the Golf Course Superintendents Association of Western Australia is collaborating with the Department of Water to assist golf courses to become more water efficient.

Table 1. Key characteristics of the seven golf courses.

No.	Name of Golf Course	Size (ha)	Linkage to Other Vegetation	Golf Course Type	Number of Holes
1	Joondalup Resort Golf Club	108.93	No	Semi-private	36
2	Wembley Golf Course	128.96	Yes	Public	36
3	Cottesloe Golf Club	61.26	Yes	Private	18
4	Lake Claremont Golf Course	4.13	Yes	Public	9
5	Sea View Golf Club	18.1	No	Private	9
6	Nedlands Golf Club	18	No	Private	9
7	Mosman Park Golf Course	24.8	Yes	Semi-private	9

Figure 2. Typical scene in a golf course in the study region in summer. The irrigated fairway is bordered by corridors of vegetation containing tall trees (*Eucalyptus*), mid-story trees (*Acacia*, *Agonis*), small shrubs, and grasstrees (*Xanthorrhoea*) (photo by Paul Barber).

2.2. Data Sources and Geospatial Analysis

To provide accurate information about the daytime LST in relation to urban land use, land cover, and vegetation characteristics, the study used multispectral, high-resolution airborne imagery, acquired at ~11:00 to 13:00 h on two typical hot late summer days (10 and 11 March 2020) with a daily maximum temperature at Perth Metro station (number: 009225) of 35.1 °C for both days [49] and calm conditions.

High-resolution RGB, seven-band multispectral, and long wave thermal radiation were acquired concurrently using the custom ArborCam^{TM} vegetation monitoring system (ArborCarbon Pty Ltd., Perth, Australia). Imagery was acquired on dedicated flights using a customized Piper PA-28 aircraft with specifications as described in Table 2.

Table 2. Acquisition parameters and resulting image Ground Sample Distance (GSD) for each of the imaging sensors for the two study areas.

	WESROC	Joondalup
Acquisition date	10 March 2020	10 and 11 March 2020
Acquisition height	2440 m	3048 m
High-resolution RGB:GSD	0.08 m	0.1 m
Multispectral: GSD	0.24 m	0.3 m
Thermal: GSD	1.0 m	1.25 m

The ArborCam sensor captures seven distinct narrow multispectral bands strategically located between 450 and 780 nm of the electromagnetic spectrum [50]. Long-wave thermal Infra-red radiation (Thermal IR 7500–14,000 nm) was converted to LST in degrees Celsius by applying a standard emissivity correction across the scene of 0.95 to produce a single-band 32-bit raster, with each pixel representing land surface temperature.

All imagery was orthorectified and radiometrically corrected using a series of propriety image processing workflows. A Digital Surface Model (DSM) was generated using a Structure from Motion processing technique during orthorectification. This DSM was further classified to identify ground surface pixels, which were then interpolated to produce a Digital Terrain Model (DTM). The difference between the DSM and DTM was calculated to determine the Feature Height Model. Final imagery was converted to units of surface reflectance using radiometric targets placed throughout the scene. Finally, vegetation within the scene was classified using a segmentation and supervised classification approach. The Arbor-Cam thermal imaging sensor uses a microbolometer with a spectral range of 7.5–14 μm, and a resolution of 640×480 with a $15° \times 11°$ field of view. Thermal radiance is corrected and converted to LST on board the camera using a standard emissivity correction of 0.95, and relative humidity of 50% at 20 °C. Linear temperature data are recorded in 16 bits, with a sensitivity of 0.05 K and a stated accuracy of ± 2 °C or ± 2%. This is a standard approach for studies of urban land surface temperature. More precise methods of emissivity correction for individual surface materials require the classification of surface materials, which is beyond the scope of the current study. The current study is concerned primarily with the relative differences in LST; therefore, the validation of the reading vs. actual LST is of lesser value.

2.2.1. NDVI Calculation

Normalized Difference Vegetation Index (NDVI) maps were developed by calculating the ratio between the red (R) and near-infrared (NIR) using Equation (1) [51]:

$$NDVI = (NIR - red)/(NIR + red) \tag{1}$$

2.2.2. Morphological Spatial Pattern Analysis (MSPA)

Morphological Spatial Pattern Analysis (MSPA) was employed in this research for analysis of spatial configuration of vegetation cover (turf and all other vegetation types) as described previously [52–54]. In order to undertake the MSPA analysis, the input data (foreground class) were defined. The binary maps (vegetation and non-vegetation) obtained from the classification of PlanetScope 3B images were used as input data with the vegetation being defined as the foreground pixels (green landscape) in the MSPA approach using the MSPA-Toolbox for ArcGIS. The output of the MSPA analysis includes the seven structural categories belonging to two groups: (1) urban vegetation patches (Cores, Edge, Perforation, and Islets) and (2) urban vegetation paths (Bridges, Branches, and Loops) [52–54]. Each of these categories was described at the pixel level [52–54] and described in ecological meaning terms based on the concept of "habitat availability" and "graphic theory" [52–54], and this can be briefly described as:

- Core—The availability of interior forest habitat;
- Islet—The isolated non-Core habitat, or potential stepping stone;
- Edge—The Edge habitat and Edge effects on interior forest habitat;
- Perforation—Edge on forest interior;
- Bridge—The structural connectivity among Core areas;
- Loop—The structural connectivity within a Core area;
- Branch—The structural connectivity that departs from a Core area and arrives at a connector, to an Edge, or a Perforation.

2.2.3. Land-Use Data

We selected eight land-use categories representing the main components in the urban landscape, as follows: conservation land (1); golf course (2); green space (3); commercial (4); industrial (5); residential (6); main road (7); and other land-use (8). This selection of land-use categories is comparable to those used in ecological research in other urban landscapes [55,56] and is described in Table 3.

Table 3. Description of land-use categories.

No.	Land-Use Category	Description
1	Conservation	Land of Bush Forever areas (described by Department of Planning [57]); areas of biodiversity conservation significance within National Parks and State and other, conservation reserves, and all classified environmental conditions, special control areas, which are of conservation concerns as dedicated in the Regional Scheme map, Regional Special Area map, and Local Scheme map as well as small-parks from the OSM platform.
2	Golf course	Golf courses are a special type of urban green space used for recreational and commercial purposes. In this study golf courses are separated for comparison with all other urban green spaces. There are 7 golf courses distributed in the study area, which are described in Figure 1 and Table 1.
3	Green space	The urban parks and other land used as set aside areas for public open space, provide for a range of active and passive recreation uses.
4	Commercial	The land is used to provide for a range of shops, offices, restaurants, and other commercial outlets in defined townsites or activity centers, a wide variety of active uses on a street level; a mix of varied but compatible land uses such as offices, showrooms, amusement centers, eating establishments, and appropriate industrial activities.
5	Industrial	Land of industrial activities to provide a broad range of industrial uses, service and storage activities.
6	Residential	Land-use areas provide for a range of housing and a choice of residential densities to meet the needs of the community by facilitating and encouraging high-quality design, built form, and streetscapes throughout residential areas.
7	Main road	The planned road network of the Western Australian Road (under the Main Roads Act 1930), and the planning responsibilities are shared by the Western Australian Planning Commission and local governments.
8	Other land-use	The land-use categories that are not classified as those above. They include designated land for future industrial development, urban development, transitional zone following the lifting of an urban deferred zoning, land of educational institutions, a broad range of essential public facilities such as halls, theatres, art galleries, educational, health and social care facilities, accommodation for the aged, other services, and other mixed land-use.

2.3. Statistical Analysis

To address the first objective (variation in LST among land-use categories and among golf courses), LST mean values were derived for each of the eight land-use categories using vector data analysis (zonal statistics) in ArcGIS 10.3 and descriptive statistics in R 3.6.1. The land-use layer (Figure 1) was used to define zonal boundaries.

To address the second objective (factors influencing the cooling effects of golf courses), the variation and correlation of LST with each driving factor were derived. We randomly generated more than 500 random points within the seven golf courses and the study area. Values for each independent variable were assigned to each point using Extract Multi Values to Points tool in ArcGIS 10.3. All geographical analyses were conducted using ArcGis version 10.3 and statistical analyses were performed in R 3.6.1 [58]. Based on the initial description of the relationship between LST and the variables, a multiple linear regression model was built with the F-statistic in R 3.6.1 to describe the effects of vegetation characteristics and geographic location that drive LST.

The explanatory variables examined are listed in Table 4. These variables were subdivided into four groups: vegetation height class, MSPA class, NDVI, and distance to water resources (Table 4). Previous studies have explored the distance to the coast as a factor impacting urban temperature [59]. However, the study area has a network of water bodies, the ocean, the estuary of the Swan River, and groundwater-derived lakes. These water bodies are likely to influence the LST, and thus the distance to the water resources (ocean, lakes, and river) was measured using the near tool in ArcGIS 10.3. The values were added to random points as an independent variable in the regression analysis. The multiple linear regression model was performed in R 3.6.1 [58] to determine the relationship between the dependent variable (LST) and its driving factors (Table 4).

Table 4. Independent variables considered in the multivariate model of LST within golf courses and other land use categories. The MSPA descriptors are the same as published by the original workers [52–54].

Variable	Description
Vegetation height class	
Turf	The top layer of a grassy area
0–3 m	Vegetation of 0–3 m height
3–10 m	Vegetation of 3–10 m height
10–15 m	Vegetation of 10–15 m height
>15 m	Vegetation of >15 m height
MSPA Class	
Bridge	The ecological vegetation that connects two Cores, which is equivalent to the connecting corridor
Core	Large-scale natural patches with high connectivity
Edge	The transition zone between vegetation and non-vegetation areas
Islet	Small natural patches that are isolated and do not connect to each other
Loop	Connecting corridor inside a large natural patch
Perforation	Unnatural patch inside the Core area
Distance to water resource	The shortest distance from the sample point to the water resources (lake, river, and coast)
NDVI	Normalized Difference Vegetation Index: NDVI = (NIR − RED)/NIR + RED) NIR—reflectance in the near-infrared spectrum RED—reflectance in the red range of the spectrum

In terms of vegetation variables, previous studies have focused on vegetation cover [59] and vegetation type (grass, shrubs, and trees) [22]. In this study, more details of vertical vegetation structure were explored where vegetation was classified into height classes (Table 4), and spatial vegetation configuration where vegetation was classified into habitat type (MSPA classes) based on the patch size and their connectivity to other vegetation areas and they were added to random points for the regression analysis.

3. Results

3.1. Variation in LST among Land-Use Categories and among Golf Courses

Overall, the conservation land was the coolest land-use category (mean LST of approximately 30 °C). Golf courses had the second lowest mean LST (around 31 °C) in the study area (Figure 3A), and thus golf courses in high-density urban areas play a role as cool islands (Figure 3B). Joondalup Resort Golf Club is shown as an example (Figure 4). The average LST for industrial, residential, and main road land uses were high, ranging from approximately 35 to 37 °C. The land use types in the order of highest to lowest temperatures were main roads, residential, industrial, other, commercial, green space, golf course, and conservation (Figure 3A).

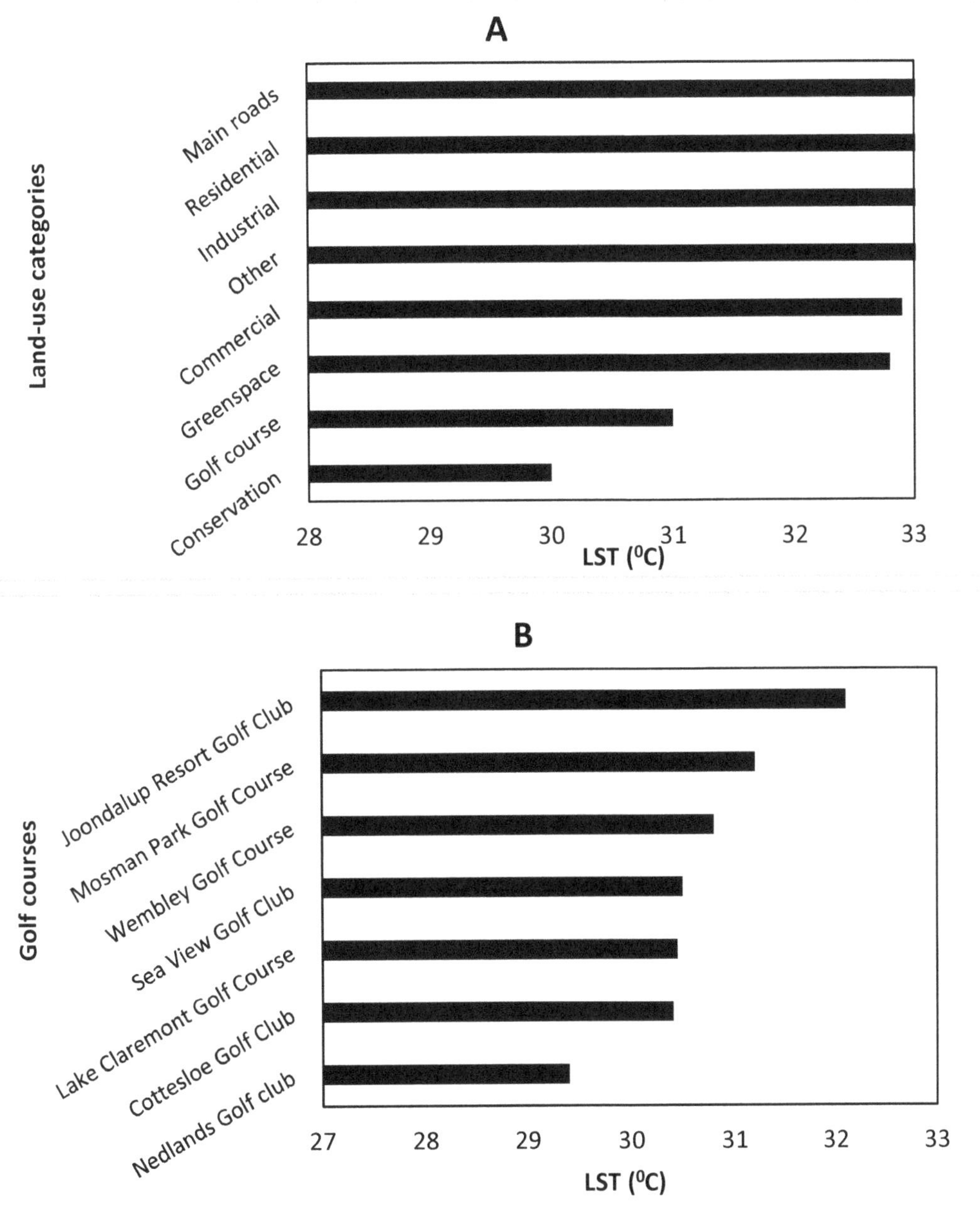

Figure 3. Variation in the mean values of LST among (**A**) land-use categories and (**B**) the seven golf courses.

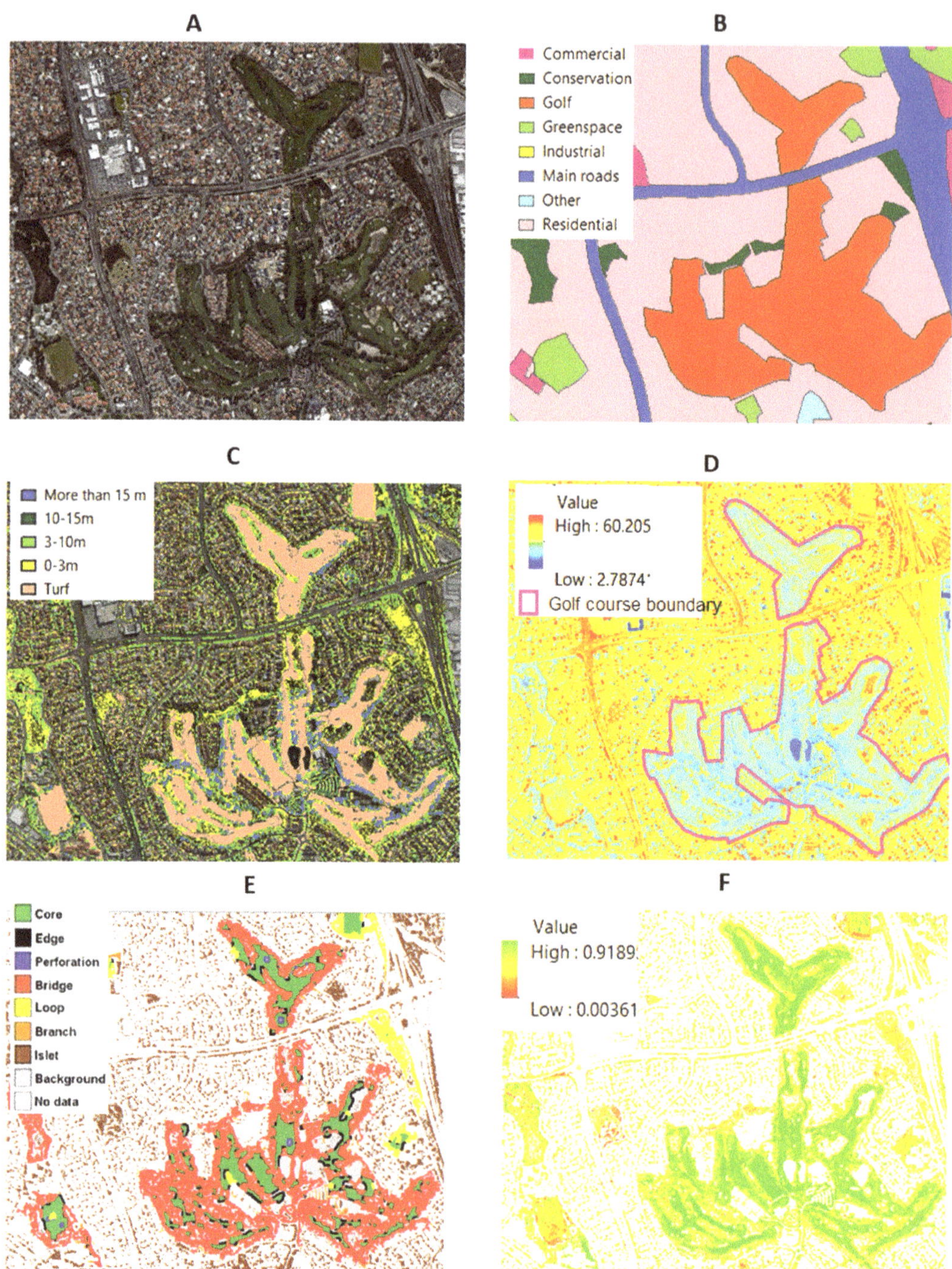

Figure 4. Joondalup Resort Golf Club and surrounds: (**A**) True color orthomosaic; (**B**) Land-use categories; (**C**) Vegetation height-strata; (**D**) Day time LST; (**E**) MSPA classes; (**F**) NDVI map. Data were collected in late summer (maximum temperature 35.1 °C) on 10 and 11 March 2020.

Notably, the LST differed markedly between golf courses. The highest mean LST occurred within Joondalup Resort Golf Club at about 32 °C (Figure 3B). Nedlands Golf Club had the lowest mean LST among studied golf courses (around 29 °C). Golf courses located nearby the coast (Cottesloe Golf Club, Sea View Golf Club, Lake Claremont Golf Course, and Wembley Golf Course) had similar mean LSTs at about 30 °C. Mosman Park Golf Course had the second highest mean LST (around 31°C) (Figure 3B). These results indicate that the golf courses have different capacities to cool their local environments, and that there may be underlying drivers leading to this variation.

3.2. Factors Influencing Cooling Effects of Golf Courses

There was a positive relationship between LST and distance to water resources (Figure 5B), which indicates the availability of water bodies can be beneficial on hot summer days. Moreover, vegetation characteristics can impact cooling. Figure 5C shows the LST in non-vegetated areas was much higher than any type of vegetation (LST median around 35 °C) indicating the role of vegetation cover in providing a cooling effect. Within areas of vegetation cover, NDVI values reflect vegetation health, and these showed a strong negative relationship with LST (R = 0.77). This means that green, healthy vegetation has a good capacity to cool urban areas (Figure 5A) with the example of Joondalup Resort Golf Club and surrounds illustrated (Figure 4D,F).

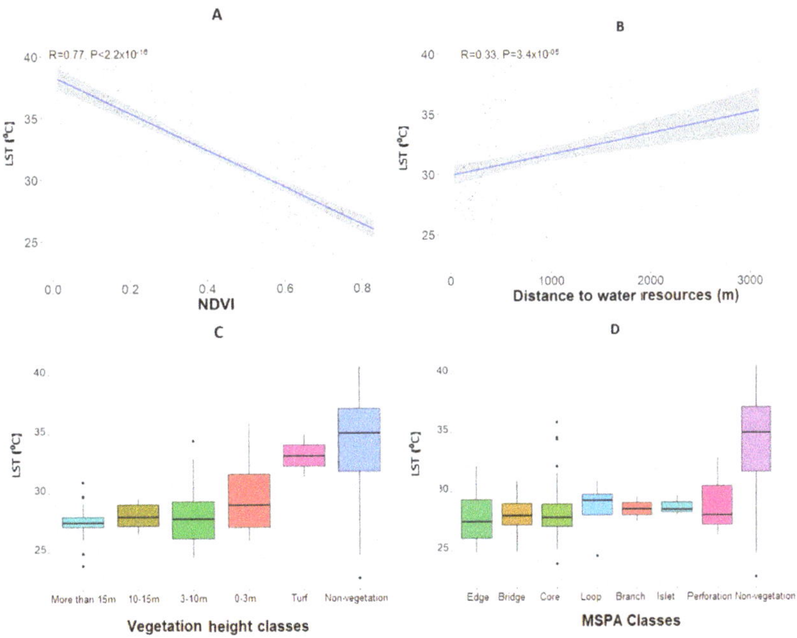

Figure 5. Factors affecting Land Surface Temperature: (**A**) Relationship between LST and NDVI; (**B**) Relationship between LST and distance to the water resources; (**C**) Variation in LST among vegetation height classes; (**D**) Variation in LST among MSPA classes.

However, not all types of vegetation have the same cooling effect. The vertical structure and horizontal configuration of vegetation further determine the capacity of vegetation for cooling. In general, the taller the vegetation the cooler the LST (Figure 5C and Table 5). Vegetation of >10 m height had a median LST of around 27 °C, 0–3 m high vegetation had a mean LST of around 29 °C, and for turf, the LST was around 33 °C (Figure 5C).

Table 5. Estimates for each independent variable from the multivariate model for predicting the LST.

Variable	Coefficient	Std. Error	z Value	Pr (> | Z |)
Intercept	3.830×10^1	2.006×10^0	19.094	$<2 \times 10^{-16}$ ***
Vegetation strata				
Non-vegetation	1.640×10^{-2}	1.177×10^0	0.014	0.98889
Turf	1.501×10^0	2.033×10^0	0.738	0.46142
3–10 m	-1.353×10^0	7.696×10^{-1}	-1.758	0.08057
10–15 m	-2.068×10^0	8.442×10^{-1}	-2.450	0.01531 *
>15 m	-1.953×10^0	8.410×10^{-1}	-2.322	0.02140 *
MSPA Class				
Bridge	-1.408×10^0	1.815×10^0	-0.776	0.43904
Core	-2.774×10^0	1.741×10^0	-1.593	0.011305 *
Edge	-4.151×10^0	1.862×10^0	-2.230	0.02709 *
Islet	2.656×10^{-1}	2.166×10^0	0.123	0.90253
Loop	-8.312×10^{-1}	2.106×10^0	0.395	0.69363
Perforation	2.535×10^0	2.206×10^0	1.149	0.25228
NDVI	-1.109×10^1	1.011×10^0	-10.968	$<2 \times 10^{-16}$ ***
Distance to water resource	8.072×10^{-4}	2.525×10^{-4}	3.196	0.00166 **

*** $p < 0.001$; ** $p < 0.01$; * $p < 0.05$.

The size of vegetation patches and the linkages among them also influence LST (Figure 5D). A large vegetation patch, comprised of the outer Edge and Core categories, had a low LST of 27–28 °C. Moreover, the Bridge class that connects two Cores also had a similar low LST (Figure 5D). The Islet representing small, isolated patches, and the Perforation representing the inner Edge had the highest LST (Figure 5D).

The multiple linear regression model for predicting LST (Thermal = Vegetation Strata + MSPA classes + Distance to water resource + NDVI) had an R-squared value of 0.72 and an adjusted R-squared value of 0.7 with $p < 0.0001$ indicating that LST was closely related to the vegetation variables and distance to water resources. However, in the subset of vegetation height class variables, only the coefficient of vegetation 10–15 m and >15 m had p-values < 0.05 (Table 5).

Similarly, among MSPA variables, only the classes representing large patches (Core, Edge) had p-values of <0.05 (Table 4). This means that taller vegetation (10–15 m and >15 m) and large patches of vegetation that combine the outer Edge and Core areas had significant effects on LST. Moreover, the multi-regression model further determined that the important factors influencing LST were the health of vegetation indicated by the NDVI value and the distance to water (coefficient p-values < 0.01) (Table 5). This suggests that these six independent variables are statistically significant predictors of the LST.

4. Discussion

4.1. Golf Courses as Cooling Islands in Urban Environments

The study revealed that urban golf courses had lower day-time land surface temperatures than other urban land-use categories, except for conservation land. This means that in a warming climate, golf courses, with most of their surface area covered with a combination of vegetation (shrubs and trees), water bodies, and turf, will be cool-islands and natural havens in cities where most of the surface is dominated by building structures and heat-absorbing hard surfaces. Green spaces in golf courses include irrigated fairways and out-of-play areas, but the cooling effects of golf courses are strongest for woodlands with complex multiple-tiered biomass structures. A similar finding was made for urban green space in Hong Kong [31,60].

In industrial and commercial land uses, the buildings often have flat concrete or metallic roofs, as seen in the aerial imagery. Concrete surfaces have low albedo from 0.1 to 0.35 [61]. In contrast, the vegetation acts as a buffer between the ground and solar radiation, and this helps to reduce the LST [62]. The similarity in LST of golf courses and conservation land can be explained by similar surface characteristics related to their vegetation cover.

Previous studies from Sydney (Australia) and Toronto (Canada) showed that mean temperatures are significantly lower for parks and recreational land uses than for highly intensive urban land-use such as industrial and commercial areas [41,63]. Thus, future changes in land-use from forest and grasslands to new urban developments (industrial, commercial, and residential) are projected to further enhance temperature increases caused by climate change [63]. The UHI problem is serious in Australia's hotter cities such as Perth, Adelaide, and Alice Springs [64]. This is due to the large proportion of impervious surfaces as a result of urbanization. Other Mediterranean-climate cities are predicted to have increases in average minimum temperatures compared to other rural areas [65]. Therefore, in hot dry climates, urban planning should pay more attention to designing cool islands to mitigate the UHI effect and its impact on city residents. With the cooling effects of golf courses identified in our study, we propose that urban golf courses should be considered as a type of cooling island in urban planning within urban heat mitigation strategies.

4.2. Vegetation Characteristics Influence Cooling Effects of Urban Green Spaces

Previous studies have confirmed the role of vegetation cover in mitigating urban heat [22,47,59], which can help to explain why urban areas without vegetation heat up easily and retain heat [59,66]. The cooling effects of vegetation within urban golf courses have not been well investigated. For example, the study of microclimate at a sub-tropical golf course in Hong Kong only investigated the differential cooling abilities of woodland, a rough grass area, and a bare-concrete rooftop control site within the golf course [31,60]. However, the role of vertical vegetation structure (vegetation height classes) and the spatial arrangement of vegetation patches in cooling urban environments is largely unexplored [67].

By using the high-resolution (0.3×0.3 m) airborne imagery, our study provides new insights into the cooling capacities of vegetation of different high classes within golf courses and other green spaces. This study suggests that tall urban forests (>10 m tall) will have the strongest effects in reducing urban heat islands, while shorter vegetation, including turf grass, will be less effective. This provides a new understanding of the relationship between vegetation and urban heat and indicates that urban heat mitigation strategies using green spaces should not solely focus on the extent of vegetation coverage but should also consider the height and vertical structure of the vegetation. Due to the limited space for vegetation in urban areas, it is necessary to maximize the effects of green space by maintaining and increasing the number of big trees to regulate temperature and improve the urban microclimate.

Moreover, our study also explored how vegetation complexity in terms of spatial configuration and arrangement might facilitate the management of urban heat. The results showed stronger cooling effects of large vegetation patches (Core area and their outer Edge) as well as the vegetation paths that link Cores (Bridge) as being more effective vegetation structures than small, isolated patches (Islet). This finding supports previous studies [68–71] where large patch sizes reduced LST due to greater shading of the periphery. Furthermore, larger vegetation patches have more interior areas, which are less affected by the ambient environment, whereas smaller, isolated patches (Islets) tend to have a greater proportion of edge areas, and thus are vulnerable to disturbance from the peripheral region [67].

In addition, our study revealed that vegetation connectivity and patch size are important when designing urban green space. The connectivity of vegetation cover can contribute to decreasing surface temperature in urban areas [72]. Therefore, increasing urban vegetation, maintaining large patches of vegetation, and reducing insolation can help to decrease urban LST. Hence, urban planning should consider the size and configuration of green spaces to operate as cooling islands without becoming masked by surrounding buildings.

Vegetation health is an important factor influencing its effectiveness in cooling urban areas. Healthy vegetation patches with NDVI values from 0.6 to 0.8 had the strongest cooling effect (Figure 5A). Therefore, together with maintaining water bodies in combination with protecting and restoring big trees in large patches, caring for vegetation health is vital to ensure the cooling effects of green spaces are optimized. Several abiotic and biotic disorders pose a threat to urban tree health with some of these, such as Phytophthora, thriving in irrigated urban parklands [73].

The correlation of LST with impact factors may explain why LST varied among land-use categories. For example, the proportion of vegetation (without turf) was highest on conservation land, and most of the vegetation existing as Core or Bridge areas in this land-use category may explain why the LST was lowest. High vegetation connectivity appears to be an important factor for conservation land having the strongest capacity to reduce urban LST. Vegetation within golf courses that is healthy with a high proportion of tall trees also contributes to the low values of LST in this land-use category. Furthermore, the golf courses were close to or contained water bodies and this helped to mitigate LST in summer in our study.

4.3. Implication for Vegetation Management and Urban Planning

This study suggests that urban vegetation management approaches are required to mitigate urban heat. Golf courses can contribute significantly to the mitigation of LST in urban landscapes. As large trees play an important role in reducing LST, golf course managers and designers should pay attention to the conservation of these trees. It is recommended that golf course managers should not only increase the natural tree canopy by planting more trees, but also actively protect tall trees and large vegetation patches to improve the cooling capacity of golf courses. It is important though to always consider the conflict between turf health and trees when designing or re-designing golf courses. Large trees provide large amounts of shade with potential negative impacts on turf health when insufficient light is received.

Urban expansion on undeveloped land containing large patches of native vegetation that involves tree clearing should embrace tree planting for future cooling effects. This study suggests that increasing the urban tree canopy should benefit heat mitigation. However, it will be difficult to reach targets without promoting planting on private land. Nowhere is this more pressing than in urban environments where there is a scarcity of available land with native vegetation. The regression equation in this study also provides an indication of how temperature can be reduced in other urban land-use categories (e.g., residential, commercial areas) by tree planting and vegetation patch maintenance. Because increasing vegetation coverage is difficult in some dense urban landscapes, measures to improve the quality of existing vegetation patches, such as tall tree conservation and irrigation, are important for mitigating temperature for improving the well-being of city dwellers.

Novel approaches for heat reduction and livable neighborhood policies should embrace the importance of developing incentives that promote multipurpose use of land and that stimulate cooperation among people and different societal sectors to support urban green space maintenance. Golf courses are an example of commercial land that can contribute to urban heat reduction that should be integrated into livable neighborhood policies to improve the life quality of urban citizens.

5. Conclusions

This study develops and demonstrates a robust and objective approach to quantify and compare variation in LST among urban land-use categories. The research used high-resolution multispectral airborne imagery to classify vegetation height classes and this helped to fill gaps in the current literature that compares the LST of different vegetation types. From our study, it is clear that vertical vegetation structure and horizontal vegetation configuration and arrangement are important in urban heat reduction. It is also evident that the vegetation of golf courses can play a beneficial role in helping to reduce urban heating during hot summer days. Effective management of vegetation for urban heat reduction and livable neighborhoods should consider the maintenance of big trees and large patches of vegetation across the urban landscape. Our study is significant because it provides insight into the ecological benefits of recreational green spaces such as golf courses in urban landscapes where such ecological roles are often valued in conservation land. The findings from this study suggest that planning for further urbanization of peri-urban land should consider opportunities for the co-planning of golf course development in conjunction with the retention of functional vegetation corridors.

Author Contributions: T.T.N.: Conceptualization, Methodology, Formal analysis, Investigation, Writing—Original draft, Writing—Review and editing. H.E.: Methodology, Resources, Review and editing. P.B.: Supervision, Conceptualization, Methodology, Resources, Review and editing, Funding acquisition. R.H.: Supervision, Conceptualization, Review and editing. B.D.: Supervision, Conceptualization, Methodology, Review and editing, Funding acquisition. All authors have read and agreed to the published version of the manuscript.

Funding: This work was part of a PhD study by Thi Thu Nguyen made possible by a joint PhD scholarship between Vietnam International Education Development (VIED) and Murdoch University. The costs of the ArborCam Imagery were borne by ArborCarbon Pty Ltd., Perth, Australia.

Data Availability Statement: Not applicable.

Acknowledgments: Naviin Hardy from ArborCarbon Pty Ltd. provided technical support with the high-quality datasets of the airborne imagery. Thanh Trang Pham from the Vietnam National University of Forestry gave advice on data analysis.

Conflicts of Interest: The authors declare no conflict of interest.

References

1. Al-Manni, A.A.A.; Abdu, A.S.A.; Mohammed, N.A.; Al-Sheeb, A.E. Urban growth and land use change detection using remote sensing and geographic information system techniques in Doha City, State of Qatar. *Arab Gulf J. Sci. Res.* **2007**, *25*, 190–198.
2. Anderson, E.C.; Avolio, M.L.; Sonti, N.F.; LaDeau, S.L. More than green: Tree structure and biodiversity patterns differ across canopy change regimes in Baltimore's urban forest. *Urban For. Urban Green.* **2021**, *65*, 127365. [CrossRef]
3. Lin, S.; Sun, J.; Marinova, D.; Zhao, D. Effects of population and land urbanization on China's environmental impact: Empirical analysis based on the extended STIRPAT model. *Sustainability* **2017**, *9*, 825. [CrossRef]
4. O'Malley, C.; Piroozfarb, P.A.E.; Farr, E.R.P.; Gates, J. An Investigation into minimizing urban heat island (UHI) effects: A UK perspective. *Energy Procedia* **2014**, *62*, 72–80. [CrossRef]
5. He, B.J.; Wang, J.; Zhu, J.; Qi, J. Beating the urban heat: Situation, background, impacts and the way forward in China. *Renew. Sustain. Energy Rev.* **2022**, *161*, 112350. [CrossRef]
6. Ramamurthy, P.; Bou-Zeid, E. Heatwaves and urban heat islands: A comparative analysis of multiple cities. *J. Geophys. Res. Atmos.* **2016**, *122*, 2. [CrossRef]
7. Tan, J.; Zheng, Y.; Tang, X.; Guo, C.; Li, L.; Song, G.; Zhen, X.; Yuan, D.; Kalkstein, A.J.; Li, F. The urban heat island and its impact on heat waves and human health in Shanghai. *Int. J. Biometeorol.* **2010**, *54*, 75–84. [CrossRef]
8. IPCC. *Climate Change 2021: The Physical Science Basis. Contribution of Working Group I to the Sixth Assessment Report of the Intergovernmental Panel on Climate Change*; Cambridge University Press: Cambridge, UK, 2021.
9. Baccini, M.; Biggeri, A.; Accetta, G.; Kosatsky, T.; Katsouyanni, K.; Analitis, A.; Anderson, H.R.; Bisanti, L.; D'Ippoliti, D.; Danova, J. Heat effects on mortality in 15 European cities. *Epidemiology* **2008**, *19*, 711–719. [CrossRef]
10. Gasparrini, A.; Guo, Y.; Hashizume, M.; Lavigne, E.; Zanobetti, A.; Schwartz, J.; Tobias, A.; Tong, S.; Rocklöv, J.; Forsberg, B. Mortality risk attributable to high and low ambient temperature: A multicountry observational study. *Lancet* **2015**, *386*, 369–375. [CrossRef]
11. Heaviside, C.; Macintyre, H.; Vardoulakis, S. The urban heat island: Implications for health in a changing environment. *Curr. Environ. Health Rep.* **2017**, *4*, 296–305. [CrossRef]
12. Lundgren-Kownacki, K.; Hornyanszky, E.D.; Chu, T.A.; Olsson, J.A.; Becker, P. Challenges of using air conditioning in an increasingly hot climate. *Int. J. Biometeorol.* **2018**, *62*, 401–412. [CrossRef] [PubMed]
13. Jain, S.; Sannigrahi, S.; Sen, S.; Bhatt, S.; Chakraborti, S.; Rahmat, S. Urban heat island intensity and its mitigation strategies in the fast-growing urban area. *J. Urban Manag.* **2020**, *9*, 54–66. [CrossRef]
14. He, B.J.; Wang, J.; Liu, H.; Ulpiani, G. Localized synergies between heat waves and urban heat islands: Implications on human thermal comfort and urban heat management. *Environ. Res.* **2021**, *193*, 110584. [CrossRef] [PubMed]
15. Aram, F.; García, E.H.; Solgi, E.; Mansournia, S. Urban green space cooling effect in cities. *Heliyon* **2019**, *5*, e01339. [CrossRef]
16. Giridharan, R.; Emmanuel, R. The impact of urban compactness, comfort strategies and energy consumption on tropical urban heat island intensity: A review. *Sustain. Cities Soc.* **2018**, *40*, 677–687. [CrossRef]
17. Oke, T.R. The energetic basis of the urban heat island. *Q. J. R. Meteorol. Soc.* **1982**, *108*, 1–24. [CrossRef]

18. Gunawardena, K.R.; Wells, M.J.; Kershaw, T. Utilising green and bluespace to mitigate urban heat island intensity. *Sci. Total Environ.* **2017**, *584–585*, 1040–1055. [CrossRef]

19. Wang, R.; Min, J.; Li, Y.; Hu, Y.; Yang, S. Analysis on seasonal variation and influencing mechanism of land surface thermal environment: A case study of Chongqing. *Remote Sens.* **2022**, *14*, 2022. [CrossRef]

20. Akbari, H.; Pomerantz, M.; Taha, H. Cool surfaces and shade trees to reduce energy use and improve air quality in urban areas. *Sol. Energy* **2001**, *70*, 295–310. [CrossRef]

21. Akbari, H.; Rosenfeld, A.H.; Taha, H. Summer heat islands, urban trees, and white surfaces. In Proceedings of the 1990 ASHRAE Winter Conference, Atlanta, GA, USA, 10–14 February 2022.

22. Duncan, J.M.A.; Boruff, B.; Saunders, A.; Sun, Q.; Hurley, J.; Amati, M. Turning down the heat: An enhanced understanding of the relationship between urban vegetation and surface temperature at the city scale. *Sci. Total Environ.* **2019**, *656*, 118–128. [CrossRef] [PubMed]

23. Wong, N.H.; Yu, C. Study of green areas and urban heat island in a tropical city. *Habitat Int.* **2005**, *29*, 547–558. [CrossRef]

24. Zhou, H.X.; Tao, G.X.; Yan, X.Y.; Sun, J.; Wu, Y. A review of research on the urban thermal environment effects of green quantity. *J. Appl. Ecol.* **2020**, *31*, 2804–2816. [CrossRef]

25. Baqa, M.F.; Lu, L.; Chen, F.; Nawaz-ul-Huda, S.; Pan, L.; Tariq, A.; Qureshi, S.; Li, B.; Li, Q. Characterizing spatiotemporal variations in the urban thermal environment related to land cover changes in Karachi, Pakistan, from 2000 to 2020. *Remote Sens.* **2022**, *14*, 2164. [CrossRef]

26. Alavipanah, S.; Wegmann, M.; Qureshi, S.; Weng, Q.; Koellner, T. The role of vegetation in mitigating urban land surface temperatures: A case study of Munich, Germany during the warm season. *Sustainability* **2015**, *7*, 4689–4706. [CrossRef]

27. Guzmán, C.A.P.; Fernández, D.J.M. Environmental impacts by golf courses and strategies to minimize them: State of the art. *Int. J. Arts Sci.* **2014**, *7*, 403.

28. Colding, J.; Folke, C. The role of golf courses in biodiversity conservation and ecosystem management. *Ecosystems* **2009**, *12*, 191–206. [CrossRef]

29. Terman, M.R. Natural links: Naturalistic golf courses as wildlife habitat. *Landsc. Urban Plan.* **1997**, *38*, 183–197. [CrossRef]

30. Nguyen, T.T.; Barber, P.; Harper, R.; Linh, T.V.K.; Dell, B. Vegetation trends associated with urban development: The role of golf courses. *PLoS ONE* **2020**, *15*, e0228090. [CrossRef]

31. Fung, C.K.W.; Jim, C.Y. Assessing the cooling effects of different vegetation settings in a Hong Kong golf course. *Procedia Environ. Sci.* **2017**, *37*, 626–636. [CrossRef]

32. Deilami, K.; Kamruzzaman, M.; Hayes, J.F. Correlation or causality between land cover patterns and the urban heat island effect? Evidence from Brisbane, Australia. *Remote Sens.* **2016**, *8*, 716. [CrossRef]

33. Shandas, V.; Voelkel, J.; Williams, J.; Hoffman, J. Integrating satellite and ground measurements for predicting locations of extreme urban heat. *Climate* **2019**, *7*, 5. [CrossRef]

34. Weng, Q. Thermal infrared remote sensing for urban climate and environmental studies: Methods, applications, and trends. *ISPRS J. Photogramm. Remote Sens.* **2009**, *64*, 335–344. [CrossRef]

35. Mushore, T.D.; Mutanga, O.; Odindi, J. Determining the influence of long term urban growth on surface urban heat islands using local climate zones and intensity analysis techniques. *Remote Sens.* **2022**, *14*, 2060. [CrossRef]

36. Wang, R.; Gao, W.; Peng, W. Downscale MODIS land surface temperature based on three different models to analyze surface urban heat island: A case study of Hangzhou. *Remote Sens.* **2020**, *12*, 2134. [CrossRef]

37. Zhang, X.; Friedl, M.A.; Schaaf, C.B.; Strahler, A.H. Climate controls on vegetation phenological patterns in northern mid-and high latitudes inferred from MODIS data. *Glob. Change Biol.* **2004**, *10*, 1133–1145. [CrossRef]

38. Evans, B.; Lyons, T.J.; Barber, P.A.; Stone, C.; Hardy, G. Dieback classification modelling using high-resolution digital multispectral imagery and in situ assessments of crown condition. *Remote Sens. Lett.* **2012**, *3*, 541–550. [CrossRef]

39. Bokaie, M.; Zarkesh, M.K.; Arasteh, P.D.; Hosseini, A. Assessment of urban heat island based on the relationship between land surface temperature and land use/land cover in Tehran. *Sustain. Cities Soc.* **2016**, *23*, 94–104. [CrossRef]

40. Fu, P.; Weng, Q. A time series analysis of urbanization induced land use and land cover change and its impact on land surface temperature with Landsat imagery. *Remote Sens. Environ.* **2016**, *175*, 205–214. [CrossRef]

41. Rinner, C.; Hussain, M. Toronto's urban heat island—Exploring the relationship between land use and surface temperature. *Remote Sens.* **2011**, *3*, 1251–1265. [CrossRef]

42. Mittermeier, R.; Gil, P.; Hoffmann, M.; Pilgrim, J.; Brooks, T.; Mittermeier, C.; Lamoreux, J.; Fonseca, G. Hotspots Revisited: Earth's Biologically Richest and Most Endangered Terrestrial Ecoregions. 2004, Volume 392. Available online: https://www.academia.edu/1438756/Hotspots_revisited_Earths_biologically_richest_and_most_endangered_terrestrial_ecoregions (accessed on 5 June 2021).

43. Myers, N.; Mittermeier, R.A.; Mittermeier, C.G.; da Fonseca, G.A.B.; Kent, J. Biodiversity hotspots for conservation priorities. *Nature* **2000**, *403*, 853–858. [CrossRef]

44. Lawrence, J.; Mackey, B.; Chiew, F.; Costello, M.J.; Hennessy, K.; Lansbury, N.; Nidumolu, U.B.; Pecl, G.; Rickards, L.; Tapper, N.; et al. Australasia. In *Climate Change 2022: Impacts, Adaptation, and Vulnerability*; Cambridge University Press: Cambridge, UK, 2022.

45. Environmental Protection Authority. *Perth and Peel@ 3.5 Million Environmental Impacts, Risks and Remedies*; Interim Strategic Advice of the Environmental Protection Authority to the Minister for Environment under Section 16e of the Environmental Protection Act; Environmental Protection Authority: Perth, Australia, 2015.

46. Subas, P.D. Glimpses of sustainability in Perth, Western Australia: Capturing and communicating the adaptive capacity of an activist group. *Cons. J. Sustain. Dev.* **2014**, *11*, 167–182.

47. Whetton, P.H.; Grose, M.R.; Hennessy, K.J. A short history of the future: Australian climate projections 1987–2015. *Clim. Serv.* **2016**, *2*, 1–14. [CrossRef]

48. The WA Local Government Association (WALGA). *2021/22 Local Government Urban Canopy Grant Program*; The WA Local Government Association: Perth, Australia, 2021.

49. Bureau of Meteorology. Daily Maximum Temperature: Perth Metro Station. Available online: http://www.bom.gov.au/jsp/ncc/cdio/weatherData/av?p_nccObsCode=122&p_display_type=dailyDataFile&p_startYear=2019&p_c=-17117203&p_stn_num=009225 (accessed on 20 January 2022).

50. Warwick-Champion, E.; Davies, K.P.; Paul, B.; Hardy, N.; Bruce, E. Characterising the Aboveground Carbon Content of Saltmarsh in Jervis Bay, NSW, Using ArborCam and PlanetScope. *Remote Sens.* **2022**, *14*, 1782. [CrossRef]

51. Tucker, C.J. Red and photographic infrared linear combinations for monitoring vegetation. *Remote Sens. Environ.* **1979**, *8*, 127–150. [CrossRef]

52. Pascual-Hortal, L.; Saura, S. Comparison and development of new graph-based landscape connectivity indices: Towards the priorization of habitat patches and corridors for conservation. *Landsc. Ecol.* **2006**, *21*, 959–967. [CrossRef]

53. Velázquez, J.; Gutiérrez, J.; García-Abril, A.; Hernando, A.; Aparicio, M.; Sánchez, B. Structural connectivity as an indicator of species richness and landscape diversity in Castilla y León (Spain). *For. Ecol. Manag.* **2019**, *432*, 286–297. [CrossRef]

54. Saura, S.; Vogt, P.; Velázquez, J.; Hernando, A.; Tejera, R. Key structural forest connectors can be identified by combining landscape spatial pattern and network analyses. *For. Ecol. Manag.* **2011**, *262*, 150–160. [CrossRef]

55. Blair, R.B. Land Use and Avian Species Diversity along an Urban Gradient. *Ecol. Appl.* **1996**, *6*, 506–519. [CrossRef]

56. Threlfall, C.G.; Ossola, A.; Hahs, A.K.; Williams, N.S.G.; Wilson, L.; Livesley, S.J. Variation in vegetation structure and composition across urban green space types. *Front. Ecol. Evol.* **2016**, *4*, 66. [CrossRef]

57. Department of Planning, Lands and Heritage. *State Planning Policy 2.8 Bushland Policy for the Perth Metropolitan Region*; Department of Planning, Lands and Heritage: Perth, WA, USA, 2010.

58. R Core Team. *R: A Language and Environment for Statistical Computing Computer Program, Version 3.6.1*; R Core Team: Vienna, Austria, 2019; Available online: http://www.R-project.org/ (accessed on 1 May 2021).

59. Ossola, A.; Jenerette, G.D.; McGrath, A.; Chow, W.; Hughes, L.; Leishman, M.R. Small vegetated patches greatly reduce urban surface temperature during a summer heatwave in Adelaide, Australia. *Landsc. Urban Plan.* **2021**, *209*, 104046. [CrossRef]

60. Fung, C.K.W.; Jim, C.Y. Microclimatic resilience of subtropical woodlands and urban-forest benefits. *Urban For. Urban Green.* **2019**, *42*, 100–112. [CrossRef]

61. Taha, H. Urban climates and heat islands: Albedo, evapotranspiration, and anthropogenic heat. *Energy Build.* **1997**, *25*, 99–103. [CrossRef]

62. Akbari, H.; Bell, R.; Brazel, T.; Cole, D.; Estes, M.; Heisler, G.; Hitchcock, D.; Johnson, B.; Lewis, M.; McPherson, G. *Reducing Urban Heat Islands: Compendium of Strategies—Urban Heat Island Basics*; Environmental Protection Agency: Washington, DC, USA, 2008; pp. 1–22.

63. Matthew, A.; Hiep, D.; Toan, T. *Impacts of Land-Use Change on Sydney's Future Temperatures*; State of NSW and Office of Environment and Heritage: Sydney, NSW, Australia, 2015.

64. Yenneti, K.; Ding, L.; Prasad, D.; Ulpiani, G.; Paolini, R.; Haddad, S.; Santamouris, M. Urban overheating and cooling potential in Australia: An evidence-based review. *Climate* **2020**, *8*, 126. [CrossRef]

65. Polydoros, A.; Mavrakou, T.; Cartalis, C. Quantifying the trends in land surface temperature and surface urban heat island intensity in Mediterranean cities in view of smart urbanization. *Urban Sci.* **2018**, *2*, 16. [CrossRef]

66. Oliveira, S.; Andrade, H.; Vaz, T. The cooling effect of green spaces as a contribution to the mitigation of urban heat: A case study in Lisbon. *Build. Environ.* **2011**, *46*, 2186–2194. [CrossRef]

67. Jiao, M.; Zhou, W.; Zheng, Z.; Wang, J.; Qian, Y. Patch size of trees affects its cooling effectiveness: A perspective from shading and transpiration processes. *Agric. For. Meteorol.* **2017**, *247*, 293–299. [CrossRef]

68. Bao, T.; Li, X.; Zhang, J.; Zhang, Y.; Tian, S. Assessing the distribution of urban green spaces and its anisotropic cooling distance on urban heat island pattern in Baotou, China. *ISPRS Int. J. Geo-Inf.* **2016**, *5*, 12. [CrossRef]

69. Kong, F.; Yin, H.; James, P.; Hutyra, L.R.; He, H.S. Effects of spatial pattern of greenspace on urban cooling in a large metropolitan area of eastern China. *Landsc. Urban Plan.* **2014**, *128*, 35–47. [CrossRef]

70. Kowe, P.; Mutanga, O.; Odindi, J.; Dube, T. Effect of landscape pattern and spatial configuration of vegetation patches on urban warming and cooling in Harare metropolitan city, Zimbabwe. *GIScience Remote Sens.* **2021**, *58*, 1–20. [CrossRef]

71. Zhang, X.; Zhong, T.; Feng, X.; Wang, K. Estimation of the relationship between vegetation patches and urban land surface temperature with remote sensing. *Int. J. Remote Sens.* **2009**, *30*, 2105–2118. [CrossRef]

72. Xie, M.; Gao, Y.; Cao, Y.; Breuste, J.; Fu, M.; Tong, D. Dynamics and temperature regulation function of urban green connectivity. *J. Urban Plan. Dev.* **2015**, *141*, A5014008. [CrossRef]
73. Barber, P.A.; Paap, T.; Burgess, T.I.; Dunstan, W.; Hardy, G.E.S.J. A diverse range of Phytophthora species are associated with dying urban trees. *Urban For. Urban Green.* **2013**, *12*, 569–575. [CrossRef]

MDPI

St. Alban-Anlage 66

4052 Basel

Switzerland

Tel. +41 61 683 77 34

Fax +41 61 302 89 18

www.mdpi.com

Remote Sensing Editorial Office

E-mail: remotesensing@mdpi.com

www.mdpi.com/journal/remotesensing